(*continued on back*)

DATE DUE

DEC 1 1 1988			
;II 3.24.05			

GAYLORD   PRINTED IN U.S.A.

10/11/88

Robust Regression
and Outlier Detection

Robust Regression and Outlier Detection

PETER J. ROUSSEEUW
University of Fribourg, Switzerland

ANNICK M. LEROY
Vrije Universiteit Brussel, Belgium

JOHN WILEY & SONS
New York • Chichester • Brisbane • Toronto • Singapore

Library of Congress Cataloging in Publication Data:

Rousseeuw, Peter J.
  Robust regression and outlier detection.

  (Wiley series in probability and mathematical
statistics. Applied probability and statistics,
ISSN 0271-6356)
  Bibliography: p.
  Includes index.
  1. Regression analysis. 2. Outliers (Statistics)
3. Least squares. I. Leroy, Annick M. II. Title.
III. Series.
QA278.2.R68 1987   519.5'36   87-8234
ISBN 0-471-85233-3

Printed in the United States of America

10 9 8 7 6 5 4 3 2 1

If among these errors are some which appear too large to be admissible, then those observations which produced these errors will be rejected, as coming from too faulty experiments, and the unknowns will be determined by means of the other observations, which will then give much smaller errors.

Legendre in 1805, in the first publication on least squares

This idea, however, from its nature, involves something vague . . . and clearly innumerable different principles can be proposed . . . . But of all these principles ours is the most simple; by the others we shall be led into the most complicated calculations.

Gauss in 1809, on the least squares criterion

The method of Least Squares is seen to be our best course when we have thrown overboard a certain portion of our data—a sort of sacrifice which has often to be made by those who sail upon the stormy seas of Probability.

Edgeworth in 1887

(Citations collected by Plackett 1972 and Stigler 1973, 1977.)

Preface

Regression analysis is an important statistical tool that is routinely applied in most sciences. Out of many possible regression techniques, the least squares (LS) method has been generally adopted because of tradition and ease of computation. However, there is presently a widespread awareness of the dangers posed by the occurrence of outliers, which may be a result of keypunch errors, misplaced decimal points, recording or transmission errors, exceptional phenomena such as earthquakes or strikes, or members of a different population slipping into the sample. Outliers occur very frequently in real data, and they often go unnoticed because nowdays much data is processed by computers, without careful inspection or screening. Not only the response variable can be outlying, but also the explanatory part, leading to so-called *leverage points*. Both types of outliers may totally spoil an ordinary LS analysis. Often, such influential points remain hidden to the user, because they do not always show up in the usual LS residual plots.

To remedy this problem, new statistical techniques have been developed that are not so easily affected by outliers. These are the robust (or resistant) methods, the results of which remain trustworthy even if a certain amount of data is contaminated. Some people think that robust regression techniques hide the outliers, but the opposite is true because the outliers are far away from the robust fit and hence can be detected by their large residuals from it, whereas the standardized residuals from ordinary LS may not expose outliers at all. The main message of this book is that robust regression is extremely useful in identifying outliers, and many examples are given where all the outliers are detected in a single blow by simply running a robust estimator.

An alternative approach to dealing with outliers in regression analysis is to construct outlier diagnostics. These are quantities computed from the data with the purpose of pinpointing influential observations, which can then be studied and corrected or deleted, followed by an LS analysis on the remaining cases. Diagnostics and robust regression have the same goal, but they proceed in the opposite order: In a diagnostic setting, one first wants to identify the outliers and then fit the good data in the classical way, whereas the robust approach first fits a regression that does justice to the majority of the data and then discovers the outliers as those points having large residuals from the robust equation. In some applications, both approaches yield exactly the same result, and then the difference is mostly subjective. Indeed, some people feel happy when switching to a more robust criterion, but they cannot accept the deletion of "true" observations (although many robust methods will, in effect, give the outliers zero influence), whereas others feel that it is all right to delete outliers, but they maintain that robust regression is "arbitrary" (although the combination of deleting outliers and then applying LS is itself a robust method). We are not sure whether this philosophical debate serves a useful purpose. Fortunately, some positive interaction between followers of both schools is emerging, and we hope that the gap will close. Personally we do not take an "ideological" stand, but we propose to judge each particular technique on the basis of its reliability by counting how many outliers it can deal with. For instance, we note that certain robust methods can withstand leverage points, whereas others cannot, and that some diagnostics allow us to detect multiple outliers, whereas others are easily masked.

In this book we consider methods with high breakdown point, which are able to cope with a large fraction of outliers. The "high breakdown" objective could be considered a kind of third generation in robustness theory, coming after minimax variance (Huber 1964) and the influence function (Hampel 1974). Naturally, the emphasis is on the methods we have worked on ourselves, although many other estimators are also discussed. We advocate the least median of squares method (Rousseeuw 1984) because it appeals to the intuition and is easy to use. No background knowledge or choice of tuning constants are needed: You just enter the data and interpret the results. It is hoped that robust methods of this type will be incorporated into major statistical packages, which would make them easily accessible. As long as this is not yet the case, you may contact the first author (PJR) to obtain an updated version of the program PROGRESS (Program for RObust reGRESSion) used in this book. PROGRESS runs on IBM-PC and compatible machines, but the structured source code is also available to enable you to include it in your own software (Recently, it was integrated in the ROBETH library in Lausanne and in the workstation package S-PLUS of Statistical Sciences,

Inc., P.O. Box 85625, Seattle WA 98145-1625.) The computation time is substantially higher than that of ordinary LS, but this is compensated by a much more important gain of the statistician's time, because he or she receives the outliers on a "silver platter." And anyway, the computation time is no greater than that of other multivariate techniques that are commonly used, such as cluster analysis or multidimensional scaling.

The primary aim of our work is to make robust regression available for everyday statistical practice. The book has been written from an applied perspective, and the technical material is concentrated in a few sections (marked with *), which may be skipped without loss of understanding. No specific prerequisites are assumed. The material has been organized for use as a textbook and has been tried out as such. Chapter 1 introduces outliers and robustness in regression. Chapter 2 is confined to simple regression for didactic reasons and to make it possible to include robustness considerations in an introductory statistics course not going beyond the simple regression model. Chapter 3 deals with robust multiple regression, Chapter 4 covers the special case of one-dimensional location, and Chapter 5 discusses the algorithms used. Outlier diagnostics are described in Chapter 6, and Chapter 7 is about robustness in related fields such as time series analysis and the estimation of multivariate location and covariance matrices. Chapters 1–3 and 6 could be easily incorporated in a modern course on applied regression, together with any other sections one would like to cover. It is also quite feasible to use parts of the book in courses on multivariate data analysis or time series. Every chapter contains exercises, ranging from simple questions to small data sets with clues to their analysis.

PETER J. ROUSSEEUW
ANNICK M. LEROY

October 1986

Acknowledgments

We are grateful to David Donoho, Frank Hampel, Werner Stahel, and many other colleagues for helpful suggestions and stimulating discussions on topics covered in this book. Parts of the manuscript were read by Leon Kaufman, Doug Martin, Elvezio Ronchetti, Jos van Soest, and Bert van Zomeren. We would also like to thank Beatrice Shube and the editorial staff at John Wiley & Sons for their assistance.

We thank the authors and publishers of figures, tables, and data for permission to use the following material:

Table 1, Chapter 1; Tables 6 and 13, Chapter 2: Rousseeuw et al. (1984a), © by North-Holland Publishing Company, Amsterdam.

Tables 1, 7, 9, and 11, Chapter 2; Tables 1, 10, 13, 16, 23, Chapter 3; Table 8, Chapter 6: © Reprinted by permission of John Wiley & Sons, New York.

Tables 2 and 3, Chapter 2; Tables 19 and 20, Chapter 3: Rousseeuw and Yohai (1984), by permission of Springer-Verlag, New York.

Tables 4 and 10, Chapter 2: © 1967, 1979, by permission of Academic Press, Orlando, Florida.

Table 6, Chapter 2; data of (3.1), Chapter 4: © Reprinted by permission of John Wiley & Sons, Chichester.

Figure 11, Chapter 2; Tables 5, 9, 22, 24, Chapter 3; data of (2.2) and Exercise 11, Chapter 4; Table 8, Chapter 6; first citation in Epigraph: By permission of the American Statistical Association, Washington, DC.

Figure 12 and Tables 5 and 12, Chapter 2: Rousseeuw et al. (1984b), © by D. Reidel Publishing Company, Dordrecht, Holland.

Table 2, Chapter 3: © 1977, Addison-Wesley Publishing Company, Inc., Reading, MA.

Table 6, Chapter 3: © 1983, Wadsworth Publishing Co., Belmont, CA
Table 22, Chapter 3: Office of Naval Research, Arlington, VA.
Data of Exercise 10, Chapter 4; second citation in Epigraph: © 1908, 1972, by permission of the Biometrika Trustees.
Third citation in Epigraph: Stigler (1977), by permission of the Institute of Mathematical Statistics.
Figure 3, Chapter 5: By permission of B. van Zomeren (1986).
Table 1, Chapter 7: By permission of Prof. Arnold Zellner, Bureau of the Census.
Data in Exercise 15, Chapter 7: Collett (1980), by permission of the Royal Statistical Society.

Contents

Robust Regression
and Outlier Detection

CHAPTER 1

Introduction

1. OUTLIERS IN REGRESSION ANALYSIS

The purpose of regression analysis is to fit equations to observed variables. The classical linear model assumes a relation of the type

$$y_i = x_{i1}\theta_1 + \cdots + x_{ip}\theta_p + e_i \quad \text{for } i = 1, \ldots, n, \tag{1.1}$$

where n is the sample size (number of cases). The variables $x_{i1}, \ldots, x_{ip}$ are called the *explanatory variables* or *carriers*, whereas the variable y_i is called the *response variable*. In classical theory, the *error term* e_i is assumed to be normally distributed with mean zero and unknown standard deviation σ. One then tries to estimate the vector of unknown parameters

$$\boldsymbol{\theta} = \begin{bmatrix} \theta_1 \\ \vdots \\ \theta_p \end{bmatrix} \tag{1.2}$$

from the data:

$$\text{Cases} \underset{\text{Variables}}{\begin{bmatrix} x_{11} & \cdots & x_{1p} & y_1 \\ \vdots & & \vdots & \vdots \\ x_{i1} & \cdots & x_{ip} & y_i \\ \vdots & & \vdots & \vdots \\ x_{n1} & \cdots & x_{np} & y_n \end{bmatrix}}. \tag{1.3}$$

Applying a regression estimator to such a data set yields

$$\hat{\boldsymbol{\theta}} = \begin{bmatrix} \hat{\theta}_1 \\ \vdots \\ \hat{\theta}_p \end{bmatrix}, \tag{1.4}$$

where the estimates $\hat{\theta}_j$ are called the *regression coefficients*. (Vectors and matrices will be denoted by boldface throughout.) Although the actual θ_j are unknown, one can multiply the explanatory variables with these $\hat{\theta}_j$ and obtain

$$\hat{y}_i = x_{i1}\hat{\theta}_1 + \cdots + x_{ip}\hat{\theta}_p, \tag{1.5}$$

where $\hat{y}_i$ is called the *predicted* or *estimated* value of y_i. The *residual* r_i of the ith case is the difference between what is actually observed and what is estimated:

$$r_i = y_i - \hat{y}_i. \tag{1.6}$$

The most popular regression estimator dates back to Gauss and Legendre (see Plackett 1972 and Stigler 1981 for some historical discussions) and corresponds to

$$\underset{\hat{\boldsymbol{\theta}}}{\text{Minimize}} \sum_{i=1}^{n} r_i^2. \tag{1.7}$$

The basic idea was to optimize the fit by making the residuals very small, which is accomplished by (1.7). This is the well-known *least squares* (LS) method, which has become the cornerstone of classical statistics. The reasons for its popularity are easy to understand: At the time of its invention (around 1800) there were no computers, and the fact that the LS estimator could be computed *explicitly* from the data (by means of some matrix algebra) made it the only feasible approach. Even now, most statistical packages still use the same technique because of tradition and computation speed. Also, in the one-dimensional situation the LS criterion (1.7) yields the arithmetic mean of the observations, which at that time seemed to be the most reasonable location estimator. Afterwards, Gauss introduced the normal (or Gaussian) distribution as the error distribution for which LS is optimal (see the citations in Huber 1972, p. 1042, and Le Cam 1986, p. 79), yielding a beautiful mathematical theory. Since then, the combination of Gaussian assumptions and LS has become a standard mechanism for the generation of statistical techniques (e.g., multivariate location, analysis of variance, and minimum variance clustering).

More recently, some people began to realize that real data usually do not completely satisfy the classical assumptions, often with dramatic effects on the quality of the statistical analysis (see, e.g., Student 1927, Pearson 1931, Box 1953, and Tukey 1960).

As an illustration, let us look at the effect of outliers in the simple regression model

$$y_i = \theta_1 x_i + \theta_2 + e_i \qquad (1.8)$$

in which the slope θ_1 and the intercept θ_2 are to be estimated. This is indeed a special case of (1.1) with $p = 2$ because one can put $x_{i1} := x_i$ and $x_{i2} := 1$ for all $i = 1, \ldots, n$. (In general, taking a carrier identical to 1 is a standard trick used to obtain regression with a constant term.) In the simple regression model, one can make a plot of the (x_i, y_i), which is sometimes called a *scatterplot*, in order to visualize the data structure. In the general multiple regression model (1.1) with large p, this would no longer be possible, so it is better to use simple regression for illustrative purposes.

Figure 1a is the scatterplot of five points, $(x_1, y_1), \ldots, (x_5, y_5)$, which almost lie on a straight line. Therefore, the LS solution fits the data very well, as can be seen from the LS line $\hat{y} = \hat{\theta}_1 x + \hat{\theta}_2$ in the plot. However, suppose that someone gets a wrong value of y_4 because of a copying or transmission error, thus affecting, for instance, the place of the decimal point. Then (x_4, y_4) may be rather far away from the "ideal" line. Figure 1b displays such a situation, where the fourth point has moved up and away from its original position (indicated by the dashed circle). This point is called an *outlier in the y-direction*, and it has a rather large influence on the LS line, which is quite different from the LS line in Figure 1a. This phenomenon has received some attention in the literature because one usually considers the y_i as observations and the $x_{i1}, \ldots, x_{ip}$ as fixed numbers (which is only true when the design has been given in advance) and because such "vertical" outliers often possess large positive or large negative residuals. Indeed, in this example the fourth point lies farthest away from the straight line, so its r_i given by (1.6) is suspiciously large. Even in general multiple regression (1.1) with large p, where one cannot visualize the data, such outliers can often be discovered from the list of residuals or from so-called *residual plots* (to be discussed in Section 4 of Chapter 2 and Section 1 of Chapter 3).

However, usually also the explanatory variables $x_{i1}, \ldots, x_{ip}$ are observed quantities subject to random variability. (Indeed, in many applications, one receives a list of variables from which one then has to choose a

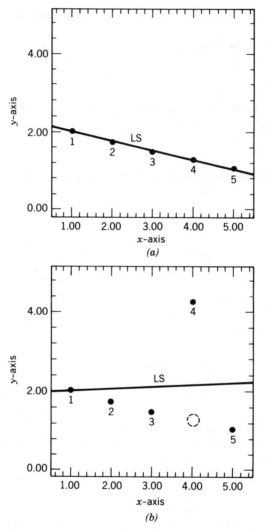

Figure 1. (*a*) Original data with five points and their least squares regression line. (*b*) Same data as in part (*a*), but with one outlier in the *y*-direction.

response variable and some explanatory variables.) Therefore, there is no reason why gross errors would only occur in the response variable y_i. In a certain sense it is even more likely to have an outlier in one of the explanatory variables $x_{i1}, \ldots, x_{ip}$ because usually p is greater than 1, and hence there are more opportunities for something to go wrong. For the effect of such an outlier, let us look at an example of simple regression in Figure 2.

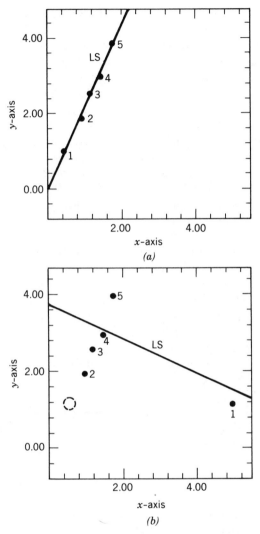

Figure 2. (*a*) Original data with five points and their least squares regression line. (*b*) Same data as in part (*a*), but with one outlier in the *x*-direction ("leverage point").

Figure 2*a* contains five points, $(x_1, y_1), \ldots, (x_5, y_5)$, with a well-fitting LS line. If we now make an error in recording x_1, we obtain Figure 2*b*. The resulting point is called an *outlier in the x-direction*, and its effect on the least squares estimator is very large because it actually tilts the LS line. Therefore the point (x_1, y_1) in Figure 2*b* is called a *leverage point*, in analogy to the notion of leverage in mechanics. This large "pull" on the

LS estimator can be explained as follows. Because x_1 lies far away, the residual r_1 from the original line (as shown in Figure 2a) becomes a very large (negative) value, contributing an enormous amount to $\Sigma_{i=1}^{5} r_i^2$ for that line. Therefore the original line cannot be selected from a least squares perspective, and indeed the line of Figure 2b possesses the smallest $\Sigma_{i=1}^{5} r_i^2$ because it has tilted to reduce that large r_1^2, even if the other four terms, $r_2^2, \ldots, r_5^2$, have increased somewhat.

In general, we call an observation (x_k, y_k) a leverage point whenever x_k lies far away from the bulk of the observed x_i in the sample. Note that this does not take y_k·into account, so the point (x_k, y_k) does not necessarily have to be a regression outlier. When (x_k, y_k) lies close to the regression line determined by the majority of the data, then it can be considered a "good" leverage point, as in Figure 3. Therefore, to say that (x_k, y_k) is a leverage point refers only to its *potential* for strongly affecting the regression coefficients $\hat{\theta}_1$ and $\hat{\theta}_2$ (due to its outlying component x_k), but it does not necessarily mean that (x_k, y_k) will actually have a large influence on $\hat{\theta}_1$ and $\hat{\theta}_2$, because it may be perfectly in line with the trend set by the other data. (In such a situation, a leverage point is even quite beneficial because it will shrink certain confidence regions.)

In multiple regression, the $(x_{i1}, \ldots, x_{ip})$ lie in a space with p dimensions (which is sometimes called the *factor space*). A leverage point is then still defined as a point $(x_{k1}, \ldots, x_{kp}, y_k)$ for which $(x_{k1}, \ldots, x_{kp})$ is outlying with respect to the $(x_{i1}, \ldots, x_{ip})$ in the data set. As before, such

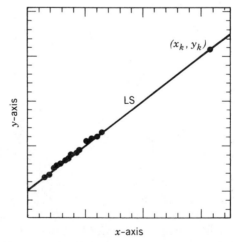

Figure 3. The point (x_k, y_k) is a leverage point because x_k is outlying. However, (x_k, y_k) is not a regression outlier because it matches the linear pattern set by the other data points.

leverage points have a potentially large influence on the LS regression coefficients, depending on the actual value of y_k. However, in this situation it is much more difficult to *identify* leverage points, because of the higher dimensionality. Indeed, it may be very difficult to discover such a point when there are 10 explanatory variables, which we can no longer visualize. A simple illustration of the problem is given in Figure 4, which plots x_{i1} versus x_{i2} for some data set. In this plot we easily see two leverage points, which are, however, invisible when the variables x_{i1} and x_{i2} are considered separately. (Indeed, the one-dimensional sample $\{x_{11}, x_{21}, \ldots, x_{n1}\}$ does not contain outliers, and neither does $\{x_{12}, x_{22}, \ldots, x_{n2}\}$.) In general, it is not sufficient to look at each variable separately or even at all plots of pairs of variables. The identification of outlying $(x_{i1}, \ldots, x_{ip})$ is a difficult problem, which will be treated in Subsection 1d of Chapter 7. However, in this book we are mostly concerned with *regression outliers*, that is, cases for which $(x_{i1}, \ldots, x_{ip}, y_i)$ deviates from the linear relation followed by the majority of the data, taking into account both the explanatory variables and the response variable simultaneously.

Many people will argue that regression outliers can be discovered by looking at the least squares residuals. Unfortunately, this is not true when the outliers are leverage points. For example, consider again Figure 2b. Case 1, being a leverage point, has tilted the LS line so much that it is now quite close to that line. Consequently, the residual $r_1 = y_1 - \hat{y}_1$ is a

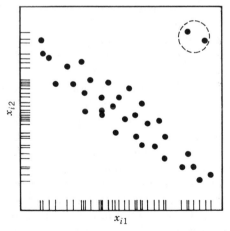

Figure 4. Plot of the explanatory variables (x_{i1}, x_{i2}) of a regression data set. There are two leverage points (indicated by the dashed circle), which are not outlying in either of the coordinates.

small (negative) number. On the other hand, the residuals r_2 and r_5 have much larger absolute values, although they correspond to "good" points. If one would apply a rule like "delete the points with largest LS residuals," then the "good" points would have to be deleted first! Of course, in such a bivariate data set there is really no problem at all because one can actually look at the data, but there are many multi-variate data sets (like those of Chapter 3) where the outliers remain invisible even through a careful analysis of the LS residuals.

To conclude, regression outliers (either in x or in y) pose a serious threat to standard least squares analysis. Basically, there are two ways out of this problem. The first, and probably most well-known, approach is to construct so-called *regression diagnostics*. A survey of these techniques is provided in Chapter 6. Diagnostics are certain quantities computed from the data with the purpose of pinpointing influential points, after which these outliers can be removed or corrected, followed by an LS analysis on the remaining cases. When there is only a single outlier, some of these methods work quite well by looking at the effect of deleting one point at a time. Unfortunately, it is much more difficult to diagnose outliers when there are several of them, and diagnostics for such multiple outliers are quite involved and often give rise to extensive computations (e.g., the number of all possible subsets is gigantic). Section 5 of Chapter 6 reports on recent developments in this direction, and in Section 6 of Chapter 6 a new diagnostic is proposed which can even cope with large fractions of outliers.

The other approach is *robust regression*, which tries to devise es-timators that are not so strongly affected by outliers. Many statisticians who have vaguely heard of robustness believe that its purpose is to simply ignore the outliers, but this is not true. On the contrary, it is by looking at the residuals from a robust (or "resistant") regression that outliers may be identified, which usually cannot be done by means of the LS residuals. Therefore, diagnostics and robust regression really have the same goals, only in the opposite order: When using diagnostic tools, one first tries to delete the outliers and then to fit the "good" data by least squares, whereas a robust analysis first wants to fit a regression to the majority of the data and then to discover the outliers as those points which possess large residuals from that robust solution.

The following step is to think about the structure that has been uncovered. For instance, one may go back to the original data set and use subject-matter knowledge to study the outliers and explain their origin. Also, one should investigate if the deviations are not a symptom for model failure, which could, for instance, be repaired by adding a quadratic term or performing some transformation.

There are almost as many robust estimators as there are diagnostics, and it is necessary to measure their effectiveness in order to differentiate between them. In Section 2, some robust methods will be compared, essentially by counting the number of outliers that they can deal with. In subsequent chapters, the emphasis will be on the application of very robust methods, which can be used to analyze extremely messy data sets as well as clean ones.

2. THE BREAKDOWN POINT AND ROBUST ESTIMATORS

In Section 1 we saw that even a single regression outlier can totally offset the least squares estimator (provided it is far away). On the other hand, we will see that there exist estimators that can deal with data containing a certain percentage of outliers. In order to formalize this aspect, the *breakdown point* was introduced. Its oldest definition (Hodges 1967) was restricted to one-dimensional estimation of location, whereas Hampel (1971) gave a much more general formulation. Unfortunately, the latter definition was asymptotic and rather mathematical in nature, which may have restricted its dissemination. We prefer to work with a simple finite-sample version of the breakdown point, introduced by Donoho and Huber (1983). Take any sample of n data points,

$$Z = \{(x_{11}, \ldots, x_{1p}, y_1), \ldots, (x_{n1}, \ldots, x_{np}, y_n)\}, \qquad (2.1)$$

and let T be a regression estimator. This means that applying T to such a sample Z yields a vector of regression coefficients as in (1.4):

$$T(Z) = \hat{\boldsymbol{\theta}}. \qquad (2.2)$$

Now consider all possible corrupted samples Z' that are obtained by replacing any m of the original data points by arbitrary values (this allows for very bad outliers). Let us denote by bias$(m; T, Z)$ the maximum bias that can be caused by such a contamination:

$$\text{bias}\,(m; T, Z) = \sup_{Z'} \| T(Z') - T(Z) \|, \qquad (2.3)$$

where the supremum is over all possible Z'. If bias$(m; T, Z)$ is infinite, this means that m outliers can have an arbitrarily large effect on T, which may be expressed by saying that the estimator "breaks down." Therefore, the (finite-sample) breakdown point of the estimator T at the sample Z is defined as

$$\varepsilon_n^*(T, Z) = \min \left\{ \frac{m}{n} ; \text{ bias } (m; T, Z) \text{ is infinite} \right\}. \qquad (2.4)$$

In other words, it is the smallest fraction of contamination that can cause the estimator T to take on values arbitrarily far from $T(Z)$. Note that this definition contains no probability distributions!

For least squares, we have seen that one outlier is sufficient to carry T over all bounds. Therefore, its breakdown point equals

$$\varepsilon_n^*(T, Z) = 1/n \qquad (2.5)$$

which tends to zero for increasing sample size n, so it can be said that LS has a breakdown point of 0%. This again reflects the extreme sensitivity of the LS method to outliers.

A first step toward a more robust regression estimator came from Edgeworth (1887), improving a proposal of Boscovich. He argued that outliers have a very large influence on LS because the residuals r_i are squared in (1.7). Therefore, he proposed the *least absolute values* regression estimator, which is determined by

$$\underset{\hat{\theta}}{\text{Minimize}} \sum_{i=1}^{n} |r_i| . \qquad (2.6)$$

(This technique is often referred to as L_1 regression, whereas least squares is L_2.) Before that time, Laplace had already used the same criterion (2.6) in the context of one-dimensional observations, obtaining the sample median (and the corresponding error law, which is now called the *double exponential* or *Laplace distribution*). The L_1 regression estimator, like the median, is not completely unique (see, e.g., Harter 1977 and Gentle et al. 1977). But whereas the breakdown point of the univariate median is as high as 50%, unfortunately the breakdown point of L_1 regression is still no better than 0%. To see why, let us look at Figure 5.

Figure 5 gives a schematic summary of the effect of outliers on L_1 regression. Figure 5a shows the effect of an outlier in the y-direction, in the same situation as Figure 1. Unlike least squares, the L_1 regression line is robust with respect to such an outlier, in the sense that it (approximately) remains where it was when observation 4 was still correct, and still fits the remaining points nicely. Therefore, L_1 protects us against outlying y_i and is quite preferable over LS in this respect. In recent years, the L_1 approach to statistics appears to have gained some ground (Bloomfield

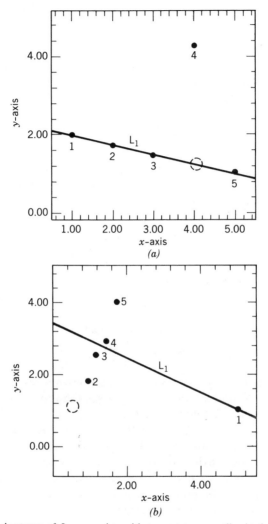

Figure 5. (*a*) Robustness of L_1 regression with respect to an outlier in the *y*-direction. (*b*) Sensitivity of L_1 regression to an outlier in the *x*-direction ("leverage point").

and Steiger 1980, 1983; Narula and Wellington 1982; Devroye and Gyorfi 1984). However, L_1 regression does not protect against outlying *x*, as we can see from Figure 5*b*, where the effect of the leverage point is even stronger than on the LS line in Figure 2. It turns out that when the leverage point lies far enough away, the L_1 line passes right through it (see exercise 10 below). Therefore, a single erroneous observation can totally offset the L_1 estimator, so its finite-sample breakdown point is also equal to $1/n$.

The next step in this direction was the use of *M-estimators* (Huber 1973, p. 800; for a recent survey see Huber 1981). They are based on the idea of replacing the squared residuals r_i^2 in (1.7) by another function of the residuals, yielding

$$\text{Minimize} \sum_{i=1}^{n} \rho(r_i), \qquad (2.7)$$

where ρ is a symmetric function [i.e., $\rho(-t) = \rho(t)$ for all t] with a unique minimum at zero. Differentiating this expression with respect to the regression coefficients $\hat{\theta}_j$ yields

$$\sum_{i=1}^{n} \psi(r_i)\mathbf{x}_i = \mathbf{0}, \qquad (2.8)$$

where ψ is the derivative of ρ, and $\mathbf{x}_i$ is the row vector of explanatory variables of the ith case:

$$\mathbf{x}_i = (x_{i1}, \ldots, x_{ip})$$
$$\mathbf{0} = (0, \ldots, 0). \qquad (2.9)$$

Therefore (2.8) is really a system of p equations, the solution of which is not always easy to find: In practice, one uses iteration schemes based on reweighted LS (Holland and Welsch 1977) or the so-called *H-algorithm* (Huber and Dutter 1974, Dutter 1977, Marazzi 1980). Unlike (1.7) or (2.6), however, the solution of (2.8) is not equivariant with respect to a magnification of the y-axis. (We use the word "equivariant" for statistics that transform properly, and we reserve "invariant" for quantities that remain unchanged.) Therefore, one has to standardize the residuals by means of some estimate of σ, yielding

$$\sum_{i=1}^{n} \psi(r_i/\hat{\sigma})\mathbf{x}_i = \mathbf{0}, \qquad (2.10)$$

where $\hat{\sigma}$ must be estimated simultaneously. Motivated by minimax asymptotic variance arguments, Huber proposed to use the function

$$\psi(t) = \min(c, \max(t, -c)). \qquad (2.11)$$

M-estimators with (2.11) are statistically more efficient (at a model with

Gaussian errors) than L_1 regression, while at the same time they are still robust with respect to outlying y_i. However, their breakdown point is again $1/n$ because of the effect of outlying $\mathbf{x}_i$.

Because of this vulnerability to leverage points, *generalized M-estimators* (GM-estimators) were introduced, with the basic purpose of bounding the influence of outlying $\mathbf{x}_i$ by means of some weight function w. Mallows (1975) proposed to replace (2.10) by

$$\sum_{i=1}^{n} w(\mathbf{x}_i)\psi(r_i/\hat{\sigma})\mathbf{x}_i = \mathbf{0} , \qquad (2.12)$$

whereas Schweppe (see Hill 1977) suggested using

$$\sum_{i=1}^{n} w(\mathbf{x}_i)\psi(r_i/(w(\mathbf{x}_i)\hat{\sigma}))\mathbf{x}_i = \mathbf{0} . \qquad (2.13)$$

These estimators were constructed in the hope of bounding the influence of a single outlying observation, the effect of which can be measured by means of the so-called *influence function* (Hampel 1974). Based on such criteria, optimal choices of ψ and w were made (Hampel 1978, Krasker 1980, Krasker and Welsch 1982, Ronchetti and Rousseeuw 1985, and Samarov 1985; for a recent survey see Chapter 6 of Hampel et al. 1986). Therefore, the corresponding GM-estimators are now generally called *bounded-influence estimators*. It turns out, however, that the breakdown point of all GM-estimators can be no better than a certain value that decreases as a function of p, where p is again the number of regression coefficients (Maronna, Bustos, and Yohai 1979). This is very unsatisfactory, because it means that the breakdown point diminishes with increasing dimension, where there are more opportunities for outliers to occur. Furthermore, it is not clear whether the Maronna–Bustos–Yohai upper bound can actually be attained, and if it can, it is not clear as to which GM-estimator can be used to achieve this goal. In Section 7 of Chapter 2, a small comparative study will be performed in the case of simple regression ($p = 2$), indicating that not all GM-estimators achieve the same breakdown point. But, of course, the real problem is with higher dimensions.

Various other estimators have been proposed, such as the methods of Wald (1940), Nair and Shrivastava (1942), Bartlett (1949), and Brown and Mood (1951); the median of pairwise slopes (Theil 1950, Adichie 1967, Sen 1968); the resistant line (Tukey 1970/1971, Velleman and Hoaglin 1981, Johnstone and Velleman 1985b); R-estimators (Jurecková 1971, Jaeckel 1972); L-estimators (Bickel 1973, Koenker and Bassett

1978); and the method of Andrews (1974). Unfortunately, in simple regression, none of these methods achieves a breakdown point of 30%. Moreover, some of them are not even defined for $p > 2$.

All this raises the question as to whether robust regression with a high breakdown point is at all possible. The affirmative answer was given by Siegel (1982), who proposed the *repeated median* estimator with a 50% breakdown point. Indeed, 50% is the best that can be expected (for larger amounts of contamination, it becomes impossible to distinguish between the "good" and the "bad" parts of the sample, as will be proven in Theorem 4 of Chapter 3). Siegel's estimator is defined as follows: For any p observations

$$(\mathbf{x}_{i_1}, y_{i_1}), \ldots, (\mathbf{x}_{i_p}, y_{i_p})$$

one computes the parameter vector which fits these points exactly. The jth coordinate of this vector is denoted by $\theta_j(i_1, \ldots, i_p)$. The repeated median regression estimator is then defined coordinatewise as

$$\hat{\theta}_j = \operatorname*{med}_{i_1} (\ldots (\operatorname*{med}_{i_{p-1}} (\operatorname*{med}_{i_p} \theta_j(i_1, \ldots, i_p))) \ldots). \qquad (2.14)$$

This estimator can be computed explicitly, but requires consideration of all subsets of p observations, which may cost a lot of time. It has been successfully applied to problems with small p. But unlike other regression estimators, the repeated median is not equivariant for linear transformations of the $\mathbf{x}_i$, which is due to its coordinatewise construction.

Let us now consider the equivariant and high-breakdown regression methods that form the core of this book. To introduce them, let us return to (1.7). A more complete name for the LS method would be *least sum of squares*, but apparently few people have objected to the deletion of the word "sum"—as if the only sensible thing to do with n positive numbers would be to add them. Perhaps as a consequence of its historical name, several people have tried to make this estimator robust by replacing the square by something else, not touching the summation sign. Why not, however, replace the sum by a median, which is very robust? This yields the *least median of squares* (LMS) estimator, given by

$$\operatorname*{Minimize}_{\hat{\theta}} \operatorname*{med}_{i} r_i^2 \qquad (2.15)$$

(Rousseeuw 1984). This proposal was essentially based on an idea of Hampel (1975, p. 380). It turns out that this estimator is very robust with respect to outliers in y as well as outliers in $\mathbf{x}$. It will be shown in Section

4 of Chapter 3 that its breakdown point is 50%, the highest possible value. The LMS is clearly equivariant with respect to linear transformations on the explanatory variables, because (2.15) only makes use of the residuals. In Section 5 of Chapter 3, we will show that the LMS is related to projection pursuit ideas, whereas the most useful algorithm for its computation (Section 1 of Chapter 5) is reminiscent of the bootstrap (Diaconis and Efron 1983). Unfortunately, the LMS performs poorly from the point of view of asymptotic efficiency (in Section 4 of Chapter 4, we will prove it has an abnormally slow convergence rate).

To repair this, Rousseeuw (1983, 1984) introduced the *least trimmed squares* (LTS) estimator, given by

$$\underset{\hat{\theta}}{\text{Minimize}} \sum_{i=1}^{h} (r^2)_{i:n} , \tag{2.16}$$

where $(r^2)_{1:n} \leq \cdots \leq (r^2)_{n:n}$ are the ordered squared residuals (note that the residuals are first squared and then ordered). Formula (2.16) is very similar to LS, the only difference being that the largest squared residuals are not used in the summation, thereby allowing the fit to stay away from the outliers. In Section 4 of Chapter 4, we will show that the LTS converges at the usual rate and compute its asymptotic efficiency. Like the LMS, this estimator is also equivariant for linear transformations on the x_i and is related to projection pursuit. The best robustness properties are achieved when h is approximately $n/2$, in which case the breakdown point attains 50%. (The exact optimal value of h will be discussed in Section 4 of Chapter 3.)

Both the LMS and the LTS are defined by minimizing a robust measure of the scatter of the residuals. Generalizing this, Rousseeuw and Yohai (1984) introduced so-called *S-estimators*, corresponding to

$$\underset{\hat{\theta}}{\text{Minimize}} \; S(\theta) , \tag{2.17}$$

where $S(\theta)$ is a certain type of robust M-estimate of the scale of the residuals $r_1(\theta), \ldots, r_n(\theta)$. The technical definition of S-estimators will be given in Section 4 of Chapter 3, where it is shown that their breakdown point can also attain 50% by a suitable choice of the constants involved. Moreover, it turns out that S-estimators have essentially the same asymptotic performance as regression M-estimators (see Section 4 of Chapter 3). However, for reasons of simplicity we will concentrate primarily on the LMS and the LTS.

Figure 6 illustrates the robustness of these new regression estimators with respect to an outlier in y or in x. Because of their high breakdown

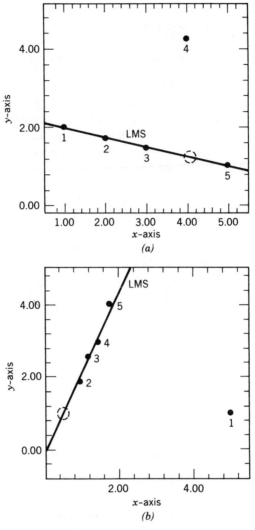

Figure 6. Robustness of LMS regression with respect to (*a*) an outlier in the *y*-direction, and (*b*) an outlier in the *x*-direction.

point, these estimators can even cope with several outliers at the same time (up to about $n/2$ of them, although, of course, this will rarely be needed in practice). This resistance is also independent of p, the number of explanatory variables, so LMS and LTS are reliable data analytic tools that may be used to discover regression outliers in such multivariate situations. The basic principle of LMS and LTS is to fit the *majority* of the data, after which outliers may be identified as those points that lie far away from the robust fit, that is, the cases with large positive or large negative residuals. In Figure 6a, the 4th case possesses a considerable residual, and that of case 1 in Figure 6b is even more apparent.

However, in general the y_i (and hence the residuals) may be in any unit of measurement, so in order to decide if a residual r_i is "large" we need to compare it to an estimate $\hat{\sigma}$ of the error scale. Of course, this scale estimate $\hat{\sigma}$ has to be robust itself, so it depends only on the "good" data and does not get blown up by the outlier(s). For LMS regression, one could use an estimator such as

$$\hat{\sigma} = C_1 \sqrt{\mathop{\mathrm{med}}_i r_i^2} \, , \qquad (2.18)$$

where r_i is the residual of case i with respect to the LMS fit. The constant C_1 is merely a factor used to achieve consistency at Gaussian error distributions. (In Section 1 of Chapter 5, a more refined version of $\hat{\sigma}$ will be discussed, which makes a correction for small samples.) For the LTS, one could use a rule such as

$$\hat{\sigma} = C_2 \sqrt{\frac{1}{n} \sum_{i=1}^{h} (r^2)_{i:n}} \, , \qquad (2.19)$$

where C_2 is another correction factor. In either case, we shall identify case i as an outlier if and only if $|r_i/\hat{\sigma}|$ is large. (Note that this ratio does not depend on the original measurement units!)

This brings us to another idea. In order to improve on the crude LMS and LTS solutions, and in order to obtain standard quantities like t-values, confidence intervals, and the like, we can apply a *weighted least squares* analysis based on the identification of the outliers. For instance, we could make use of the following weights:

$$w_i = \begin{cases} 1 & \text{if } |r_i/\hat{\sigma}| \leq 2.5 \\ 0 & \text{if } |r_i/\hat{\sigma}| > 2.5 \, . \end{cases} \qquad (2.20)$$

This means simply that case i will be retained in the weighted LS if its LMS residual is small to moderate, but disregarded if it is an outlier. The bound 2.5 is, of course, arbitrary, but quite reasonable because in a Gaussian situation there will be very few residuals larger than $2.5\hat{\sigma}$. Instead of 'hard" rejection of outliers as in (2.20), one could also apply "smooth" rejection, for instance by using continuous functions of $|r_i/\hat{\sigma}|$, thereby allowing for a region of doubt (e.g., points with $2 \leq |r_i/\hat{\sigma}| \leq 3$ could be given weights between 1 and 0). Anyway, we then apply weighted least squares defined by

$$\text{Minimize}_{\hat{\theta}} \sum_{i=1}^{n} w_i r_i^2 , \qquad (2.21)$$

which can be computed fast. The resulting estimator still possesses the high breakdown point, but is more efficient in a statistical sense (under Gaussian assumptions) and yields all the standard output one has become accustomed to when working with least squares, such as the coefficient of determination (R^2) and a variance–covariance matrix of the regression coefficients.

The next chapter treats the application of robust methods to simple regression. It explains how to use the program PROGRESS, which yields least squares, least median of squares, and reweighted least squares, together with some related material. Several examples are given with their interpretation, as well as a small comparative study. Chapter 3 deals with multiple regression, illustrating the use of PROGRESS in identifying outliers. Chapter 3 also contains the theoretical robustness results. In Chapter 4 the special case of estimation of one-dimensional location is covered. Chapter 5 discusses several algorithms for high-breakdown regression, and also reports on a small simulation study. Chapter 6 gives a survey of outlier diagnostics, including recent developments. Chapter 7 treats the extension of the LMS, LTS, and S-estimators to related areas such as multivariate location and covariance estimation, time series, and orthogonal regression.

EXERCISES AND PROBLEMS

Section 1

1. Take a real data set with two variables (or just construct one artificially, e.g., by taking 10 points on some line). Calculate the corresponding LS fit by means of a standard program. Now replace the y-value of one of the points by an arbitrary value, and recompute the LS fit. Repeat this experiment by changing an x-value, and try to interpret the residuals.

2. Why is one-dimensional location a special case of regression? Show that the LS criterion, when applied to a one-dimensional sample $y_1, \ldots, y_n$, becomes

$$\text{Minimize}_{\hat{\theta}} \sum_{i=1}^{n} (y_i - \hat{\theta})^2$$

and yields the arithmetic mean $\hat{\theta} = (1/n) \sum_{i=1}^{n} y_i$.

3. Find (or construct) an example of a good leverage point and a bad leverage point. Are these points easy to identify by means of their LS residuals?

Section 2

4. Show that least squares and least absolute deviations are M-estimators.

5. When all y_i are multiplied by a nonzero constant, show that the least squares (1.7) and least absolute deviations (2.6) estimates, as well as the LMS (2.15) and LTS (2.16) estimates, are multiplied by the same factor.

6. Obtain formula (2.8) by differentiating (2.7) with respect to the coefficients $\hat{\theta}_j$, keeping in mind that r_i is given by (1.6).

7. In the special case of simple regression, write down Siegel's repeated median estimates of slope and intercept, making use of (2.14). What does this estimator reduce to in a one-dimensional location setting?

8. The following real data come from a large Belgian insurance company. Table 1 shows the monthly payments in 1979, made as a result of the end of period of life-insurance contracts. (Because of company rules, the payments are given as a percentage of the total amount in that year.) In December a very large sum was paid, mainly because of one extremely high supplementary pension.

Table 1. Monthly Payments in 1979

Month (x)	Payment (y)
1	3.22
2	9.62
3	4.50
4	4.94
5	4.02
6	4.20
7	11.24
8	4.53
9	3.05
10	3.76
11	4.23
12	42.69

Source: Rousseeuw et al. (1984a)

(a) Make a scatterplot of the data. Is the December value an outlier in the x-direction or in the y-direction? Are there any other outliers?

(b) What is the trend of the good points (of the majority of the data)? Fit a robust line by eye. For this line, compute $\sum_{i=1}^{n} r_i^2$, $\sum_{i=1}^{n} |r_i|$, and med$_i$ r_i^2.

(c) Compute the LS line (e.g., by means of a standard statistical program) and plot it in the same figure. Also compute $\sum_{i=1}^{n} r_i^2$, $\sum_{i=1}^{n} |r_i|$, and med$_i$ r_i^2 for this line, and explain why the lines are so different.

(d) Compute both the Pearson product-moment correlation coefficient and the Spearman rank correlation coefficient, and relate them to (b) and (c).

9. Show that the weighted least squares estimate defined by (2.21) can be computed by replacing all $(\mathbf{x}_i, y_i)$ by $(w_i^{1/2}\mathbf{x}_i, w_i^{1/2}y_i)$ and then applying ordinary least squares.

10. (E. Ronchetti) Let $\bar{x}$ be the average of all x_i in the data set $\{(x_1, y_1), \ldots, (x_n, y_n)\}$. Suppose that x_1 is an outlier, which is so far away that all remaining x_i lie on the other side of $\bar{x}$ (as in Figure 5b). Then show that the L_1 regression line goes right through the leverage point (x_1, y_1). (Hint: assume that the L_1 line does not go through (x_1, y_1) and show that $\sum_{i=1}^{n} |r_i|$ will decrease when the line is tilted about $\bar{x}$ to reduce $|r_1|$.)

CHAPTER 2

Simple Regression

1. MOTIVATION

The simple regression model

$$y_i = \theta_1 x_i + \theta_2 + e_i \qquad (i = 1, \ldots, n) \tag{1.1}$$

has been used in Chapter 1 for illustrating some problems that occur when fitting a straight line to a two-dimensional data set. With the aid of some scatterplots, we showed the effect of outliers in the y-direction and of outliers in the x-direction on the ordinary least squares (LS) estimates (see Figures 1 and 2 of Chapter 1). In this chapter we would like to apply high-breakdown regression techniques that can cope with these problems. We treat simple regression separately for didactic reasons, because in this situation it is easy to see the outliers. In Chapter 3, the methods will be generalized to the multiple regression model.

The phrase "simple regression" is also sometimes used for a linear model of the type

$$y_i = \theta_1 x_i + e_i \qquad (i = 1, \ldots, n), \tag{1.2}$$

which does not have a constant term. This model can be used in applications where it is natural to assume that the response should become zero when the explanatory variable takes on the value zero. Graphically, it corresponds to a straight line passing through the origin. Some examples will be given in Section 6.

The following example illustrates the need for a robust regression technique. We have resorted to the so-called *Pilot-Plant data* (Table 1) from Daniel and Wood (1971). The response variable corresponds to the

21

Table 1. Pilot-Plant Data Set

Observation (i)	Extraction (x_i)	Titration (y_i)
1	123	76
2	109	70
3	62	55
4	104	71
5	57	55
6	37	48
7	44	50
8	100	66
9	16	41
10	28	43
11	138	82
12	105	68
13	159	88
14	75	58
15	88	64
16	164	88
17	169	89
18	167	88
19	149	84
20	167	88

Source: Daniel and Wood (1971).

acid content determined by titration, and the explanatory variable is the organic acid content determined by extraction and weighing. Yale and Forsythe (1976) also analyzed this data set.

The scatterplot (Figure 1) suggests a strong statistical relationship between the response and the explanatory variable. The tentative assumption of a linear model such as (1.1) appears to be reasonable.

The LS fit is

$$\hat{y} = 0.322x + 35.458 \qquad \text{(dashed line)}.$$

The least median of squares (LMS) line, defined by formula (2.15) of Chapter 1, corresponds to

$$\hat{y} = 0.311x + 36.519 \qquad \text{(solid line)}.$$

In examining the plot, we see no outliers. As could be expected in such

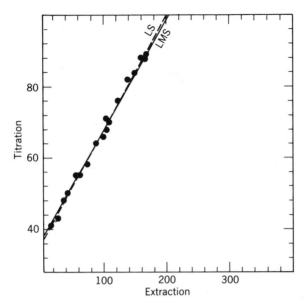

Figure 1. Pilot-Plant data with LS fit (dashed line) and LMS fit (solid line).

a case, only marginal differences exist between the robust estimates and those based on least squares.

Suppose now that one of the observations has been wrongly recorded. For example, the x-value of the 6th observation might have been registered as 370 instead of 37. This error produces an outlier in the x-direction, which is surrounded by a dashed circle in the scatterplot in Figure 2.

What will happen with the regression coefficients for this contaminated sample? The least squares result

$$\hat{y} = 0.081x + 58.939$$

corresponds to the dashed line in Figure 2. It has been attracted very strongly by this single outlier, and therefore fits the other points very badly. On the other hand, the solid line was obtained by applying least median of squares, yielding

$$\hat{y} = 0.314x + 36.343 .$$

This robust method has succeeded in staying away from the outlier, and

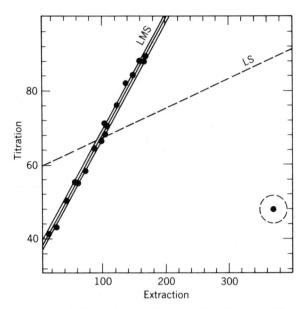

Figure 2. Same data set as in Figure 1, but with one outlier. The dashed line corresponds to the LS fit. The solid LMS line is surrounded by the narrowest strip containing half of the points.

yields a good fit to the majority of the data. Moreover, it lies close to the LS estimate applied to the original uncontaminated data. It would be wrong to say that the robust technique ignores the outlier. On the contrary, the LMS fit exposes the presence of such points.

The LMS solution for simple regression with intercept is given by

$$\underset{\hat{\theta}_1, \hat{\theta}_2}{\text{Minimize}} \ \underset{i}{\text{med}} \ (y_i - \hat{\theta}_1 x_i - \hat{\theta}_2)^2 \ . \tag{1.3}$$

Geometrically, it corresponds to *finding the narrowest strip covering half of the observations.* (To be precise, by "half" we mean $[n/2] + 1$, where $[n/2]$ denotes the integer part of $n/2$. Moreover, the thickness of this strip is measured in the vertical direction.) The LMS line lies exactly at the middle of this band. (We will prove this fact in Theorem 1 of Chapter 4, Section 2.) Note that this notion is actually much easier to explain to most people than the classical LS definition. For the contaminated Pilot-Plant data, this strip is drawn in Figure 2.

The outlier in this example was artificial. However, it is important to realize that this kind of mistake appears frequently in real data. Outlying data points can be present in a sample because of errors in recording observations, errors in transcription or transmission, or an exceptional occurrence in the investigated phenomenon. In the two-dimensional case (such as the example above), it is rather easy to detect atypical points just by plotting the observations. This visual tracing is no longer possible for higher dimensions. So in practice, one needs a procedure that is able to lessen the impact of outliers, thereby exposing them in the residual plots (examples of this are given in Section 3). In addition, when no outliers occur, the result of the alternative procedure should hardly differ from the LS solution. It turns out that LMS regression does meet these requirements.

Let us now look at some real data examples with outliers. In the Belgian Statistical Survey (published by the Ministry of Economy), we found a data set containing the total number (in tens of millions) of international phone calls made. These data are listed in Table 2 and plotted in Figure 3.

The plot seems to show an upward trend over the years. However, this time series contains heavy contamination from 1964 to 1969. Upon inquiring, it turned out that from 1964 to 1969, another recording system

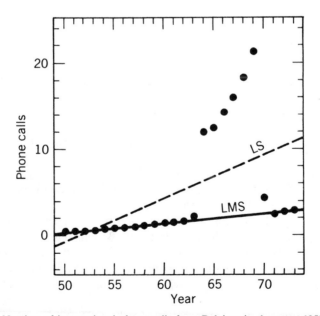

Figure 3. Number of international phone calls from Belgium in the years 1950–1973 with the LS (dashed line) and LMS fit (solid line).

Table 2. Number of International Calls from Belgium

Year (x_i)	Number of Calls[a] (y_i)
50	0.44
51	0.47
52	0.47
53	0.59
54	0.66
55	0.73
56	0.81
57	0.88
58	1.06
59	1.20
60	1.35
61	1.49
62	1.61
63	2.12
64	11.90
65	12.40
66	14.20
67	15.90
68	18.20
69	21.20
70	4.30
71	2.40
72	2.70
73	2.90

[a]In tens of millions.

was used, giving the total number of *minutes* of these calls. (The years 1963 and 1970 are also partially affected because the transitions did not happen exactly on New Year's Day, so the number of calls of some months were added to the number of minutes registered in the remaining months!) This caused a large fraction of outliers in the y-direction.

The ordinary LS solution for these data is given by $\hat{y} = 0.504x - 26.01$ and corresponds to the dashed line in Figure 3. This dashed line has been affected very much by the y values associated with the years 1964–1969. As a consequence, the LS line has a large slope and does not fit the good or the bad data points. This is what one would obtain by not looking critically at these data and by applying the LS method in a routine way. In fact, some of the good observations (such as the 1972 one) yield even larger LS residuals than some of the bad values! Now let us apply the

LMS method. This yields $\hat{y} = 0.115x - 5.610$ (plotted as a solid line in Figure 3), which avoids the outliers. It corresponds to the pattern one sees emerging when simply looking at the plotted data points. Clearly, this line fits the majority of the data. (This is not meant to imply that a linear fit is necessarily the best model, because collecting more data might reveal a more complicated kind of relationship.)

Another example comes from astronomy. The data in Table 3 form the Hertzsprung–Russell diagram of the star cluster CYG OB1, which contains 47 stars in the direction of Cygnus. Here x is the logarithm of the effective temperature at the surface of the star (T_e), and y is the logarithm of its light intensity (L/L_0). These numbers were given to us by C. Doom (personal communication), who extracted the raw data from Humphreys (1978) and performed the calibration according to Vansina and De Greve (1982).

Table 3. Data for the Hertzsprung–Russell Diagram of the Star Cluster CYG OB1

Index of Star (i)	$\log T_e$ (x_i)	$\log [L/L_0]$ (y_i)	Index of Star (i)	$\log T_e$ (x_i)	$\log [L/L_0]$ (y_i)
1	4.37	5.23	25	4.38	5.02
2	4.56	5.74	26	4.42	4.66
3	4.26	4.93	27	4.29	4.66
4	4.56	5.74	28	4.38	4.90
5	4.30	5.19	29	4.22	4.39
6	4.46	5.46	30	3.48	6.05
7	3.84	4.65	31	4.38	4.42
8	4.57	5.27	32	4.56	5.10
9	4.26	5.57	33	4.45	5.22
10	4.37	5.12	34	3.49	6.29
11	3.49	5.73	35	4.23	4.34
12	4.43	5.45	36	4.62	5.62
13	4.48	5.42	37	4.53	5.10
14	4.01	4.05	38	4.45	5.22
15	4.29	4.26	39	4.53	5.18
16	4.42	4.58	40	4.43	5.57
17	4.23	3.94	41	4.38	4.62
18	4.42	4.18	42	4.45	5.06
19	4.23	4.18	43	4.50	5.34
20	3.49	5.89	44	4.45	5.34
21	4.29	4.38	45	4.55	5.54
22	4.29	4.22	46	4.45	4.98
23	4.42	4.42	47	4.42	4.50
24	4.49	4.85			

The Hertzsprung–Russell diagram itself is shown in Figure 4. It is the scatterplot of these points, where the log temperature x is plotted from right to left. In the plot, one sees two groups of points: the majority, which seems to follow a steep band, and the four stars in the upper right corner. These parts of the diagram are well known in astronomy: The 43 stars are said to lie on the main sequence, whereas the four remaining stars are called giants. (The giants are the points with indices 11, 20, 30, and 34.)

Application of our LMS estimator to these data yields the solid line $\hat{y} = 3.898x - 12.298$, which fits the main sequence nicely. On the other hand, the LS solution $\hat{y} = -0.409x + 6.78$ corresponds to the dashed line in Figure 4, which has been pulled away by the four giant stars (which it does not fit well either). These outliers are leverage points, but they are not errors: It would be more appropriate to say that the data come from two different populations. These two groups can easily be distinguished on the basis of the LMS residuals (the large residuals correspond to the giant stars), whereas the LS residuals are rather homogeneous and do not allow us to separate the giants from the main-sequence stars.

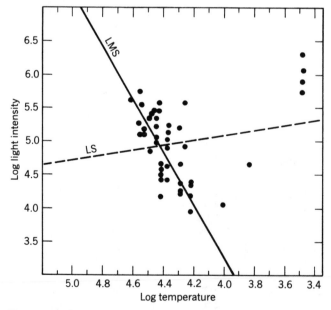

Figure 4. Hertzsprung–Russell diagram of the star cluster CYG OB1 with the LS (dashed line) and LMS fit (solid line).

2. COMPUTATION OF THE LEAST MEDIAN OF SQUARES LINE

The present section describes the use of PROGRESS, a program implementing LMS regression. (Its name comes from Program for RObust reGRESSion.) The algorithm itself is explained in detail in Chapter 5. Without the aid of a computer, it would never have been possible to calculate high-breakdown regression estimates. Indeed, one does not have an explicit formula, such as the one used for LS. It appears there are deep reasons why high-breakdown regression cannot be computed cheaply. [We are led to this assertion by means of partial results from our own research and because of some arguments provided by Donoho (1984) and Steele and Steiger (1986).] Fortunately, the present evolution of computers has made robust regression quite feasible.

PROGRESS is designed to run on an IBM-PC or a compatible microcomputer. At least 256K RAM must be available. The boundaries of the arrays in the program allow regression analysis with at most 300 cases and 10 coefficients. PROGRESS starts by asking the data specifications and the options for treatment and output. This happens in a fully interactive way, which makes it very easy to use the program. The user only has to answer the questions appearing on the screen. No knowledge of informatics or computer techniques is required. Nevertheless, we will devote this section to the input. [The mainframe version described in Leroy and Rousseeuw (1984) was written in very portable FORTRAN, so it was not yet interactive.] We will treat the Pilot-Plant example (with outlier) of the preceding section. The words typed by the user are printed in italics to distinguish them from the words or lines coming from PROGRESS.

The first thing to do, of course, is to insert the diskette containing the program. In order to run PROGRESS, the user only has to type A:PROGRESS in case the program is on drive A. (Other possibilities would be drive B or hard disk C.) Then the user has to press the ENTER key. Having done this, the program generates the following screen:

```
* * * * * * * * * * * * * * * *
*   P R O G R E S S   *
* * * * * * * * * * * * * * * *
```

ENTER THE NUMBER OF CASES PLEASE: *20*

The user now has to enter the number of cases he or she wants to handle in the analysis. Note that there are limits to the size of the data sets that can be treated. (This restriction is because of central memory limitations of the computer.) Therefore, PROGRESS gives a warning when the number of cases entered by the user is greater than this limit. When PROGRESS has accepted the number of cases, the following question appears:

```
DO YOU WANT A CONSTANT TERM IN THE REGRESSION?
PLEASE ANSWER YES OR NO: YES
```

When the user answers YES to this question, PROGRESS performs a regression with a constant term. Otherwise, the program yields a regression through the origin. The general models for regression with and without a constant are, respectively,

$$y_i = x_{i1}\theta_1 + \cdots + x_{i,p-1}\theta_{p-1} + \theta_p + e_i \qquad (i = 1, \ldots, n) \qquad (2.1)$$

and

$$y_i = x_{i1}\theta_1 + \cdots + x_{i,p-1}\theta_{p-1} + x_{ip}\theta_p + e_i \qquad (i = 1, \ldots, n). \qquad (2.2)$$

In (2.2) the estimate of y_i is equal to zero when all x_{ij} ($j = 1, \ldots, p$) are zero. [Note that (2.1) is a special case of (2.2), obtained by putting the last explanatory variable x_{ip} equal to 1 for all cases.]

It may happen that the user has a large data set, consisting of many more variables than those he or she wishes to insert in a regression model. PROGRESS allows the user to select some variables out of the entire set. Furthermore, for each variable in the regression, PROGRESS asks for a label in order to facilitate the interpretation of the output. Therefore the user has to answer the following questions:

```
WHAT IS THE TOTAL NUMBER OF VARIABLES IN YOUR DATA SET?
------------------------------------------------
PLEASE GIVE A NUMBER BETWEEN 1 AND 50: 5

WHICH VARIABLE DO YOU CHOOSE AS RESPONSE VARIABLE?
-------------------------------------------
OUT OF THESE 5 GIVE ITS POSITION: 4

GIVE A LABEL FOR THIS VARIABLE (AT MOST 10 CHARACTERS): TITRATION

HOW MANY EXPLANATORY VARIABLES DO YOU WANT TO USE IN THE ANALYSIS?
-----------------------------------------------------------
(AT MOST 4): 1
```

The answer to each question is verified by PROGRESS. This means that a message is given when an answer is not allowed. For example, when the user answers 12 to the question

```
WHICH VARIABLE DO YOU CHOOSE AS RESPONSE VARIABLE?
----------------------------------------
OUT OF THESE 5 GIVE ITS POSITION: 12
```

the following prompt will appear:

```
NOT ALLOWED! ENTER YOUR CHOICE AGAIN: 4
```

Also, the program checks whether the number of cases is more than twice the number of regression coefficients (including the constant term if there is one). If there are fewer cases, the program stops.

The question

```
HOW MANY EXPLANATORY VARIABLES DO YOU WANT TO USE IN THE ANALYSIS?
------------------------------------------------------------
(AT MOST 4):
```

may be answered with 0. In that situation the response variable is analyzed in a one-dimensional way, yielding robust estimates of its location and scale. (More details on this can be found in Chapter 4.)

When the number of explanatory variables is equal to the number of remaining variables (this means, all but the response variable) in the data set, the user has to fill up a table containing one line for each explanatory variable. Each of these variables is identified by means of a label of at most 10 characters. These characters have to be typed below the arrows. On the other hand, when the number of explanatory variables is less, the user also has to give the position of the selected variable in the data set together with the corresponding label. For our example, this table would be

```
EXPLANATORY VARIABLES     :   POSITION  LABEL (AT MOST 10 CHARACTERS)
---------------------------↓↓↓↓-- ↓↓↓↓↓↓↓↓↓↓ -------------
NUMBER 1                  :    2          EXTRACTION
```

An option concerning the amount of output can be chosen in the following question:

```
HOW MUCH OUTPUT DO YOU WANT?
----------------------
0=SMALL OUTPUT          :  LIMITED TO BASIC RESULTS
1=MEDIUM-SIZED OUTPUT:  ALSO INCLUDES A TABLE WITH THE OBSERVED VALUES OF Y,
                           THE ESTIMATES OF Y, THE RESIDUALS AND THE WEIGHTS
2=LARGE OUTPUT          :  ALSO INCLUDES THE DATA ITSELF
ENTER YOUR CHOICE: 2
```

If the user types 0, the output is limited to the basic results, namely the LS, the LMS, and the reweighted least squares (RLS) estimates, with their standard deviations (in order to construct confidence intervals around the estimated regression coefficients) and t-values. The scale estimates are also given. In the case of regression with one explanatory variable, a plot of y versus x is produced. This permits us to detect a pattern in the data.

Setting the print option at 1 yields more information: a table with the observed values of y, the estimated values of y, the residuals, and the residuals divided by the scale estimate (which are called *standardized residuals*); and for reweighted least squares, an additional column with the weight (resulting from LMS) of each observation. Apart from the output produced with print option 1, print option 2 also lists the data itself.

A careful analysis of residuals is an important part of applied regression. Therefore we have added a plot option that permits us to obtain a plot of the standardized residuals versus the estimated value of y (this is performed when the plot option is set at 1) or a plot of the standardized residuals versus the index i of the observation (which is executed when the plot option is set at 2). If the plot option is set at 3, both types of plots are given. If the plot option is set at 0, the output contains no residual plots. The plot option is selected by means of the following question:

```
DO YOU WANT TO LOOK AT THE RESIDUALS?
-----------------------------
0=NO RESIDUAL PLOTS
1=PLOT OF THE STANDARDIZED RESIDUALS VERSUS THE ESTIMATED VALUE OF Y
2=PLOT OF THE STANDARDIZED RESIDUALS VERSUS THE INDEX OF THE OBSERVATION
3=PERFORMS BOTH TYPES OF RESIDUAL PLOTS
ENTER YOUR CHOICE: 0
```

When the following question is answered with YES, the program yields some outlier diagnostics, which will be described in Chapter 6.

```
DO YOU WANT TO COMPUTE OUTLIER DIAGNOSTICS?
PLEASE ANSWER YES OR NO: NO
```

When the data set has already been stored in a file, the user only has to give the name of that file in response to the following question. If such a file does not already exist, the user still has the option of entering his or her data by keyboard in an interactive way during a PROGRESS session. In that case the user has to answer KEY. The entered data set has to contain as many variables as mentioned in the third question of the interactive input. The program then picks out the response and the explanatory variables for the analysis.

```
GIVE THE NAME OF THE FILE CONTAINING THE DATA (e.g. TYPE A:EXAMPLE.DAT),
or TYPE KEY IF YOU PREFER TO ENTER THE DATA BY KEYBOARD.
WHAT DO YOU CHOOSE? KEY
```

Moreover, PROGRESS enables the user to store the data (in case KEY has been answered) by means of the following dialogue:

```
DO YOU WANT TO SAVE YOUR DATA IN A FILE?
PLEASE ANSWER YES OR NO: YES

IN WHICH FILE DO YOU WANT TO SAVE YOUR DATA?
(WARNING: IF THERE ALREADY EXISTS A FILE WITH THE SAME NAME,
         THEN THE OLD FILE WILL BE OVERWRITTEN.)
TYPE e.g. B:SAVE.DAT : B:PILOT.DAT
```

The whole data set will be stored, even those variables that are not used right now. This enables the user to perform another analysis afterwards, with a different combination of variables.

Depending on the answer to the following question, the output provided by PROGRESS will be written on the screen, on paper, or in a file.

```
WHERE DO YOU WANT YOUR OUTPUT?
----------------------
    TYPE CON IF YOU WANT IT ON THE SCREEN
or TYPE PRN IF YOU WANT IT ON THE PRINTER
or TYPE THE NAME OF A FILE (e.g. B:EXAMPLE.OUT)
(WARNING: IF THERE ALREADY EXISTS A FILE WITH THE SAME NAME,
         THEN THE OLD FILE WILL BE OVERWRITTEN.)
WHAT DO YOU CHOOSE? PRN
```

We would like to give the user a warning concerning the latter two questions. The name of a DOS file is unique. This means that if the user enters a name of a file that already exists on the diskette, the old file will be overwritten by the new file.

The plots constructed by PROGRESS are intended for a printer using 8 lines per inch. (Consequently, on the screen these plots are slightly stretched out.) It is therefore recommended to adapt the printer to 8 lines per inch. For instance, this can be achieved by typing the DOS command "MODE LPT1:80,8" before running PROGRESS.

Next, PROGRESS requests a title, which will be reproduced on the output. This title should consist of at most 60 characters. When the user enters more characters, only the first 60 will be read:

```
PLEASE ENTER A TITLE FOR THE OUTPUT (AT MOST 60 CHARACTERS):
------------------------------------------------
PILOT-PLANT DATA SET WITH ONE LEVERAGE POINT
```

The answer to the following question tells PROGRESS the way in which the data has to be read. Two possibilities are available. The first consists of reading the data in free format, which is performed by answering YES to:

```
DO YOU WANT TO READ THE DATA IN FREE FORMAT?
----------------------------------
THIS MEANS THAT YOU ONLY HAVE TO INSERT BLANK(S) BETWEEN NUMBERS.
(WE ADVISE USERS WITHOUT KNOWLEDGE OF FORTRAN FORMATS TO ANSWER YES.)
MAKE YOUR CHOICE (YES/NO): YES
```

In order to use the free format, it suffices that the variables for each case be separated by at least one blank. On the other hand, when the user answers NO to the above question, PROGRESS requests the FORTRAN format to be used to input the data. The program expects the format necessary for reading the total number of variables of the data set (in this case 5). The program will then select the variables for actual use by means of the positions chosen above. The FORTRAN format has to be set between brackets, and it should be described in at most 60 characters (including the brackets). The observations are to be processed as real numbers, so they should be read in F-formats and/or E-formats. The formats X and / are also allowed.

Because the execution time for large data sets may be quite long, the user has the option of choosing a faster version of the algorithm in that case. In other cases it is recommended to use the extensive search version because of its greater precision. (More details about the algorithm will be provided in Chapter 5.)

```
WHICH VERSION OF THE ALGORITHM WOULD YOU LIKE TO USE?
----------------------------------------
Q=QUICK VERSION
E=EXTENSIVE SEARCH
ENTER YOUR CHOICE PLEASE (Q OR E): E
```

PROGRESS also allows the user to deal with missing values. However, we shall postpone the discussion of these options until Section 2 of Chapter 3.

```
CHOOSE AN OPTION FOR THE TREATMENT OF MISSING VALUES
----------------------------------------
0=THERE ARE NO MISSING VALUES IN THE DATA
1=ELIMINATION OF THE CASES FOR WHICH AT LEAST ONE VARIABLE IS MISSING
2=ESTIMATES ARE FILLED IN FOR UNOBSERVED VALUES
ENTER YOUR CHOICE: 0
```

Finally, PROGRESS gives a survey of the options that were selected.

```
* * * * * * * * * * * * * * * * * * * * * * * * * * * * * * * * * * * * * * * * * *
*   P R O G R E S S  WILL PERFORM A REGRESSION WITH CONSTANT TERM   *
* * * * * * * * * * * * * * * * * * * * * * * * * * * * * * * * * * * * * * * * * *

THE NUMBER OF CASES EQUALS                        20
THE NUMBER OF EXPLANATORY VARIABLES EQUALS       1
TITRATION IS THE RESPONSE VARIABLE.
THE DATA WILL BE READ FROM THE KEYBOARD.
THE DATA WILL BE SAVED IN FILE: B:PILOT.DAT
TITLE FOR OUTPUT: PILOT-PLANT DATA SET WITH ONE LEVERAGE POINT
THE DATA WILL BE READ IN FREE FORMAT.
LARGE OUTPUT IS WANTED.
NO RESIDUAL PLOTS ARE WANTED.
THE EXTENSIVE SEARCH VERSION WILL BE USED.
THERE ARE NO MISSING VALUES.
YOUR OUTPUT WILL BE WRITTEN ON: PRN

ARE ALL THESE OPTIONS OK? YES OR NO: YES
```

When the data have to be read from the keyboard, the user has to type the measurements for each case. For the example we are working with, this would look as follows:

```
ENTER YOUR DATA FOR EACH CASE.

THE DATA FOR CASE NUMBER    1:   1 123 0 76 28
THE DATA FOR CASE NUMBER    2:   2 109 0 70 23
THE DATA FOR CASE NUMBER    3:   3 62 1 55 29
THE DATA FOR CASE NUMBER    4:   4 104 1 71 28
THE DATA FOR CASE NUMBER    5:   5 57 0 55 27
THE DATA FOR CASE NUMBER    6:   6 370 0 48 35
THE DATA FOR CASE NUMBER    7:   7 44 1 50 24
THE DATA FOR CASE NUMBER    8:   8 100 1 66 23
THE DATA FOR CASE NUMBER    9:   9 16 0 41 27
THE DATA FOR CASE NUMBER   10:   10 28 1 43 29
THE DATA FOR CASE NUMBER   11:   11 138 0 82 21
THE DATA FOR CASE NUMBER   12:   12 105 0 68 28
THE DATA FOR CASE NUMBER   13:   13 159 1 88 24
THE DATA FOR CASE NUMBER   14:   14 75 1 58 26
THE DATA FOR CASE NUMBER   15:   15 88 0 64 26
THE DATA FOR CASE NUMBER   16:   16 164 0 88 26
THE DATA FOR CASE NUMBER   17:   17 169 1 89 23
THE DATA FOR CASE NUMBER   18:   18 167 1 88 36
THE DATA FOR CASE NUMBER   19:   19 149 0 84 24
THE DATA FOR CASE NUMBER   20:   20 167 1 88 21
```

Out of these five variables, we only use two. (In fact, in this example the other three variables are artificial, because they were added to illustrate the program.)

The output file corresponding to this example is printed below, and is followed by a discussion of the results in Section 3.

```
***********************************************
* ROBUST REGRESSION WITH A CONSTANT TERM.  *
***********************************************

NUMBER OF CASES =      20
NUMBER OF COEFFICIENTS (INCLUDING CONSTANT TERM) =      2

THE EXTENSIVE SEARCH VERSION WILL BE USED.
DATA SET =          PILOT-PLANT DATA SET WITH ONE LEVERAGE POINT

THERE ARE NO MISSING VALUES.

LARGE OUTPUT IS WANTED.

THE OBSERVATIONS ARE:

          EXTRACTION    TITRATION
     1      123.0000      76.0000
     2      109.0000      70.0000
     3       62.0000      55.0000
     4      104.0000      71.0000
     5       57.0000      55.0000
     6      370.0000      48.0000
     7       44.0000      50.0000
     8      100.0000      66.0000
     9       16.0000      41.0000
    10       28.0000      43.0000
    11      138.0000      82.0000
    12      105.0000      68.0000
    13      159.0000      88.0000
    14       75.0000      58.0000
    15       88.0000      64.0000
    16      164.0000      88.0000
    17      169.0000      89.0000
    18      167.0000      88.0000
    19      149.0000      84.0000
    20      167.0000      88.0000
```

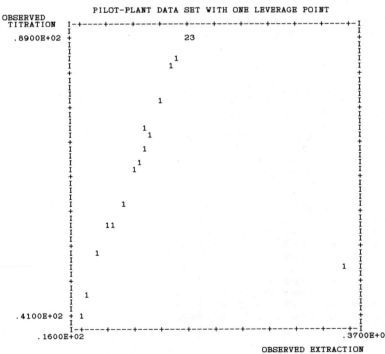

PILOT-PLANT DATA SET WITH ONE LEVERAGE POINT

```
MEDIANS  =
         EXTRACTION   TITRATION
         107.0000     69.0000

DISPERSIONS =
         EXTRACTION   TITRATION
         70.4235      21.4977

THE STANDARDIZED OBSERVATIONS ARE:

         EXTRACTION   TITRATION
    1       .2272       .3256
    2       .0284       .0465
    3      -.6390      -.6512
    4      -.0426       .0930
    5      -.7100      -.6512
    6      3.7345      -.9768
    7      -.8946      -.8838
    8      -.0994      -.1395
    9     -1.2922     -1.3025
   10     -1.1218     -1.2094
   11       .4402       .6047
   12      -.0284      -.0465
   13       .7384       .8838
   14      -.4544      -.5117
   15      -.2698      -.2326
   16       .8094       .8838
   17       .8804       .9303
   18       .8520       .8838
   19       .5964       .6977
   20       .8520       .8838
```

PEARSON CORRELATION COEFFICIENTS BETWEEN THE VARIABLES
(TITRATION IS THE RESPONSE VARIABLE)

```
EXTRACTION   1.00
TITRATION     .38  1.00
```

SPEARMAN RANK CORRELATION COEFFICIENTS BETWEEN THE VARIABLES
(TITRATION IS THE RESPONSE VARIABLE)

```
EXTRACTION   1.00
TITRATION     .76  1.00
```

****

LEAST SQUARES REGRESSION
****

VARIABLE	COEFFICIENT	STAND. ERROR	T - VALUE	P - VALUE
EXTRACTION	.08071	.04695	1.71914	.10274
CONSTANT	58.93883	6.61420	8.91096	.00000

```
SUM OF SQUARES        =     4379.69300

DEGREES OF FREEDOM    =     18

SCALE ESTIMATE        =     15.59860
```

VARIANCE - COVARIANCE MATRIX =

```
    .2204D-02
   -.2638D+00    .4375D+02
```

COEFFICIENT OF DETERMINATION (R SQUARED) =   .14103

THE F-VALUE =   2.955 (WITH   1 AND   18 DF)   P - VALUE = .10274

OBSERVED TITRATION	ESTIMATED TITRATION	RESIDUAL	NO	RES/SC
76.00000	68.86635	7.13365	1	.46
70.00000	67.73638	2.26362	2	.15
55.00000	63.94294	-8.94294	3	-.57
71.00000	67.33282	3.66718	4	.24
55.00000	63.53938	-8.53938	5	-.55
48.00000	88.80209	-40.80209	6	-2.62
50.00000	62.49014	-12.49014	7	-.80
66.00000	67.00998	-1.00998	8	-.06
41.00000	60.23021	-19.23021	9	-1.23
43.00000	61.19875	-18.19875	10	-1.17
82.00000	70.07702	11.92298	11	.76
68.00000	67.41354	.58646	12	.04
88.00000	71.77196	16.22804	13	1.04
58.00000	64.99220	-6.99220	14	-.45
64.00000	66.04144	-2.04144	15	-.13
88.00000	72.17552	15.82448	16	1.01
89.00000	72.57908	16.42092	17	1.05
88.00000	72.41766	15.58234	18	1.00
84.00000	70.96484	13.03516	19	.84
88.00000	72.41766	15.58234	20	1.00

****
```
```

LEAST MEDIAN OF SQUARES REGRESSION
****

VARIABLE	COEFFICIENT
EXTRACTION	.31429
CONSTANT	36.34286

FINAL SCALE ESTIMATE   =   1.33279

COEFFICIENT OF DETERMINATION =   .99641

OBSERVED TITRATION	ESTIMATED TITRATION	RESIDUAL	NO	RES/SC
76.00000	75.00000	1.00000	1	.75
70.00000	70.60000	-.60000	2	-.45
55.00000	55.82857	-.82857	3	-.62
71.00000	69.02856	1.97144	4	1.48
55.00000	54.25714	.74286	5	.56
48.00000	152.62860	-104.62860	6	-78.50
50.00000	50.17143	-.17143	7	-.13
66.00000	67.77142	-1.77142	8	-1.33
41.00000	41.37143	-.37143	9	-.28
43.00000	45.14286	-2.14286	10	-1.61
82.00000	79.71428	2.28572	11	1.71
68.00000	69.34285	-1.34285	12	-1.01
88.00000	86.31429	1.68571	13	1.26
58.00000	59.91428	-1.91428	14	-1.44
64.00000	64.00000	.00000	15	.00
88.00000	87.88571	.11429	16	.09
89.00000	89.45714	-.45714	17	-.34
88.00000	88.82857	-.82857	18	-.62
84.00000	83.17142	.82858	19	.62
88.00000	88.82857	-.82857	20	-.62

****

REWEIGHTED LEAST SQUARES BASED ON THE LMS
****

VARIABLE	COEFFICIENT	STAND. ERROR	T - VALUE	P - VALUE
EXTRACTION	.32261	.00595	54.21467	.00000
CONSTANT	35.31744	.69617	50.73091	.00000

WEIGHTED SUM OF SQUARES =   26.75224

DEGREES OF FREEDOM   =   17

SCALE ESTIMATE   =   1.25446

VARIANCE - COVARIANCE MATRIX =

   .3541D-04
  -.3772D-02   .4847D+00

COEFFICIENT OF DETERMINATION (R SQUARED) =   .99425

THE F-VALUE =   2939.231 (WITH   1 AND   17 DF)   P - VALUE =   .00000

THERE ARE   19 POINTS WITH NON-ZERO WEIGHT.

AVERAGE WEIGHT   =   .95000

OBSERVED TITRATION	ESTIMATED TITRATION	RESIDUAL	NO	RES/SC	WEIGHT
76.00000	74.99884	1.00116	1	.80	1.0
70.00000	70.48225	-.48225	2	-.38	1.0
55.00000	55.31945	-.31945	3	-.25	1.0
71.00000	68.86919	2.13081	4	1.70	1.0
55.00000	53.70638	1.29362	5	1.03	1.0
48.00000	154.68420	-106.68420	6	-85.04	.0
50.00000	49.51241	.48759	7	.39	1.0
66.00000	67.57874	-1.57874	8	-1.26	1.0
41.00000	40.47925	.52075	9	.42	1.0
43.00000	44.35061	-1.35061	10	-1.08	1.0
82.00000	79.83803	2.16197	11	1.72	1.0
68.00000	69.19180	-1.19180	12	-.95	1.0
88.00000	86.61290	1.38710	13	1.11	1.0
58.00000	59.51341	-1.51341	14	-1.21	1.0
64.00000	63.70738	.29262	15	.23	1.0
88.00000	88.22597	-.22597	16	-.18	1.0
89.00000	89.83904	-.83904	17	-.67	1.0
88.00000	89.19380	-1.19380	18	-.95	1.0
84.00000	83.38677	.61323	19	.49	1.0
88.00000	89.19380	-1.19380	20	-.95	1.0

3. INTERPRETATION OF THE RESULTS

The output provided by PROGRESS starts with some general information about the data set. In the above example, the data were two-dimensional so they could be plotted. A point in the scattergram is represented by a digit. This digit corresponds to the number of points having approximately the same coordinates. When more than nine points coincide, an asterisk (∗) is printed in that position. In simple regression, such a plot reveals immediately which points may exert a strong influence on the LS estimates.

The program then prints the median m_j of each variable j. In the Pilot-Plant output displayed above, the median extraction value is 107 and the median titration value equals 69. On the next line, the dispersion s_j of each variable is given, which can be considered as a robust version of its standard deviation (the exact definition will be given in Section 1 of Chapter 4). When large output has been requested, the program then provides a list of the standardized observations, in which each measurement x_{ij} is replaced by

$$\frac{x_{ij} - m_j}{s_j} \tag{3.1}$$

(the response variable is standardized in the same way). The columns of the resulting table each have a median of 0 and a dispersion of 1. This enables us to identify outliers in any single variable. Indeed, if the absolute value of a standardized observation is large (say, larger than 2.5) then it is an outlier for that particular variable. In the Pilot-Plant output we discover in this way that case 6 has an unusually large extraction measurement, because its standardized value is 3.73. (Sometimes the standardized values can tell us that the distribution in a column is skewed, thereby suggesting data transformation.) The standardization is performed differently when there is no constant term in the regression, as we shall see in Section 6.

Next, PROGRESS provides the Pearson (product-moment) correlation coefficients between the variables, as well as the nonparametric Spearman correlation coefficients.

Before giving the robust estimates, PROGRESS starts with the classical LS results. A table is printed with the estimated coefficients, along with their standard error. The standard error of the jth LS regression coefficient is the square root of the jth diagonal element of the variance–covariance matrix of the LS regression coefficients, given by $\sigma^2(\mathbf{X}'\mathbf{X})^{-1}$, where the matrix $\mathbf{X}$ is given by

$$\mathbf{X} = \begin{bmatrix} x_{11} & \cdots & x_{1p} \\ \vdots & & \vdots \\ x_{i1} & \cdots & x_{ip} \\ \vdots & & \vdots \\ x_{n1} & \cdots & x_{np} \end{bmatrix}.$$

The ith row of $\mathbf{X}$ equals the vector $\mathbf{x}_i$ consisting of the p explanatory variables of the ith observation. The unknown σ^2 is estimated by

$$s^2 = \frac{1}{n-p} \sum_{i=1}^{n} r_i^2, \tag{3.2}$$

where $r_i = y_i - \hat{y}_i$ is the ith residual. The estimated variance–covariance matrix $s^2(\mathbf{X}'\mathbf{X})^{-1}$ is also contained in the output. (Because this matrix is symmetric, only its lower triangular part is given, including the main diagonal.) For simple regression, $p = 1$ when there is no intercept term, and $p = 2$ otherwise. In the output, the quantity indicated by SCALE ESTIMATE is $s = \sqrt{s^2}$.

To construct confidence intervals for the parameters θ_j, one has to assume that the errors e_i are independently normally distributed with mean zero and variance σ^2. Under these conditions, it is well known that each of the quantities

$$\frac{\hat{\theta}_j - \theta_j}{\sqrt{s^2((\mathbf{X}'\mathbf{X})^{-1})_{jj}}}, \qquad j = 1, \ldots, p \tag{3.3}$$

has a Student distribution with $n - p$ degrees of freedom. Let us denote the $1 - \alpha/2$ quantile of this distribution by $t_{n-p,1-\alpha/2}$. Then a $(1 - \alpha) \times 100\%$ confidence interval for θ_j is given by

$$[\hat{\theta}_j - t_{n-p,1-\alpha/2}\sqrt{s^2((\mathbf{X}'\mathbf{X})^{-1})_{jj}}, \ \hat{\theta}_j + t_{n-p,1-\alpha/2}\sqrt{s^2((\mathbf{X}'\mathbf{X})^{-1})_{jj}}] \tag{3.4}$$

for each $j = 1, \ldots, p$. The same result can also be used for testing the significance of any regression coefficient, such as $\hat{\theta}_j$. The hypotheses are

$$\begin{array}{lll} H_0: & \theta_j = 0 & \text{(null hypothesis)} \\ H_1: & \theta_j \neq 0 & \text{(alternative hypothesis)}. \end{array} \tag{3.5}$$

Such a test may be helpful in determining if the jth variable might be deleted from the model. If the null hypothesis in (3.5) is accepted (for a certain value of α), then this indicates that the jth explanatory variable does not contribute much to the explanation of the response variable. The test statistic for this hypothesis is

$$\frac{\hat{\theta}_j}{\sqrt{s^2((\mathbf{X}^t\mathbf{X})^{-1})_{jj}}}, \qquad j = 1, \ldots, p. \tag{3.6}$$

This ratio corresponds to "T-VALUE" in the output of PROGRESS. The null hypothesis will be rejected at the level α when the absolute value of "T-VALUE" is larger than $t_{n-p,1-\alpha/2}$ (in this case we will say that the jth regression coefficient is significantly different from zero).

In the present example, $n = 20$ and $p = 2$. The 97.5% quantile (since we use $\alpha = 5\%$) of the Student distribution with $n - p = 18$ degrees of freedom equals 2.101. Therefore, one can conclude that the LS slope is not significantly different from zero, because its associated t-value equals 1.719. On the other hand, the intercept t-value equals 8.911, which is quite significant.

Next to each t-value, PROGRESS also prints the corresponding (two-sided) "p-value" or "significance level." This is the probability that a Student-distributed random variable with $n - p$ degrees of freedom becomes larger in absolute value than the t-value that was actually obtained. In order to compute this probability, we used the exact formulas of Lackritz (1984) as they were implemented by van Soest and van Zomeren (1986, personal communication). When the p-value is smaller than 0.05, then the corresponding regression coefficient is significant at the 5% level. In the above output, the p-value of the LS slope equals 0.10274 so it is not significant, whereas the p-value of the LS intercept is given as 0.00000, which makes it significant (even at the 0.1% level). The printed p-values make hypothesis testing very easy, because probability tables are no longer necessary.

However, (3.6) has to be used with caution because it is not robust at all. The distribution theory of this statistic only holds when the errors really follow a Gaussian distribution, which is rarely fulfilled in practice. It is therefore advisable to look first at the robust fit in order to be aware of the possible presence of outliers. In Section 4, we will give some guidelines based on the residuals resulting from the robust fit for identifying the harmful observations. When it appears that the data set contains influential points, then one has to resort to the t-values of the RLS solution, which will be presented below.

In order to obtain an idea of the strength of the linear relationship between the response variable and the explanatory variable(s), the coefficient of determination (R^2) is displayed in the output. R^2 measures the proportion of total variability explained by the regression. For the exact formula one has to distinguish between regression with and without a constant term. For LS regression, R^2 can be calculated as

$$R^2 = 1 - \frac{\text{SSE}}{\text{SST}} \quad \text{in the model without constant term}$$

and as

$$R^2 = 1 - \frac{\text{SSE}}{\text{SST}_m} \quad \text{in the model with constant term,}$$

where

$$\text{SSE} = \text{residual error sum of squares}$$

$$= \sum_{i=1}^{n} (y_i - \hat{y}_i)^2,$$

$$\text{SST} = \text{total sum of squares}$$

$$= \sum_{i=1}^{n} y_i^2,$$

and

$$\text{SST}_m = \text{total sum of squares corrected for the mean}$$

$$= \sum_{i=1}^{n} (y_i - \bar{y})^2,$$

where

$$\bar{y} = \frac{1}{n} \sum_{i=1}^{n} y_i.$$

In the case of simple regression with constant term, the coefficient of determination equals the square of the Pearson correlation coefficient between x and y, which explains the notation R^2.

For the Pilot-Plant data (with outlier), the R^2 value corresponding to LS equals 0.141. This means that only 14.1% of the variability in y is explained by the simple regression model.

One can also consider testing the hypothesis that R^2 equals zero. Formally,

$$H_0: \quad R^2 = 0 \quad \text{(null hypothesis)}$$
$$H_1: \quad R^2 \neq 0 \quad \text{(alternative hypothesis)} . \tag{3.7}$$

These hypotheses are equivalent to testing whether the whole vector of regression coefficients (except for the constant term if the model has one) equals the zero vector, that is, (3.7) is equivalent to

$$H_0: \quad \text{All nonintercept } \theta_j \text{'s are together equal to zero}$$
$$H_1: \quad H_0 \text{ is not true} . \tag{3.8}$$

This is quite different from the above t-test on an individual regression coefficient, because here the coefficients are considered together. If the e_i are normally distributed, then the following statistics have an F-distribution:

$$\frac{R^2/(p-1)}{(1-R^2)/(n-p)} = \frac{(\text{SST}_m - \text{SSE})/(p-1)}{\text{SSE}/(n-p)} \sim F_{p-1,n-p}$$

for regression with a constant, and

$$\frac{R^2/p}{(1-R^2)/(n-p)} = \frac{(\text{SST} - \text{SSE})/p}{\text{SSE}/(n-p)} \sim F_{p,n-p} \quad \text{otherwise} .$$

If the calculated value of the appropriate statistic is less than the $(1-\alpha)$th quantile of the associated F-distribution, then H_0 can be accepted. If not, H_0 may be rejected. To facilitate this test, PROGRESS prints the p-value of the F-statistic. This p-value (again computed according to Lackritz 1984) is the probability that an F-distributed random variable with the proper degrees of freedom exceeds the actual F-value.

For the contaminated Pilot-Plant data we obtain an F-value of 2.955, with 1 and 18 degrees of freedom. The corresponding p-value is 0.103, so we have no significance at the 5% level. Consequently, one can say that it appears from the LS estimates that the explanatory variable does not really "explain" the response in a significant way, since one cannot reject H_0.

The interpretation of t-test, R^2, and F-test is still valid for the multidimensional data sets that will be considered in Chapter 3.

When intermediate or large output is requested, the program continues by listing, for each observation, the actual value of the response

(y_i), the estimated response $(\hat{y}_i)$, the residual $(r_i = y_i - \hat{y}_i)$, and the standardized residual (r_i/s). This table allows us to identify outlying residuals. However, because case 6 has attracted the LS line so strongly (see Figure 2), and has also blown up s, its standardized residual is merely -2.62, which is perhaps slightly conspicuous but in no way dramatic.

The results for LMS regression are printed below the output for LS. First the estimates of the regression parameters are given, together with a corresponding scale estimate. This scale estimate is also defined in a robust way. For that purpose a preliminary scale estimate s^0 is calculated, based on the value of the objective function, multiplied by a finite sample correction factor dependent on n and p:

$$s^0 = 1.4826\left(1 + \frac{5}{n-p}\right)\sqrt{\operatorname*{med}_i r_i^2} .$$

With this scale estimate, the standardized residuals r_i/s^0 are computed and used to determine a weight w_i for the ith observation as follows:

$$w_i = \begin{cases} 1 & \text{if } |r_i/s^0| \le 2.5 \\ 0 & \text{otherwise} . \end{cases} \tag{3.9}$$

The scale estimate for the LMS regression is then given by

$$\sigma^* = \sqrt{\frac{\displaystyle\sum_{i=1}^{n} w_i r_i^2}{\displaystyle\sum_{i=1}^{n} w_i - p}} . \tag{3.10}$$

Note that σ^* also has a 50% breakdown point, which means that it does not explode $(\sigma^* \to \infty)$ or implode $(\sigma^* \to 0)$ for less than 50% of contamination. More details on this scale estimate will be given in Chapter 5. In the present example, $\sigma^* = 1.33$, which is much smaller than the LS scale $s = 15.6$ computed in the beginning.

The LMS also possesses a measure to determine how well the fitted model explains the observed variability in y. In analogy to the classical one, we called it also R^2 or coefficient of determination. In the case of regression with constant term, it is defined by

$$R^2 = 1 - \left(\frac{\operatorname{med}|r_i|}{\operatorname{mad}(y_i)}\right)^2 \tag{3.11}$$

and by

$$R^2 = 1 - \left(\frac{\text{med} |r_i|}{\text{med} |y_i|} \right)^2 \qquad (3.12)$$

when the model has no intercept term. Here, the abbreviation "mad" stands for *median absolute deviation*, defined as

$$\text{mad}(y_i) = \underset{i}{\text{med}} \{ |y_i - \underset{j}{\text{med}} \, y_j| \} .$$

Independent of our work, formula (3.11) was also proposed by Kvalseth (1985). In the Pilot-Plant output the robust coefficient of determination equals 0.996, which means that the majority of the data fits a linear model quite nicely.

Also for the LMS, a table with observed y_i, estimated $\hat{y}_i$, residual r_i, and standardized residual r_i/σ^* is given. It now shows clearly that case 6 is an outlier, because r_6/σ^* equals the enormous value of -78.50.

The last part of the output is about reweighted least squares (RLS) regression. This corresponds to minimizing the sum of the squared residuals multiplied by a weight w_i:

$$\underset{\hat{\theta}}{\text{Minimize}} \sum_{i=1}^{n} w_i r_i^2 . \qquad (3.13)$$

The weights w_i are determined from the LMS solution as in (3.9), but with the final scale estimate σ^* (3.10) instead of s^0. The effect of the weights, which can only take the values 0 or 1, is the same as deleting the cases for which w_i equals zero. The scale estimate associated with (3.13) is given by

$$s = \sqrt{\frac{\sum_{i=1}^{n} w_i r_i^2}{\sum_{i=1}^{n} w_i - p}} . \qquad (3.14)$$

(One finds again the scale estimate for ordinary LS when putting $w_i = 1$ for all cases.) Therefore, the RLS can be seen as ordinary LS on a "reduced" data set, consisting of only those observations that received a nonzero weight. Because this reduced data set does not contain regression outliers anymore, the statistics and inferences are more trustworthy than those associated with LS on the whole data set. The underlying distribution theory is no longer entirely exact (because the weights depend on the data in a complicated way), but is still useful as a good approximation, as was confirmed by means of some Monte Carlo trials.

In the present example all w_i are equal to 1, except for case 6, which indeed was the outlier we had produced. The regression coefficients obtained by RLS strongly resemble those determined in the first step by LMS. Note that, without the outlier, the slope estimate becomes significantly different from zero.

The determination coefficient for the RLS is defined in an analogous way as for LS, but all terms are now multiplied by their weight w_i. In this example it is highly significant, because the F-value becomes very large and hence the corresponding p-value is close to zero.

To end this section, we would like to warn the reader about a common misunderstanding. When the LS and the RLS results are substantially different, the right thing to do is to identify the outliers (by means of the RLS residuals) and to study them. Instead, some people are inclined to think they have to *choose* between the LS and the RLS output, and typically they will prefer the estimates with the most significant t-values or F-value, often assuming that the highest R^2 corresponds to the "best" regression. This makes no sense, because the LS inference is very sensitive to outliers, which may affect R^2 in *both* directions. Indeed, the least squares R^2 of any data set can be made arbitrarily close to 1 by means of one or more leverage points. It often happens that RLS discards the outliers that were responsible for a high R^2 and correctly comes to the conclusion that the R^2 of the majority is not so high at all (or it may find that some θ_j are no longer significantly different from zero).

4. EXAMPLES

In this section we will further explain the results provided by PROGRESS, by means of some real-data examples appearing in the literature.

Example 1:   First Word—Gesell Adaptive Score Data

This two-dimensional data set comes from Mickey et al. (1967) and has been widely cited. The explanatory variable is the age (in months) at which a child utters its first word, and the response variable is its Gesell adaptive score. These data (for 21 children) appear in Table 4, and they are plotted in Figure 5.

Mickey et al. (1967) decided that observation 19 is an outlier, by means of a sequential approach to detect outliers via stepwise regression. Andrews and Pregibon (1978), Draper and John (1981), and Paul (1983)

Table 4. First Word—Gesell Adaptive Score Data

Child (i)	Age in Months (x_i)	Gesell Score (y_i)
1	15	95
2	26	71
3	10	83
4	9	91
5	15	102
6	20	87
7	18	93
8	11	100
9	8	104
10	20	94
11	7	113
12	9	96
13	10	83
14	11	84
15	11	102
16	10	100
17	12	105
18	42	57
19	17	121
20	11	86
21	10	100

Source: Mickey et al. (1967).

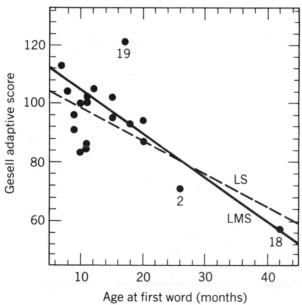

Figure 5. Scatterplot of Gesell adaptive score versus age at first word.

applied outlier diagnostics to this data set. (The use of such diagnostics will be discussed in Chapter 6.) The most important results of our own analysis are given below.

```
LEAST SQUARES REGRESSION
*************************
   VARIABLE        COEFFICIENT    STAND. ERROR    T - VALUE     P - VALUE
-------------------------------------------------------------------------
  AGE(MONTH)         -1.12699         .31017      -3.63343       .00177
  CONSTANT          109.87380        5.06780      21.68077       .00000

SUM OF SQUARES            =       2308.58600

DEGREES OF FREEDOM       =         19

SCALE ESTIMATE            =         11.02291

VARIANCE - COVARIANCE MATRIX =
    .9621D-01
   -.1384D+01    .2568D+02

COEFFICIENT OF DETERMINATION (R SQUARED) =            .40997

THE F-VALUE =     13.202 (WITH   1 AND   19 DF)   P - VALUE =   .00177

            OBSERVED             ESTIMATED            RESIDUAL    NO   RES/SC
          GESELL SC.            GESELL SC.
            95.00000             92.96901             2.03099     1     .18
            71.00000             80.57213            -9.57213     2    -.87
            83.00000             98.60395           -15.60395     3   -1.42
            91.00000             99.73094            -8.73094     4    -.79
           102.00000             92.96901             9.03099     5     .82
            87.00000             87.33406             -.33406     6    -.03
            93.00000             89.58804             3.41196     7     .31
           100.00000             97.47696             2.52304     8     .23
           104.00000            100.85790             3.14207     9     .29
            94.00000             87.33406             6.66594    10     .60
           113.00000            101.98490            11.01508    11    1.00
            96.00000             99.73094            -3.73094    12    -.34
            83.00000             98.60395           -15.60395    13   -1.42
            84.00000             97.47696           -13.47696    14   -1.22
           102.00000             97.47696             4.52304    15     .41
           100.00000             98.60395             1.39605    16     .13
           105.00000             96.34998             8.65002    17     .78
            57.00000             62.54031            -5.54031    18    -.50
           121.00000             90.71503            30.28497    19    2.75
            86.00000             97.47696           -11.47696    20   -1.04
           100.00000             98.60395             1.39605    21     .13

*****************************************************************************
LEAST MEDIAN OF SQUARES REGRESSION
**********************************
   VARIABLE        COEFFICIENT
-----------------------------------
  AGE(MONTH)         -1.50000
  CONSTANT          119.75000

FINAL SCALE ESTIMATE         =         8.83928

COEFFICIENT OF DETERMINATION =          .44460

            OBSERVED             ESTIMATED            RESIDUAL    NO   RES/SC
          GESELL SC.            GESELL SC.
            95.00000             97.25000            -2.25000     1    -.25
            71.00000             80.75000            -9.75000     2   -1.10
            83.00000            104.75000           -21.75000     3   -2.46
            91.00000            106.25000           -15.25000     4   -1.73
           102.00000             97.25000             4.75000     5     .54
            87.00000             89.75000            -2.75000     6    -.31
            93.00000             92.75000              .25000     7     .03
           100.00000            103.25000            -3.25000     8    -.37
           104.00000            107.75000            -3.75000     9    -.42
            94.00000             89.75000             4.25000    10     .48
           113.00000            109.25000             3.75000    11     .42
            96.00000            106.25000           -10.25000    12   -1.16
            83.00000            104.75000           -21.75000    13   -2.46
            84.00000            103.25000           -19.25000    14   -2.18
           102.00000            103.25000            -1.25000    15    -.14
           100.00000            104.75000            -4.75000    16    -.54
           105.00000            101.75000             3.25000    17     .37
            57.00000             56.75000              .25000    18     .03
           121.00000             94.25000            26.75000    19    3.03
            86.00000            103.25000           -17.25000    20   -1.95
           100.00000            104.75000            -4.75000    21    -.54

*****************************************************************************
```

```
REWEIGHTED LEAST SQUARES BASED ON THE LMS
*******************************************
```

VARIABLE	COEFFICIENT	STAND. ERROR	T - VALUE	P - VALUE
AGE(MONTH)	-1.19331	.24348	-4.90100	.00012
CONSTANT	109.30470	3.96997	27.53290	.00000

```
WEIGHTED SUM OF SQUARES  =        1340.02400

DEGREES OF FREEDOM       =          18

SCALE ESTIMATE           =           8.62820


VARIANCE - COVARIANCE MATRIX =
    .5928D-01
   -.8448D+00      .1576D+02

COEFFICIENT OF DETERMINATION (R SQUARED) =          .57163

THE F-VALUE =        24.020 (WITH    1 AND    18 DF)  P - VALUE =  .00012

THERE ARE    20 POINTS WITH NON-ZERO WEIGHT.

AVERAGE WEIGHT           =             .95238
```

OBSERVED GESELL SC.	ESTIMATED GESELL SC.	RESIDUAL	NO	RES/SC	WEIGHT
95.00000	91.40502	3.59498	1	.42	1.0
71.00000	78.27860	-7.27860	2	-.84	1.0
83.00000	97.37157	-14.37157	3	-1.67	1.0
91.00000	98.56488	-7.56488	4	-.88	1.0
102.00000	91.40502	10.59498	5	1.23	1.0
87.00000	85.43846	1.56154	6	.18	1.0
93.00000	87.82509	5.17491	7	.60	1.0
100.00000	96.17826	3.82174	8	.44	1.0
104.00000	99.75819	4.24181	9	.49	1.0
94.00000	85.43846	8.56154	10	.99	1.0
113.00000	100.95150	12.04849	11	1.40	1.0
96.00000	98.56488	-2.56488	12	-.30	1.0
83.00000	97.37157	-14.37157	13	-1.67	1.0
84.00000	96.17826	-12.17826	14	-1.41	1.0
102.00000	96.17826	5.82174	15	.67	1.0
100.00000	97.37157	2.62843	16	.30	1.0
105.00000	94.98495	10.01505	17	1.16	1.0
57.00000	59.18563	-2.18563	18	-.25	1.0
121.00000	89.01840	31.98160	19	3.71	.0
86.00000	96.17826	-10.17826	20	-1.18	1.0
100.00000	97.37157	2.62843	21	.30	1.0

Because medium-sized output was requested, PROGRESS gives for each estimator a table with the observed response variable (y_i), its estimated value ($\hat{y}_i$), the residual (r_i), and the standardized residual (denoted by "RES/SC"). The standardized residuals for each regression are obtained by dividing the raw residuals by the scale estimate of the fit. A supplementary column with weights is added to the table for RLS regression. (These weights are determined from the LMS solution, as described in Section 3.) In this example, the case with index 19 received a zero weight. Indeed, this case has been identified as outlying because it has a large residual from the LMS fit. The equations of the LS and the RLS do not differ very much for this data set. The pair of points 2 and 18 has pulled the LS in the "good" direction. These points are good leverage points and possess small LMS residuals as well as small LS residuals. (The deletion of one or both of these points would have a considerable effect on the size of the confidence intervals.) One might even say that this data set is not a very good example of linear regression because deleting the leverage points (2 and 18) would not leave much of a linear relationship between x and y. (We will return to this at the end of Section 6.)

Example 2: *Number of Fires in 1976–1980*

This data set (listed in Table 5) shows the trend from 1976 to 1980 of the number of reported claims of Belgian fire-insurance companies (from the annual report of the Belgian Association of Insurance Companies). It is included here to have an example with very few points.

When looking at the scatterplot in Figure 6, one notices a slight upward trend over the years. However, the number for 1976 is extraordinarily high. The reason lies in the fact that in that year the summer was extremely hot and dry (compared to Belgian standards), causing trees and

Table 5. Number of Fire Claims in Belgium from 1976 to 1980

Year (x_i)	Number of Fires (y_i)
76	16,694
77	12,271
78	12,904
79	14,036
80	13,874

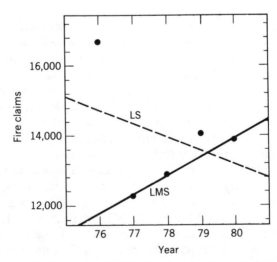

Figure 6. Number of fire claims in Belgium for the years 1976–1980.

bushes to catch fire spontaneously. It is striking that the LS $\hat{y} = -387.5x + 44180.8$ (dashed line) and the LMS $\hat{y} = 534.3x - 28823.3$ (solid line) are very different. The LS fits the data with a decreasing trend, whereas the LMS line increases. The outlier lies on the outside of the x-range and causes LS to grossly misbehave. It does not even possess the largest LS residual. This example shows again that examination of the LS residuals is not sufficient to identify the outlier(s).

Of course, in such a small data set one cannot draw any strong statistical conclusions, but it does show that one should think carefully about the data whenever LS and LMS yield substantially different results.

Example 3: Annual Rates of Growth of Prices in China

Table 6 contains the annual rates of growth of the average prices in the main cities of Free China from 1940 to 1948 (Simkin 1978). For instance, in 1940 prices went up 1.62% as compared to the previous year. In 1948 a huge jump occurred as a result of enormous government spending, the budget deficit, and the war, leading to what is called *hyperinflation*.

The LMS regression equation is given by

$$\hat{y} = 0.102x - 2.468 ,$$

whereas the LS estimate corresponds to

$$\hat{y} = 24.845x - 1049.468 ,$$

Table 6. Annual Rates of Growth of Average Prices in the Main Cities of Free China from 1940 to 1948

Year (x_i)	Growth of Prices (y_i)	Estimated Growth	
		By LMS	By LS
40	1.62	1.61	−55.67
41	1.63	1.71	−30.82
42	1.90	1.82	−5.98
43	2.64	1.92	18.87
44	2.05	2.02	43.71
45	2.13	2.12	68.56
46	1.94	2.22	93.40
47	15.50	2.33	118.25
48	364.00	2.43	143.09

Source: Simkin (1978).

which is totally different. To show which of these lines yields the better fit, Table 6 lists the estimated values by both methods. The LMS provides a fair approximation to the majority of the data, except of course for the last two years, where the observed y_i go astray. On the other hand, the LS fit is bad everywhere: The estimated $\hat{y}_i$ is even negative for the first three years, after which it becomes much too large, except for the 1948 value, which it cannot match either. Least squares smears out the effect (of nonlinearity of the original data) over the whole column, whereas LMS fits the majority of the data (where it is indeed linear) and allows the discrepancy to show up in those two years where actually something went wrong. Applying PROGRESS to this data set yields (among other things) the following output:

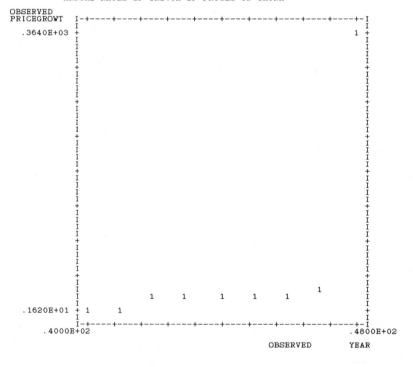

```
**********************************************
* ROBUST REGRESSION WITH A CONSTANT TERM.  *
**********************************************

NUMBER OF CASES =        9
NUMBER OF COEFFICIENTS (INCLUDING CONSTANT TERM) =     2

THE EXTENSIVE SEARCH VERSION WILL BE USED.
DATA SET = ANNUAL RATES OF GROWTH OF PRICES IN CHINA
THERE ARE NO MISSING VALUES.

              ANNUAL RATES OF GROWTH OF PRICES IN CHINA

      OBSERVED
      PRICEGROWT I-+----+----+----+----+----+----+----+----+----+-I
               I                                                  I
      .3640E+03 +                                              1 +
               I                                                  I
               I                                                  I
               I                                                  I
               +                                                  +
               I                                                  I
               I                                                  I
               I                                                  I
               +                                                  +
               I                                                  I
               I                                                  I
               I                                                  I
               +                                                  +
               I                                                  I
               I                                                  I
               I                                                  I
               +                                                  +
               I                                                  I
               I                                                  I
               I                                                  I
               +                                                  +
               I                                                  I
               I                                                  I
               I                                                  I
               +                                                  +
               I                                                  I
               I                                                  I
               I                                                  I
               +                                                  +
               I                                            1     I
               I         1    1    1    1    1                    I
      .1620E+01 + 1       1                                       +
               I                                                  I
               I-+----+----+----+----+----+----+----+----+----+-I
        .4000E+02                                          .4800E+02

                                          OBSERVED        YEAR
```

MEDIANS =

```
          YEAR PRICEGROWT
       44.0000    2.0500
```

DISPERSIONS =

```
          YEAR PRICEGROWT
        2.9652     .6227
```

THE STANDARDIZED OBSERVATIONS ARE:

```
          YEAR PRICEGROWT
   1    -1.3490     -.6906
   2    -1.0117     -.6745
   3     -.6745     -.2409
   4     -.3372      .9475
   5      .0000      .0000
   6      .3372      .1285
   7      .6745     -.1767
   8     1.0117    21.5998
   9     1.3490   581.2665
```

PEARSON CORRELATION COEFFICIENTS BETWEEN THE VARIABLES
(PRICEGROWT IS THE RESPONSE VARIABLE)

```
           YEAR    1.00
PRICEGROWT          .57  1.00
```

SPEARMAN RANK CORRELATION COEFFICIENTS BETWEEN THE VARIABLES
(PRICEGROWT IS THE RESPONSE VARIABLE)

```
           YEAR    1.00
PRICEGROWT          .85  1.00
```

****

LEAST SQUARES REGRESSION
****

VARIABLE	COEFFICIENT	STAND. ERROR	T - VALUE	P - VALUE
YEAR	24.84500	13.67404	1.81695	.11207
CONSTANT	-1049.46800	602.69280	-1.74130	.12517

```
SUM OF SQUARES          =     78531.34000

DEGREES OF FREEDOM      =         7

SCALE ESTIMATE          =       105.91870
```

VARIANCE - COVARIANCE MATRIX =

```
   .1870D+03
 -.8227D+04     .3632D+06
```

COEFFICIENT OF DETERMINATION (R SQUARED) =   .32047

THE F-VALUE =   3.301 (WITH   1 AND   7 DF)   P - VALUE = .11207

OBSERVED PRICEGROWT	ESTIMATED PRICEGROWT	RESIDUAL	NO	RES/SC
1.62000	-55.66766	57.28766	1	.54
1.63000	-30.82269	32.45269	2	.31
1.90000	-5.97766	7.87766	3	.07
2.64000	18.86731	-16.22731	4	-.15
2.05000	43.71228	-41.66228	5	-.39
2.13000	68.55737	-66.42738	6	-.63
1.94000	93.40234	-91.46234	7	-.86
15.50000	118.24730	-102.74730	8	-.97
364.00000	143.09230	220.90770	9	2.09

ANNUAL RATES OF GROWTH OF PRICES IN CHINA

--- L E A S T   S Q U A R E S ---

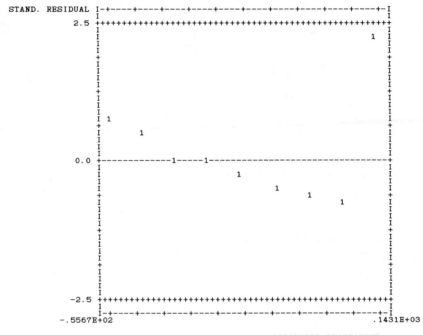

ESTIMATED PRICEGROWT

****

LEAST MEDIAN OF SQUARES REGRESSION
****
VARIABLE	COEFFICIENT
YEAR	.10200
CONSTANT	-2.46800

FINAL SCALE ESTIMATE   =   .15475

COEFFICIENT OF DETERMINATION =   .93824

OBSERVED PRICEGROWT	ESTIMATED PRICEGROWT	RESIDUAL	NO	RES/SC
1.62000	1.61200	.00800	1	.05
1.63000	1.71400	-.08400	2	-.54
1.90000	1.81600	.08400	3	.54
2.64000	1.91800	.72200	4	4.67
2.05000	2.02000	.03000	5	.19
2.13000	2.12200	.00800	6	.05
1.94000	2.22400	-.28400	7	-1.84
15.50000	2.32600	13.17400	8	85.13
364.00000	2.42800	361.57200	9	2336.42

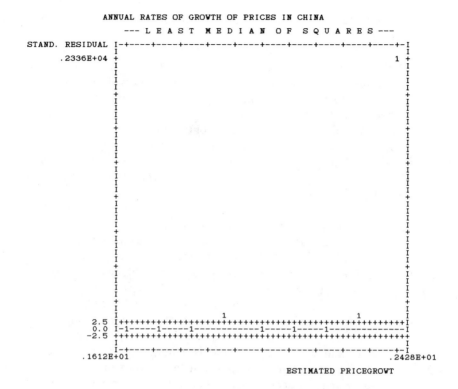

ANNUAL RATES OF GROWTH OF PRICES IN CHINA

--- L E A S T   M E D I A N   O F   S Q U A R E S ---

ESTIMATED PRICEGROWT

Note that observations 8 and 9 are outliers in the y-direction, which show up in the second column of the standardized data. Also, the Spearman rank correlation is much higher than the Pearson correlation because the outliers have disturbed the linear relation, whereas they obey the near-monotone relation between both variables.

For this example, we will discuss the residual plots (printed above) provided by PROGRESS. They display the standardized residual versus the estimated value of the response. An index plot is similar, except it contains the standardized residual plotted versus the index of the observation. (This makes it easier to find which observation corresponds to a given point in the plot. The index plot is also useful because the indices often reflect the order in which the observations were measured, so it may reveal sudden changes in the data recording process.) In both plots, a dashed line is drawn through zero, and a horizontal band on the interval $[-2.5, 2.5]$ is marked. These lines facilitate the interpretation of the results. When the observed value y_i equals the estimated value $\hat{y}_i$, then the resulting residual becomes zero. Points in the neighborhood of this zero line are best fitted by the model. If the residuals are normally distributed, then one can expect that roughly 98% of the standardized residuals will lie in the interval $[-2.5, 2.5]$. Thus, observations for which the standardized residuals are situated far from the horizontal band can be identified as outlying.

The first residual plot in the above output shows how the LS fit masks the bad point. The LS has been pulled away by this outlier, and its scale estimate has exploded. The LS residual associated with the outlier even lies within the horizontal band. Because of this effect, the interpretation of a residual plot corresponding to the LS estimator is dangerous. In the residual plot of the LMS, the outlier is very far away from the band. Residual plots corresponding to robust estimators are even more useful in problems with several variables, as will be illustrated in Chapter 3.

These graphical tools are very convenient for spotting the outlying observations. The LS result can be trusted only if the residual plots of both the robust and nonrobust regression methods agree closely.

Besides the identification of outliers, the residual plot provides a diagnostic tool for assessing the adequacy of the fit and for suggesting transformations. An ideal pattern of the residual plot, which indicates an adequate model and well-behaved data, is a horizontal cloud of points with constant vertical scatter. Anomalies in the pattern of residuals can lead to several courses of action. For instance, the plot may indicate that the variance of the residuals increases or decreases with increasing estimated y or with another variable. (See the education expenditure data in Section 3 of Chapter 3 for an illustration.) This is called *heteroscedasticity*, in contrast to the classical model where the errors have a constant variance, in which case we speak of *homoscedasticity*. This problem may be approached by applying a suitable transformation on either an explanatory or the response variable. If this heteroscedasticity appears in an index plot, then one should turn back to the origin of the data in order to look for the cause of the phenomenon. For example, it could be that a time-related variable has to be included in the model. Also, other model failures may be visible in residual plots. A pattern resembling a horseshoe may be caused by nonlinearity. (See the cloud point data in Section 3 of Chapter 3 for an example.) In such a situation, a transformation on an explanatory or on the response variable, or an additional squared term or a cross-product term in the model, or the addition of an extra explanatory variable may be required. (In many applications of regression, there is a substantial amount of prior knowledge that can be useful in choosing between these possibilities.)

Example 4: Brain and Weight Data

Table 7 presents the brain weight (in grams) and the body weight (in kilograms) of 28 animals. (This sample was taken from larger data sets in Weisberg 1980 and Jerison 1973.) It is to be investigated whether a larger brain is required to govern a heavier body.

A clear picture of the relationship between the logarithms (to the base 10) of these measurements is shown in Figure 7. This logarithmic

Table 7. Body and Brain Weight for 28 Animals

Index (i)	Species	Body Weight[a] (x_i)	Brain Weight[b] (y_i)
1	Mountain beaver	1.350	8.100
2	Cow	465.000	423.000
3	Gray wolf	36.330	119.500
4	Goat	27.660	115.000
5	Guinea pig	1.040	5.500
6	Diplodocus	11700.000	50.000
7	Asian elephant	2547.000	4603.000
8	Donkey	187.100	419.000
9	Horse	521.000	655.000
10	Potar monkey	10.000	115.000
11	Cat	3.300	25.600
12	Giraffe	529.000	680.000
13	Gorilla	207.000	406.000
14	Human	62.000	1320.000
15	African elephant	6654.000	5712.000
16	Triceratops	9400.000	70.000
17	Rhesus monkey	6.800	179.000
18	Kangaroo	35.000	56.000
19	Hamster	0.120	1.000
20	Mouse	0.023	0.400
21	Rabbit	2.500	12.100
22	Sheep	55.500	175.000
23	Jaguar	100.000	157.000
24	Chimpanzee	52.160	440.000
25	Brachiosaurus	87000.000	154.500
26	Rat	0.280	1.900
27	Mole	0.122	3.000
28	Pig	192.000	180.000

[a] In kilograms.
[b] In grams.
Source: Weisberg (1980) and Jerison (1973).

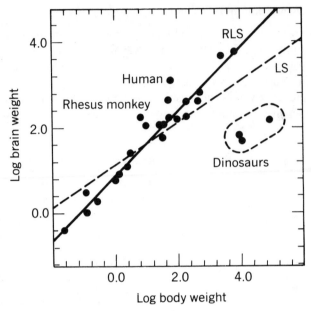

Figure 7. Logarithmic brain weight versus logarithmic body weight for 28 animals with LS (dashed line) and RLS fit (solid line).

transformation was necessary because plotting the original measurements would fail to represent either the smaller or the larger measurements. Indeed, both original variables range over several orders of magnitude. A linear fit to this transformed data would be equivalent to a relationship of the form

$$\hat{y} = \hat{\theta}'_2 x^{\hat{\theta}_1}$$

between brain weight (y) and body weight (x). Looking at Figure 7, it seems that this transformation makes things more linear. Another important advantage of the log scale is that the heteroscedasticity disappears.

The LS fit is given by

$$\log \hat{y} = 0.49601 \log x + 1.10957$$

(dashed line in Figure 7). The standard error associated with the slope equals 0.0782, and that of the intercept term is 0.1794. In Section 3, we explained how to construct a confidence interval for the unknown regression parameters. For the present example, $n = 28$ and $p = 2$, so one has to use the 97.5% quantile of the t-distribution with 26 degrees of freedom, which equals 2.0555. Using the LS results, a 95% confidence interval for the slope is given by [0.3353; 0.6567]. The RLS yields the solid line in Figure 7, which is a fit with a steeper slope:

$$\log \hat{y} = 0.75092 \log x + 0.86914 .$$

The slope estimated by the RLS technique even falls outside the 95% confidence interval associated with the LS fit! The standard error of the regression coefficients in RLS is reduced remarkably as compared with

Table 8. Standardized LS and RLS Residuals for the Brain and Body Weight Data

Index	Species	Standardized LS Residuals	Standardized RLS Residuals	w_i
1	Mountain beaver	−0.40	−0.27	1
2	Cow	0.29	−1.13	1
3	Gray wolf	0.29	0.17	1
4	Goat	0.36	0.50	1
5	Guinea pig	−0.57	−0.65	1
6	Diplodocus	−2.15	<u>−10.19</u>	0
7	Asian elephant	1.30	1.08	1
8	Donkey	0.58	0.21	1
9	Horse	0.54	−0.43	1
10	Potar monkey	0.68	2.02	1
11	Cat	0.06	0.68	1
12	Giraffe	0.56	−0.37	1
13	Gorilla	0.53	0.00	1
14	Human	1.69	<u>4.15</u>	0
15	African elephant	1.13	0.08	1
16	Triceratops	−1.86	<u>−9.20</u>	0
17	Rhesus monkey	1.10	<u>3.47</u>	0
18	Kangaroo	−0.19	−1.29	1
19	Hamster	−0.98	−0.81	1
20	Mouse	−1.04	−0.17	1
21	Rabbit	−0.34	−0.39	1
22	Sheep	0.40	0.29	1
23	Jaguar	0.14	−0.80	1
24	Chimpanzee	1.02	2.22	1
25	Brachiosaurus	−2.06	<u>−10.94</u>	0
26	Rat	−0.84	−0.80	1
27	Mole	−0.27	1.35	1
28	Pig	0.02	−1.50	1

the LS, namely 0.0318 for the slope and 0.0618 for the constant term. A 95% confidence interval for the unknown slope is now given by [0.6848; 0.8171], which is narrower than the interval coming from LS. The t-values associated with the RLS regression coefficients are very large, which implies that the slope and intercept are significantly different from zero. Moreover, the determination coefficient R^2, which is a summary measure for overall goodness of fit, increases from 0.608 for LS to 0.964 for RLS. This example shows that not only the LS regression coefficients, but also the whole LS inference, may become doubtful in the presence of outliers.

Table 8 lists the standardized LS and RLS residuals and the w_i determined on the basis of the LMS. From the RLS, it is easy to detect unusual observations and to give them special consideration. Indeed, looking at the five cases with zero w_i, one can easily understand why they have to be considered as outlying. The most severe (and highly negative) RLS residuals are those of cases 6, 16, and 25, which are responsible for the low slope of the LS fit. These are three dinosaurs, each of which possessed a small brain as compared with a heavy body. In this respect they contrast with the mammals which make up the rest of the data set. The LMS regression also produced a zero weight for cases 14 and 17, namely the human and the rhesus monkey. For them, the actual brain weight is higher than that predicted by the linear model. Unlike the dinosaurs, their residuals are therefore positive. Concluding, one could say that dinosaurs, humans, and rhesus monkeys do not obey the same trend as the one followed by the majority of the data.

5. AN ILLUSTRATION OF THE EXACT FIT PROPERTY

The phrase "exact fit" stands for situations where a large percentage of the observations fits some linear equation exactly. For example, in simple regression this happens when the majority of the data lie exactly on a straight line. In such a case a robust regression method should recover that line. At an Oberwolfach Meeting, Donoho (1984) called this the *exact fit property*. For instance, the repeated median satisfies this property (Siegel 1982), as well as the LMS (Rousseeuw 1984). When at least $n - [n/2] + 1$ of the observations lie on the same line, then the equation of this line will be the LMS solution. More details on the exact fit property and its relation to the breakdown point can be found in Section 4 of Chapter 3.

The data in Table 9 come from Emerson and Hoaglin (1983, p. 139). They were devised by A. Siegel as a counterexample for the resistant line estimator (which will be briefly discussed in Section 7). Looking at the

Table 9. Siegel's Data Set

i	x_i	y_i
1	−4	0
2	−3	0
3	−2	0
4	−1	0
5	0	0
6	1	0
7	2	−5
8	3	5
9	12	1

Source: Emerson and Hoaglin (1983).

scatterplot of these data (Figure 8), Emerson and Hoaglin suggest that a line with slope 0 would be a reasonable summary. Indeed, six out of the nine points actually lie *on* the line with zero slope and zero intercept. By running PROGRESS we see that least median of squares yields this line exactly, unlike least squares.

Exact fit situations also occur in real data. One example has to do with the use of an electron microscope in crystallography. Several discrete variables (such as the number of edges and certain symmetry properties) are observed for a large number of "cells" (consisting of molecules) in a

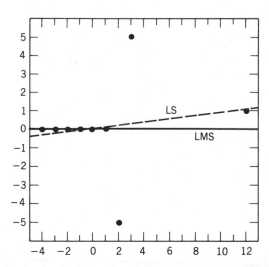

Figure 8. Example of exact fit: Scatterplot of Siegel's data set with LS (dashed line) and LMS fit (solid line).

regular lattice. In practice, most of these cells are good (and hence fit exactly), whereas a fraction are damaged by the radiation of the microscope itself.

6. SIMPLE REGRESSION THROUGH THE ORIGIN

When it is known in advance that the intercept term is zero, then one has to impose this on the model. This leads to equation (1.2) in Section 1 of this chapter.

Afifi and Azen (1979, p. 125) give an example (copied in Table 10) where such a model is suitable. The data concern the calibration of an instrument that measures lactic acid concentration in blood. One compares the true concentration x_i with the measured value y_i. The explanatory variable in this data set is designed. This means that its values are fixed in advance. Consequently, such a data set does not contain leverage points. The scatterplot in Figure 9 displays a roughly linear relationship

Table 10. Data on the Calibration of an Instrument that Measures Lactic Acid Concentration in Blood

Index (i)	True Concentration (x_i)	Instrument (y_i)
1	1.0	1.1
2	1.0	0.7
3	1.0	1.8
4	1.0	0.4
5	3.0	3.0
6	3.0	1.4
7	3.0	4.9
8	3.0	4.4
9	3.0	4.5
10	5.0	7.3
11	5.0	8.2
12	5.0	6.2
13	10.0	12.0
14	10.0	13.1
15	10.0	12.6
16	10.0	13.2
17	15.0	18.7
18	15.0	19.7
19	15.0	17.4
20	15.0	17.1

Source: Afifi and Azen (1979).

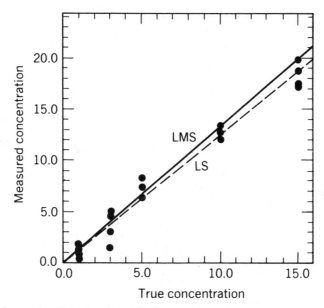

Figure 9. Scatterplot of the data in Table 10. A model without intercept term is used. The LS estimator corresponds to the dashed line and the LMS estimator corresponds to the solid line.

between both variables. Indeed, for this example the LS (dashed line) and the LMS (solid line) almost coincide. However, the literature also contains calibration data with outliers. Massart et al. (1986) apply the LMS to detect outliers and model errors in some real-data examples from analytical chemistry.

Let us now look at data used in Hampel et al. (1986, Chapter 6), which are reproduced here in Table 11. They are measurements of water flow at two different points (Libby, Montana and Newgate, British Columbia) on the Kootenay river in January, for the years 1931–1943. The original data came from Ezekiel and Fox (1959, pp. 57–58), and Hampel et al. changed the Newgate measurement for the year 1934 to 15.7 for illustrative purposes (thereby converting a "good" leverage point into a "bad" one).

For each variable j, PROGRESS will now compute another type of dispersion,

$$s_j = 1.4826 \underset{i}{\text{med}} |x_{ij}| , \qquad (6.1)$$

because it is logical to consider the deviations from zero (and not from some average or median value). For the same reason, the data are

Table 11. Water Flow Measurements in Libby and Newgate on the Kootenay River in January for the Years 1931–1943

Year	Libby (x_i)	Newgate (y_i)
31	27.1	19.7
32	20.9	18.0
33	33.4	26.1
34	77.6	15.7[a]
35	37.0	26.1
36	21.6	19.9
37	17.6	15.7
38	35.1	27.6
39	32.6	24.9
40	26.0	23.4
41	27.6	23.1
42	38.7	31.3
43	27.8	23.8

[a]The original value was 44.9.

Source: Ezekiel and Fox (1959).

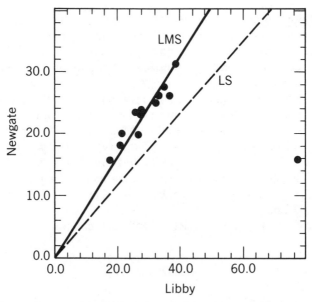

Figure 10. Scatterplot of the water flow in Libby and Newgate, with LS (dashed line) and LMS fit (solid line).

standardized by replacing each x_{ij} by

$$\frac{x_{ij}}{s_j},$$ (6.2)

which is printed when large output was requested. For the Kootenay data, this yields the following output:

```
DISPERSION OF ABSOLUTE VALUES=

    41.2163      34.6928

THE STANDARDIZED OBSERVATIONS ARE:

         LIBBY      NEWGATE
  1      0.6575      0.5678
  2      0.5071      0.5188
  3      0.8104      0.7523
  4      1.8828      0.4525
  5      0.8977      0.7523
  6      0.5241      0.5736
  7      0.4270      0.4525
  8      0.8516      0.7956
  9      0.7909      0.7177
 10      0.6308      0.6745
 11      0.6696      0.6658
 12      0.9389      0.9022
 13      0.6745      0.6860
```

This shows us that case 4 (the year 1934) is a leverage point.

Estimating the unknown slope with the LS technique and the LMS technique gives rise to the dashed and the solid lines in Figure 10.

The LS line ($\hat{y} = 0.5816x$) is attracted by the leverage point associated with the year 1934. On the other hand, this contaminated case produces a large standardized residual with respect to the LMS fit given by $\hat{y} = 0.8088x$. The coefficient of determination corresponding to this fit is rather large too, namely 0.997, whereas it was only 0.798 for the LS fit. From the scatterplot in Figure 10, it appears that the LMS fits the good points very closely, whereas the LS fit fails to give a good prediction for the response variable because its slope is too small.

*7. OTHER ROBUST TECHNIQUES FOR SIMPLE REGRESSION

During the past 50 years, many techniques have been proposed for estimating slope and intercept in a simple regression model. Most of them do not attain a breakdown point of 30%, as will be shown by means of the breakdown plot discussed below.

Emerson and Hoaglin (1983) give a historical survey that contains some explanation about the techniques of Wald (1940), Nair and Shrivastava (1942), Bartlett (1949), and Brown and Mood (1951). These regression methods are all based on the idea of splitting up the data set and then defining a summary value in each group. They testify to the concern about the dramatic lack of robustness of least squares regression, but can only be applied to simple regression.

The resistant line method of Tukey (1970/1971) is a variant of some of these first approaches to robust estimation. This "pencil-and-paper" technique for fitting a line to bivariate data starts from a partition of the data set into three parts, as nearly equal in size as possible. This allocation is performed according to the smallest, intermediate, and largest x-values. Tied x-values are assigned to the same group. Then, the resistant slope is determined such that the median of the residuals in the two outermost groups (L stands for left, and R stands for right) are equal, that is,

$$\operatorname*{med}_{i \in L} (y_i - \hat{\theta}_1 x_i) = \operatorname*{med}_{i \in R} (y_i - \hat{\theta}_1 x_i) , \qquad (7.1)$$

and the intercept is chosen to make the median residuals of both groups zero. The worst-case bound on the available protection against outliers for this technique is $1/6$ because one uses the median, which has a breakdown point of $1/2$, in both groups. Velleman and Hoaglin (1981) provide an algorithm and a portable program for finding the resistant line. Some theoretical and Monte Carlo results on this estimator are provided by Johnstone and Velleman (1985b). The Brown and Mood (1951) technique is defined in a similar way, but with two groups instead of three. As a consequence, the breakdown point for this technique increases to $1/4$.

Andrews (1974) developed another median-based method for obtaining a straight-line fit. He starts by ordering the x-values from smallest to largest. Then he eliminates a certain number of the smallest and of the largest x-values, and also a certain number of the values in the neighborhood of the median x-value. Of the two remaining subsets of x-values, he calculates the medians (med x_1 and med x_2). For the corresponding y-values, the medians are calculated too (med y_1 and med y_2). The slope of the fit is then defined as

$$\hat{\theta}_1 = \frac{\operatorname{med} y_2 - \operatorname{med} y_1}{\operatorname{med} x_2 - \operatorname{med} x_1}$$

The breakdown point of this procedure is at most 25% because half of the data in either subset can determine the fitted line. Andrews generalized this technique to multiple regression by applying a so-called "sweep" operator. This means that at each iteration the dependence of one variable on another is removed by adjusting a variable by a multiple (determined in an earlier stage) of another. The same idea has been used to generalize the resistant line (Johnstone and Velleman 1985b, Emerson and Hoaglin 1985).

Another group of simple regression estimators uses the pairwise slopes as building stones in their definition, without splitting up the data set. Theil (1950) proposed as an estimator of $\hat{\theta}_1$ the median of all C_n^2 slopes, namely

$$\hat{\theta}_1 = \operatorname*{med}_{1 \le i < j \le n} \frac{y_j - y_i}{x_j - x_i}, \tag{7.2}$$

which possesses a high asymptotic efficiency. Sen (1968) extended this estimator to handle ties among the x_i. It turns out that the breakdown point of these techniques is about 29.3%. This value can be explained by the following reasoning: Since the proportion of "good" slopes has to be at least $1/2$, one needs that $(1 - \varepsilon)^2$ be at least $1/2$, where ε is the fraction of outliers in the data. From this it follows that

$$\varepsilon \le 1 - (1/2)^{1/2} \approx 0.293 . \tag{7.3}$$

More recently, Siegel (1982) increased the breakdown point to 50% by means of a crucial improvement, leading to the repeated median estimator. In this method, one has to compute a two-stage median of the pairwise slopes instead of a single median such as in (7.2). The slope and intercept are then defined as

$$\hat{\theta}_1 = \operatorname*{med}_{i} \operatorname*{med}_{j \ne i} \frac{y_j - y_i}{x_j - x_i}$$

$$\hat{\theta}_2 = \operatorname*{med}_{i} (y_i - \hat{\theta}_1 x_i) . \tag{7.4}$$

This list of simple regression techniques is not complete. Several other methods will be dealt with in Section 6 of Chapter 3.

In order to compare the fit obtained by different regression methods in the presence of contamination, we considered an artificial example. Thirty "good" observations were generated according to the linear relation

$$y_i = 1.0x_i + 2.0 + e_i , \tag{7.5}$$

where x_i is uniformly distributed on $[1, 4]$, and e_i is normally distributed with mean zero and standard deviation 0.2. Then a cluster of 20 "bad" observations was added, possessing a spherical bivariate normal distribution with mean $(7, 2)$ and standard deviation 0.5. This yielded 40% of contamination in the pooled sample, which is very high. Actually, this amount was chosen to demonstrate what happens if one goes above the breakdown point of most estimators for simple regression. Let us now see which estimator succeeds best in describing the pattern of the majority of the data. The classical least squares method yields $\hat{\theta}_1 = -0.47$ and $\hat{\theta}_2 = 5.62$: It clearly fails because it tries to suit both the good and the bad points. Making use of the ROBETH library (Marazzi 1980), three robust estimators were applied: Huber's M-estimator [Chapter 1, equation (2.10)] with $\psi(x) = \min(1.5, \max(-1.5, x))$, Mallows' generalized M-estimator [Chapter 1, equation (2.12)] with Hampel weights, and Schweppe's generalized M-estimator [Chapter 1, equation (2.13)] with Hampel–Krasker weights (both Mallows' and Schweppe's estimators use the same Huber function). All three methods, however, gave results virtually indistinguishable from the LS solution: the four lines almost coincide in Figure 11.

The repeated median estimator (7.4) yields $\hat{\theta}_1 = 0.30$ and $\hat{\theta}_2 = 3.11$. If the cluster of "bad" points is moved further down, the repeated median line follows it a little more and then stops. Therefore, this method does not break down, although in this particular example it does not yield a

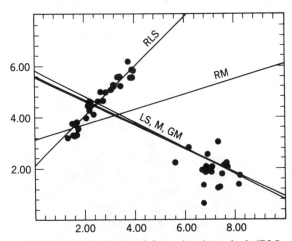

Figure 11. Regression lines for the simulated data using six methods (RLS, reweighted least squares based on LMS; LS, least squares; M, Huber's M-estimator; GM, Mallows' and Schweppe's generalized M-estimators; RM, repeated median). *Source:* Rousseeuw (1984).

very good fit. On the other hand, the LMS-based RLS yields $\hat{\theta}_1 = 0.97$ and $\hat{\theta}_2 = 2.09$, which comes quite close to the original values of θ_1 and θ_2. When the cluster of bad points is moved further away, this solution does not change any more. Moreover, the RLS method does not break down even when only 26 "good" points and 24 outliers are used.

The breakdown properties of these estimators were investigated more extensively in a larger experiment. To begin with, we generated 100 "good" observations according to the linear relation (7.5). To these data, we applied the same fitting techniques as in the above example (see Figure 11). Because the data were well behaved, all estimators yielded values of $\hat{\theta}_1$ and $\hat{\theta}_2$ which were very close to the original θ_1 and θ_2. Then we started to contaminate the data. At each step we deleted one "good" point and replaced it by a "bad" point generated according to a bivariate normal distribution with mean $(7, 2)$ and standard deviation 0.5. We repeated this until only 50 "good" points remained. The LS was immediately affected by these leverage points, so the estimated slope $\hat{\theta}_1$ became negative, moving away from the ideal value $\theta_1 = 1.0$. In Figure 12, the value of $\hat{\theta}_1$ is drawn as a function of the percentage of outliers. We call this a *breakdown plot*.

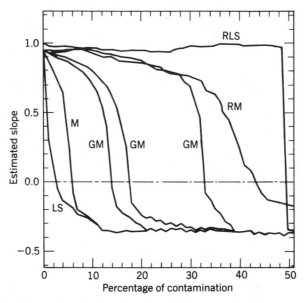

Figure 12. Breakdown plot, showing the estimated slope as a function of the percentage of contamination. The estimators are those of Figure 11, applied to a similar data configuration. *Source:* Rousseeuw et al. (1984b).

We see that LS breaks down first and is then followed by the Huber-type M-estimator and the GM-estimators. (The best of those appears to be the Mallows estimator with Hampel weights, which could tolerate slightly over 30% of outliers in this experiment. Note, however, that GM-estimators will break down much earlier in higher dimensions.) The repeated median goes down gradually and becomes negative at about 40%, whereas the RLS holds on until the very end before breaking down.

REMARK. At this point we would like to say something about the interpretation of breakdown. By definition, breakdown means that the bias $\|T(Z') - T(Z)\|$ becomes unbounded. For smaller fractions of contamination the bias remains bounded, but this does not yet imply that it is small. Although the breakdown point of LMS regression is always 50%, we have found that the effect of outliers may differ according to the quality of the "good" data. For instance, compare the data configurations in Figure 13a and b. In both situations there is a fraction $1 - \varepsilon$ of original data and a percentage ε of outliers. Because $\varepsilon < 50\%$, moving the outliers around does not make the LMS break down in either case. Nevertheless, its maximal bias becomes much larger—though not infinite!—in Figure 13b, where the majority of the data is "ball-shaped" with R^2 close to 0, than in Figure 13a, where the good data already have a strong linear structure with $R^2 \approx 1$. It seems to us that this behavior lies in the nature

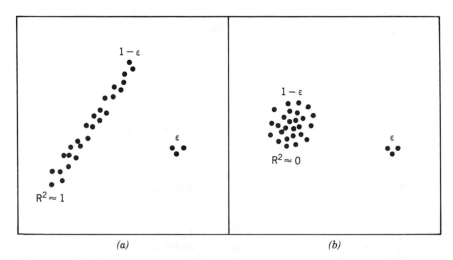

(a)   (b)

Figure 13. Sketch of two data configurations: (a) the good data possess a strong linear structure so the LMS will have only a small bias, and (b) the majority of the points have no linear structure so the effect of outliers will become much larger.

of things. For instance, the data in Figures 2, 3, 4, 6, 7, 10, and 11 are like Figure 13a, whereas the Mickey data (Figure 5) are not much better than Figure 13b.

EXERCISES AND PROBLEMS

Sections 1–4

1. Which is more robust, the Pearson product-moment correlation coefficient or the Spearman rank correlation coefficient? How does this reflect in the correlation coefficients (between extraction and titration) in the PROGRESS output reproduced at the end of Section 2?

2. In the framework of LS regression with intercept, explain why the (one-sided) p-value of the coefficient of determination equals the (one-sided) p-value of the F-statistic. If there is only one explanatory variable apart from the intercept, show that this probability is also equal to the (two-sided) p-value of the t-statistic of the slope and to the (two-sided) p-value of the Pearson correlation coefficient. What happens if we switch to RLS?

3. Making use of the definition of the LMS, can you explain heuristically why it is required that the number of cases be more than twice the number of regression coefficients (i.e., $n > 2p$)? Apart from this constraint, can you also argue why small n might lead to problems, based on the probability of linear clusters occurring merely by virtue of random fluctuations?

4. Reanalyze the Pilot-Plant data (Table 1) with an outlier in the y-direction. For instance, assume that the y-value of the 19th observation was recorded as 840 instead of 84. Apply PROGRESS to these contaminated data, and compare the effect of the outlier on LS, LMS, and RLS.

5. Find (or construct) an example where the least squares R^2 is larger with the outlier included than without it.

6. Apply PROGRESS to the monthly payments data of exercise 8 of Chapter 1. Give a detailed discussion of the results, including the standardized observations, the correlation coefficients, the p-values, the coefficients of determination, and the standardized residuals. Compare the LMS and the RLS with the eye fit. How many outliers are identified?

7. Let us look at an example on Belgian employers' liability insurance. Table 12 lists the number of reported accidents of people going to or

Table 12. Number of Accidents in 1975–1981

Year (x_i)	Number of Accidents (y_i)
75	18,031
76	18,273
77	16,660
78	15,688
79	23,805
80	15,241
81	13,295

Source: The 1981 Annual Report of the Belgian Association of Insurance Companies.

coming from work, from 1975 to 1981. Only accidents leading to at least 1 day of absence were counted. When looking at these data, we notice a downward trend over the years. However, the number for 1979 is very high. This is because during the first few months of 1979 it was extremely cold, with snow and ice on most days (one has to go back many decades to find records of a similar winter). Therefore, the number of fractures, sprains, and dislocations in January and February was overwhelming. Discuss the PROGRESS output for these data. Is 1979 an outlier in x or in y? (Look at the scatterplot.) Can the outlier be identified by means of the standardized observations? Explain the difference between the Pearson correlation and the Spearman correlation. How many outliers can be identified on the basis of the standardized residuals of the LS solution? And with the robust regressions? What is the effect of the outlier on the p-value of the LS slope?

8. Table 13 lists the total 1981 premium income of pension funds of Dutch firms, for 18 professional branches. In the next column the respective premium reserves are given. The highest amounts correspond to P.G.G.M., the pension fund for medical care workers. Its premium income is three times larger than the second largest branch (the building industry). Draw a plot of reserves versus income (e.g., by means of PROGRESS). Compare the simple regression of reserves as a function of income using LS with the LMS regression line. Is P.G.G.M. a good or a bad leverage point for this model? Does it have the largest LS residual? How does this compare with the RLS residuals, and what happens to the p-values of the coefficients? The residual plot indicates that a linear model is not very suitable here. Is it possible to transform either variable to make the relation more linear?

Table 13. Pension Funds for 18 Professional Branches

Index	Premium Income[a]	Premium Reserves[a]
1	10.4	272.2
2	15.6	212.9
3	16.2	120.7
4	17.9	163.6
5	37.8	226.1
6	46.9	622.9
7	52.4	1353.2
8	52.9	363.6
9	71.0	951.7
10	73.9	307.2
11	76.3	588.4
12	77.0	952.5
13	131.2	1157.3
14	151.7	2105.6
15	206.1	3581.4
16	314.7	3404.7
17	470.8	4095.3
18	1406.3	6802.7

[a]In millions of guilders.

Source: de Wit (1982, p. 52).

Sections 5 and 6

9. Run the exact fit example of Section 5 again as a simple regression through the origin.

10. Look for an example where a line through the origin is more appropriate (from subject-matter knowledge) than with intercept.

11. Repeat exercise 8 (on the pension funds of 18 professions) by means of simple regression without intercept. This makes sense because the RLS intercept was not significantly different from zero. Draw the corresponding lines through the origin. Which transformation do you propose to linearize the data?

Section 7

12. Apply the resistant line (7.1), Theil's estimator (7.2), and the repeated median (7.4) to the monthly payments data (Table 1 of Chapter 1).

13. In the framework of simple regression, Hampel (1975, p. 379) considered the line through ($\text{med}_j x_j$, $\text{med}_j y_j$) which has equal num-

bers of positive residuals on both sides of $\text{med}_j\, x_j$. (He then dismissed this estimator by stating that it may lead to a lousy fit.) Simon (1986) proposed the "median star" line, which also goes through $(\text{med}_j\, x_j,\ \text{med}_j\, y_j)$ and has the slope

$$\hat{\theta}_1 = \underset{i}{\text{med}} \left(\frac{y_i - \underset{j}{\text{med}}\, y_j}{x_i - \underset{j}{\text{med}}\, x_j} \right).$$

Show that this is the same estimator. What is its breakdown point? Write a small program and apply it to the situations of Figures 11 and 12. Repeat this for another version of the median star in which the intercept is chosen so that $\text{med}_i\, (r_i) = 0$.

14. (Research problem). Is it possible to construct a standardized version of the breakdown plot, which does not depend on random number generation?

CHAPTER 3

Multiple Regression

1. INTRODUCTION

In multiple regression, the response variable y_i is related to p explanatory variables $x_{i1}, \ldots, x_{ip}$ in the model

$$y_i = x_{i1}\theta_1 + \cdots + x_{ip}\theta_p + e_i \qquad (i = 1, \ldots, n) . \qquad (1.1)$$

As in simple regression, the least squares (LS) technique for estimating the unknown parameters $\theta_1, \ldots, \theta_p$ is quite sensitive to the presence of outlying points. The identification of such points becomes more difficult, because it is no longer possible to spot the influential points in a scatterplot. Therefore, it is important to have a tool for identifying such points.

In the last few decades, several statisticians have given consideration to robust regression, whereas others have directed their attention to regression diagnostics (see Chapter 6). Both approaches are closely related by two important common aims, namely, identifying outliers and pointing out inadequacies of the model. However, they proceed in a different way. Regression diagnostics first attempt to identify points that have to be deleted from the data set, before applying a regression method. Robust regression tackles these problems in the inverse order, by designing estimators that dampen the impact of points that would be highly influential otherwise. A robust procedure tries to accommodate the majority of the data. Bad points, lying far away from the pattern formed by the good ones, will consequently possess large residuals from the robust fit. So in addition to insensitivity to outliers, a robust regression estimator makes the detection of these points an easy job. Of course, the residuals from LS cannot be used for this purpose, because the outliers

75

may possess very small LS residuals as the LS fit is pulled too much in the direction of these deviating points.

Let us look at some examples to illustrate the need for a robust alternative to LS. The first example is the well-known stackloss data set presented by Brownlee (1965). We have selected this example because it is a set of real data and it has been examined by a great number of statisticians (Draper and Smith 1966, Daniel and Wood 1971, Andrews 1974, Andrews and Pregibon 1978, Cook 1979, Dempster and Gasko-Green 1981, Atkinson 1982, Carroll and Ruppert 1985, Li 1985, and many others) by means of several methods. The data describe the operation of a plant for the oxidation of ammonia to nitric acid and consist of 21 four-dimensional observations (listed in Table 1). The stackloss (y) has to be explained by the rate of operation (x_1), the cooling water inlet temperature (x_2), and the acid concentration (x_3). Summarizing the findings cited in the literature, it can be said that most

Table 1. Stackloss Data

Index (i)	Rate (x_1)	Temperature (x_2)	Acid Concentration (x_3)	Stackloss (y)
1	80	27	89	42
2	80	27	88	37
3	75	25	90	37
4	62	24	87	28
5	62	22	87	18
6	62	23	87	18
7	62	24	93	19
8	62	24	93	20
9	58	23	87	15
10	58	18	80	14
11	58	18	89	14
12	58	17	88	13
13	58	18	82	11
14	58	19	93	12
15	50	18	89	8
16	50	18	86	7
17	50	19	72	8
18	50	19	79	8
19	50	20	80	9
20	56	20	82	15
21	70	20	91	15

Source: Brownlee (1965).

people concluded that observations 1, 3, 4, and 21 were outliers. According to some people, observation 2 is reported as an outlier too. Least
squares regression yields the equation

$$\hat{y} = 0.716x_1 + 1.295x_2 - 0.152x_3 - 39.9 \ .$$

The LS index plot is shown in Figure 1. The standardization of the
residuals is performed by the division of the raw residuals ($r_i = y_i - \hat{y}_i$) by
the scale estimate corresponding to the fit. A horizontal band encloses the
standardized residuals between -2.5 and 2.5. In Figure 1, no outliers
strike the eye. From the LS index plot, one would conclude that the data
set contains no outliers at all because all the standardized LS residuals fall
nicely within the band. However, let us now look at Figure 2, the index
plot associated with the least median of squares (LMS) fit

$$\hat{y} = 0.714x_1 + 0.357x_2 + 0.000x_3 - 34.5 \ .$$

This plot is based on a robust fit, and does indeed reveal the presence of
harmful points. From this index plot it becomes immediately clear that

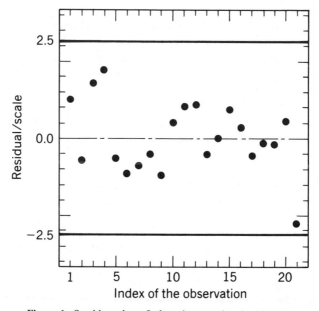

Figure 1. Stackloss data: Index plot associated with LS.

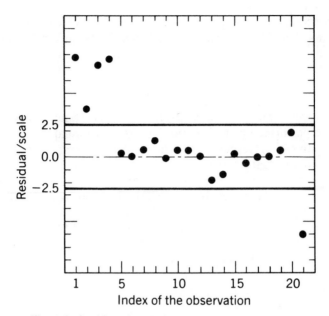

Figure 2. Stackloss data: Index plot associated with LMS.

the observations 1, 3, 4, and 21 are the most outlying, and that case 2 is intermediate because it is on the verge of the area containing the outliers. This shows how our robust regression technique is able to analyze these data in a single blow, which should be contrasted to some of the earlier analyses of the same data set, which were long and laborious.

This example once more illustrates the danger of merely looking at the LS residuals. We would like to repeat that it is necessary to compare the standardized residuals of both the LS *and* the robust method in each regression analysis. If the results of the two procedures are in substantial agreement, then the LS can be trusted. If they differ, the robust technique can be used as a reliable tool for identifying the outliers, which may then be thoroughly investigated and perhaps corrected (if one has access to the original measurements) or deleted. Another possibility is to change the model (e.g., by adding squared or cross-product terms and/or transforming the response variable). In this way, Atkinson (1985, pp. 129–136) analyzes the stackloss data by setting up models explaining $\log(y)$ by means of x_1, x_2, $x_1 x_2$, and x_1^2.

The following example comes from the social sciences. The data set contains information on 20 schools from the Mid-Atlantic and New England states, drawn from a population studied by Coleman et al.

(1966). Mosteller and Tukey (1977) analyze this sample consisting of measurements on six different variables, one of which will be treated as response. They can be described as follows:

x_1 = staff salaries per pupil

x_2 = percent of white-collar fathers

x_3 = socioeconomic status composite deviation: means for family size, family intactness, father's education, mother's education, and home items

x_4 = mean teacher's verbal test score

x_5 = mean mother's educational level (one unit is equal to two school years)

y = verbal mean test score (all sixth graders).

The data set itself is exhibited in Table 2.

Table 2. Coleman Data Set, Containing Information on 20 Schools from the Mid-Atlantic and New England States

Index	x_1	x_2	x_3	x_4	x_5	y
1	3.83	28.87	7.20	26.60	6.19	37.01
2	2.89	20.10	−11.71	24.40	5.17	26.51
3	2.86	69.05	12.32	25.70	7.04	36.51
4	2.92	65.40	14.28	25.70	7.10	40.70
5	3.06	29.59	6.31	25.40	6.15	37.10
6	2.07	44.82	6.16	21.60	6.41	33.90
7	2.52	77.37	12.70	24.90	6.86	41.80
8	2.45	24.67	−0.17	25.01	5.78	33.40
9	3.13	65.01	9.85	26.60	6.51	41.01
10	2.44	9.99	−0.05	28.01	5.57	37.20
11	2.09	12.20	−12.86	23.51	5.62	23.30
12	2.52	22.55	0.92	23.60	5.34	35.20
13	2.22	14.30	4.77	24.51	5.80	34.90
14	2.67	31.79	−0.96	25.80	6.19	33.10
15	2.71	11.60	−16.04	25.20	5.62	22.70
16	3.14	68.47	10.62	25.01	6.94	39.70
17	3.54	42.64	2.66	25.01	6.33	31.80
18	2.52	16.70	−10.99	24.80	6.01	31.70
19	2.68	86.27	15.03	25.51	7.51	43.10
20	2.37	76.73	12.77	24.51	6.96	41.01

Source: Mosteller and Tukey (1977).

The ordinary LS regression for all 20 schools is given as

$$\hat{y} = -1.79x_1 + 0.044x_2 + 0.556x_3 + 1.11x_4 - 1.81x_5 + 19.9 \ .$$

Least median of squares regression yields the fit

$$\hat{y} = 0.580x_1 + 0.058x_2 + 0.637x_3 + 0.740x_4 - 2.32x_5 + 25.1 \ .$$

The left side of Table 3 lists the LS estimates and the associated residuals. These results reveal that the LS equation slightly underestimates the response for schools 3 and 11 and overestimates it for school 18. By only examining the LS results, the conclusion would be that schools 3, 11, and 18 are furthest away from the linear model. But from the right side of Table 3, it appears that school 11 does not deviate at all from the robust fit.

The robust regression spots schools 3, 17, and 18 as outliers by assigning large standardized residuals to them. Afterwards these standar-

Table 3. Coleman Data: Estimated Verbal Mean Test Score and the Associated Residuals for the LS Fit and the LMS Fit

	LS Results		LMS Results		
Index	Estimated Response	Standardized Residuals	Estimated Response	Standardized Residuals	Weights
1	36.661	0.17	38.883	−1.59	1
2	26.860	−0.17	26.527	−0.01	1
3	40.460	−1.90	41.273	−4.04	0
4	41.174	−0.23	42.205	−1.28	1
5	36.319	0.38	37.117	−0.01	1
6	33.986	−0.04	33.917	−0.01	1
7	41.081	0.35	41.628	0.15	1
8	33.834	−0.21	32.922	0.41	1
9	40.386	0.30	41.519	−0.43	1
10	36.990	0.10	34.847	2.00	1
11	25.508	−1.06	23.169	0.11	1
12	33.454	0.84	33.514	1.43	1
13	35.949	−0.51	34.917	−0.01	1
14	33.446	−0.17	32.591	0.43	1
15	24.479	−0.86	22.717	−0.01	1
16	38.403	0.63	40.041	−0.29	1
17	33.240	−0.69	35.121	−2.82	0
18	26.698	2.41	24.918	5.76	0
19	41.977	0.54	42.662	0.37	1
20	40.747	0.13	41.027	−0.01	1

dized residuals are used for computing weights, by giving a weight of 0 to all observations that have an absolute standardized residual larger than 2.5. Doing this for Coleman's data set gives rise to the last column in Table 3. These weights are then employed for a reweighted least squares (RLS) analysis. This amounts to the same thing as performing LS regression on the reduced data set containing only the 17 points with a weight of 1.

Apart from the fitted equation and the associated residuals, outliers also affect the t-based significance levels. This is important for the construction of confidence intervals and for hypothesis testing about regression coefficients (see also Section 3 of Chapter 2). For the Coleman data set, the significance of the regression coefficients turns out to be quite different in the LS fit and the RLS fit. The t-values in Table 4 test the null hypothesis $H_0: \theta_j = 0$ against the alternative $H_1: \theta_j \neq 0$ for the LS estimates. From this table it is seen that only the variables x_3 and x_4 have LS regression coefficients that are significantly different from zero for $\alpha = 5\%$, because their t-values exceed the critical value 2.1448 of the Student distribution with 14 $(= n - p)$ degrees of freedom, and hence their p-values are below 0.05.

Let us now analyze these data with the RLS using the weights of Table 3. This gives rise to the coefficients and t-values on the right side of Table 4. It is striking that for the cleaned data set, all the explanatory variables now have regression coefficients significantly different from zero for $\alpha = 5\%$ (because the 97.5% quantile of a t-distribution with 11 degrees of freedom equals 2.2010, so all p-values are less than 0.05).

As in Chapter 2, it must be noted that the opposite can also happen. In many examples, the "significance" of certain LS regression coefficients is only caused by an outlier, and then the corresponding RLS coefficients may no longer be significantly different from zero.

The ordinary LS method is not immune to the masking effect. This means that after the deletion of one or more influential points, another

Table 4. Coleman Data: t-Values Associated with the LS and the RLS Fit

Variable	LS Results			RLS Reults		
	Coefficient	t-Value	p-Value	Coefficient	t-Value	p-Value
x_1	−1.793	−1.454	0.1680	−1.203	−2.539	0.0275
x_2	0.044	0.819	0.4267	0.082	4.471	0.0009
x_3	0.556	5.979	0.0000	0.659	19.422	0.0000
x_4	1.110	2.559	0.0227	1.098	7.289	0.0000
x_5	−1.810	−0.893	0.3868	−3.898	−5.177	0.0003
Constant	19.949	1.464	0.1652	29.750	6.095	0.0001

observation may emerge as extremely influential, which was not visible at first. Therefore, the use of a high-breakdown regression method (such as the LMS) for the determination of the weights is indispensable. As an illustration, let us consider the "Salinity data" (Table 5) that were listed by Ruppert and Carroll (1980). It is a set of measurements of water salinity (i.e., its salt concentration) and river discharge taken in North Carolina's Pamlico Sound. We will fit a linear model where the salinity is regressed against salinity lagged by two weeks (x_1), the trend, that is, the number of biweekly periods elapsed since the beginning of the spring

Table 5. Salinity Data

Index (i)	Lagged Salinity (x_1)	Trend (x_2)	Discharge (x_3)	Salinity (y)
1	8.2	4	23.005	7.6
2	7.6	5	23.873	7.7
3	4.6	0	26.417	4.3
4	4.3	1	24.868	5.9
5	5.9	2	29.895	5.0
6	5.0	3	24.200	6.5
7	6.5	4	23.215	8.3
8	8.3	5	21.862	8.2
9	10.1	0	22.274	13.2
10	13.2	1	23.830	12.6
11	12.6	2	25.144	10.4
12	10.4	3	22.430	10.8
13	10.8	4	21.785	13.1
14	13.1	5	22.380	12.3
15	13.3	0	23.927	10.4
16	10.4	1	33.443	10.5
17	10.5	2	24.859	7.7
18	7.7	3	22.686	9.5
19	10.0	0	21.789	12.0
20	12.0	1	22.041	12.6
21	12.1	4	21.033	13.6
22	13.6	5	21.005	14.1
23	15.0	0	25.865	13.5
24	13.5	1	26.290	11.5
25	11.5	2	22.932	12.0
26	12.0	3	21.313	13.0
27	13.0	4	20.769	14.1
28	14.1	5	21.393	15.1

Source: Ruppert and Carroll (1980).

season (x_2); and the volume of river discharge into the sound (x_3). Carroll and Ruppert (1985) describe the physical background of the data. They indicated that cases 5 and 16 correspond to periods of very heavy discharge. Their analysis showed that the third and sixteenth observations conspire to hide the discrepant number 5. In fact, observation 5 can be recognized as influential only after the deletion of cases 3 and 16. This is a prime example of the masking effect. On the other hand, the LMS is not affected by this phenomenon and identifies 5 and 16 in a single blow. The LS fit is given by

$$\hat{y} = 0.777x_1 - 0.026x_2 - 0.295x_3 + 9.59 \, ,$$

whereas the LMS yields the equation

$$\hat{y} = 0.356x_1 - 0.073x_2 - 1.30x_3 + 36.7 \, .$$

A residual plot associated with the LS fit is presented in Figure 3. In this figure the standardized residuals are plotted against the estimated response. In Figure 4 such a residual plot is given for the LMS regression. From Figure 3, it appears that there is nothing wrong with the fit. However, one has to keep in mind that leverage points tend to produce small LS residuals simply by virtue of their leverage. On the other hand,

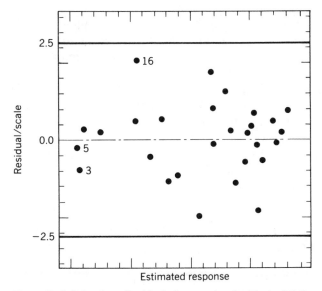

Figure 3. Salinity data: Residual plot associated with the LS fit.

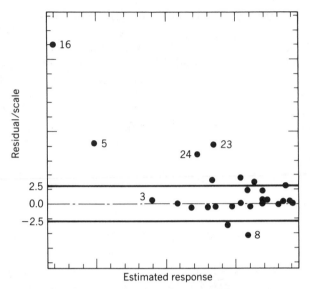

Figure 4. Salinity data: Residual plot associated with the LMS fit.

Figure 4 gives evidence of the presence of outlying observations, because some points fall far from the band. In this example, the LS residual plot cannot be trusted because it differs too much from the one associated with a robust fit.

As we said before, residual plots can also indicate possible defects in the model's functional form in the direction of the fitted values. A residual plot may, for example, display a variance pattern that is a monotone function of the response. If the functional part of the model is not misspecified, then plotting the standardized LMS residuals versus $\hat{y}_i$ gives rise to a horizontal band of points that look "structureless." Also, anomalies in the pattern might suggest a transformation of the variables in the model. For example, a curved pattern in a residual plot may lead to replacing the observed y_i by some function of y_i ($\log y_i$ or y_i raised to some power). In example 2 of Section 3 we will illustrate the use of residual plots to remedy model failures.

2. COMPUTATION OF LEAST MEDIAN OF SQUARES MULTIPLE REGRESSION

The question "How do we run program PROGRESS?" has for the most part been answered in Section 2 of Chapter 2, in the context of simple regression. The interactive session accompanying a multiple regression analysis is completely identical.

The special treatment of data sets with missing values has not yet been discussed. In that case the interactive input becomes a little bit longer. We will illustrate this situation for part of the "Air Quality" data that originated with the New York State Department of Conservation and the National Weather Service; these data were reported in Chambers et al. (1983). The whole data set consists of daily readings of air quality values from May 1, 1973 to September 30, 1973. We will use only the values for May in our example. The variables are the mean ozone concentration (in parts per billion) from 1300 to 1500 hours at Roosevelt Island (OZONE ppb), solar radiation in Longleys in the frequency band 4000–7700 Å from 0800 to 1200 hours at Central Park (SOLAR RADI), average wind speed (in miles per hour) between 0700 and 1000 hours at La Guardia Airport (WINDSPEED), and maximum daily temperature (in degrees Fahrenheit) at La Guardia Airport (TEMPERATUR). The data are exhibited in Table 6.

The aim of the analysis is to explain the ozone concentration by means of the other variables. In Table 6, one can observe that the measurements for OZONE ppb and/or SOLAR RADI are not registered for some days. PROGRESS provides two methods for handling such an incomplete data set. One or the other can be chosen by answering 1 or 2 to the question

```
CHOOSE AN OPTION FOR THE TREATMENT OF MISSING VALUES
----------------------------------------
0=THERE ARE NO MISSING VALUES IN THE DATA
1=ELIMINATION OF THE CASES FOR WHICH AT LEAST ONE VARIABLE IS MISSING
2=ESTIMATES ARE FILLED IN FOR UNOBSERVED VALUES
ENTER YOUR CHOICE:
```

When the option for missing values is not equal to zero, PROGRESS needs additional information on the missing value codes for each variable. First of all, the question

```
IS THERE A UNIQUE VALUE WHICH IS TO BE INTERPRETED
AS A MISSING MEASUREMENT FOR ANY VARIABLE?
ANSWER YES OR NO:
```

must be answered. When the answer is YES, the statement

```
PLEASE ENTER THIS VALUE:
```

appears, where the user has to give a value that will be interpreted as the

Table 6. Air Quality Data Set for May 1973[a]

Index (i)	SOLAR RADI (x_1)	WINDSPEED (x_2)	TEMPERATUR (x_3)	OZONE ppb (y)
1	190	7.4	67	41
2	118	8.0	72	36
3	149	12.6	74	12
4	313	11.5	62	18
5	9999	14.3	56	999
6	9999	14.9	66	28
7	299	8.6	65	23
8	99	13.8	59	19
9	19	20.1	61	8
10	194	8.6	69	999
11	9999	6.9	74	7
12	256	9.7	69	16
13	290	9.2	66	11
14	274	10.9	68	14
15	65	13.2	58	18
16	334	11.5	64	14
17	307	12.0	66	34
18	78	18.4	57	6
19	322	11.5	68	30
20	44	9.7	62	11
21	8	9.7	59	1
22	320	16.6	73	11
23	25	9.7	61	4
24	92	12.0	61	32
25	66	16.6	57	999
26	266	14.9	58	999
27	9999	8.0	57	999
28	13	12.0	67	23
29	252	14.9	81	45
30	223	5.7	79	115
31	279	7.4	76	37

[a]The values 9999 of x_1 (solar radiation) indicate missing measurements. Also the numbers 999 of y (ozone ppb) correspond to missing values.

Source: Chambers et al. (1983).

missing value code for all the variables in the analysis. Otherwise, the user has to answer the following question for each variable:

```
DOES VARIABLE .......... CONTAIN MISSING VALUE(S)?
ANSWER YES OR NO:
```

(Instead of the dots, the actual label of the variable concerned will be printed.) If the answer to this question is YES, then the user has to enter the missing value code for this variable:

```
ENTER THE VALUE OF THIS VARIABLE WHICH HAS TO BE INTERPRETED AS
THE MISSING VALUE CODE:
```

In both missing value options, the program makes an inventory of the cases with incomplete data, for each variable for which missing values were announced. When there are variables for which more than 80% of the cases have a missing value, the program will terminate (after giving a message). When analyzing the same data again, the user should no longer include these partially observed variables because they do not contain enough information.

Let us now look at the two missing value options provided by PROGRESS. The first avenue open to the researcher is to eliminate the cases for which at least one variable is missing. This can be achieved by setting the missing value option equal to 1. When the print option differs from 0, a complete table is given, where for each case the user of the program can see which variables were missing. The regression analysis is then performed on the remaining cases. (Each case keeps its original number.)

In fact, this option can also be used for another purpose. It may happen that one wishes to perform the analysis only on a part of the data set. For example, in a sample containing attributes of many people, one may want to fit a linear model to the women only. Then one can use the missing value option 1 for the variable associated with the sex of each person. The value corresponding to men then has to be taken as the "missing" value code, in order to eliminate the observations for men.

An alternative treatment, which corresponds to option 2, consists of filling in guesses for unobserved values. This can be necessary in circumstances where deleting all partially observed cases would result in an extremely small sample (which might even be empty). Nevertheless, even for this option, cases for which the response y is lacking will first be dropped from the data set. Also, cases for which *all* explanatory variables are missing will be removed. On this reduced data set, missing data will be replaced by the median of the corresponding variable. On the resulting data set, the previous methods of estimation can be applied. When the print option is not 0, the output of PROGRESS delivers a complete inventory of the deleted cases and of the missing values that are replaced by medians. Also here, the cases retain the numbering of the original data set.

For the Air Quality data, we encoded the missing values of variable SOLAR RADI by 9999, and those of OZONE ppb by 999. These codes are acceptable because these values have not been observed. Let us now look at a listing of the interactive input for this data set.

```
                    **********************
                    *  P R O G R E S S  *
                    **********************

ENTER THE NUMBER OF CASES PLEASE : 31

DO YOU WANT A CONSTANT TERM IN THE REGRESSION?
ANSWER YES OR NO : YES

WHAT IS THE TOTAL NUMBER OF VARIABLES IN YOUR DATA SET?
------------------------------------------------------------
PLEASE GIVE A NUMBER BETWEEN 1 AND 50 :4

WHICH VARIABLE DO YOU CHOOSE AS RESPONSE VARIABLE?
------------------------------------------------------------
OUT OF THESE    4 GIVE ITS POSITION : 4

GIVE A LABEL FOR THIS VARIABLE (AT MOST 10 CHARACTERS) : OZONE ppb

HOW MANY EXPLANATORY VARIABLES DO YOU WANT TO USE IN THE ANALYSIS?
------------------------------------------------------------------
(AT MOST    3 ) : 3

EXPLANATORY VARIABLES         :   POSITION   LABEL (AT MOST 10 CHARACTERS)
------------------------------↓↓↓↓----↓↓↓↓↓↓↓↓↓↓--------------------
NUMBER     1                  :      1       SOLAR RADI
NUMBER     2                  :      2       WINDSPEED
NUMBER     3                  :      3       TEMPERATUR

HOW MUCH OUTPUT DO YOU WANT?
---------------------------
0 = SMALL OUTPUT        : LIMITED TO BASIC RESULTS
1 = MEDIUM-SIZED OUTPUT: ALSO INCLUDES TABLE WITH THE OBSERVED VALUES OF Y,
                         THE ESTIMATES OF Y, THE RESIDUALS AND THE WEIGHTS
2 = LARGE OUTPUT        : ALSO INCLUDES THE DATA ITSELF
ENTER YOUR CHOICE : 2

DO YOU WANT TO LOOK AT THE RESIDUALS?
-------------------------------------
0 = NO RESIDUAL PLOTS
1 = PLOT OF THE STANDARDIZED RESIDUALS VERSUS THE ESTIMATED VALUE OF Y
2 = PLOT OF THE STANDARDIZED RESIDUALS VERSUS THE INDEX OF THE OBSERVATION
3 = PERFORMS BOTH TYPES OF RESIDUAL PLOTS
ENTER YOUR CHOICE : 3

DO YOU WANT TO COMPUTE OUTLIER DIAGNOSTICS ?
YES OR NO: NO

GIVE THE NAME OF THE FILE CONTAINING THE DATA (e.g. TYPE  A:EXAMPLE.DAT ),
or TYPE  KEY  IF YOU PREFER TO ENTER THE DATA BY KEYBOARD.
WHAT DO YOU CHOOSE ?   B:AIRMAY.DAT
```

```
WHERE DO YOU WANT YOUR OUTPUT?
------------------------------
    TYPE  CON  IF YOU WANT IT ON THE SCREEN
or TYPE  PRN  IF YOU WANT IT ON THE PRINTER
or TYPE THE NAME OF A FILE (e.g. B:EXAMPLE.OUT)
(WARNING : IF THERE ALREADY EXISTS A FILE WITH THE SAME NAME
           THE OLD FILE WILL BE OVERWRITTEN.)
WHAT DO YOU CHOOSE ?  B:AIRMAY.RES


PLEASE ENTER A TITLE FOR THE OUTPUT (AT MOST 60 CHARACTERS):
-----------------------------------------------------------
  AIR QUALITY MEASUREMENTS FOR NEW YORK

DO YOU WANT TO READ THE DATA IN FREE FORMAT?
--------------------------------------------
THIS MEANS THAT YOU ONLY HAVE TO INSERT BLANK(S) BETWEEN NUMBERS.
(WE ADVISE USERS WITHOUT KNOWLEDGE OF FORTRAN FORMATS TO ANSWER YES.)
MAKE YOUR CHOICE (YES/NO): YES


WHICH VERSION OF THE ALGORITHM WOULD YOU LIKE TO USE?
-----------------------------------------------------
Q = QUICK VERSION
E = EXTENSIVE SEARCH
ENTER YOUR CHOICE PLEASE (Q OR E) : E

CHOOSE AN OPTION FOR THE TREATMENT OF MISSING VALUES
----------------------------------------------------
0 = THERE ARE NO MISSING VALUES IN THE DATA
1 = ELIMINATION OF THE CASES FOR WHICH AT LEAST ONE VARIABLE IS MISSING
2 = ESTIMATES ARE FILLED IN FOR UNOBSERVED VALUES
ENTER YOUR CHOICE : 1




        *********************************************************************
        * P R O G R E S S  WILL PERFORM A REGRESSION WITH CONSTANT TERM *
        *********************************************************************

THE NUMBER OF CASES EQUALS                     31
THE NUMBER OF EXPLANATORY VARIABLES EQUALS      3
OZONE ppb  IS THE RESPONSE VARIABLE.
YOUR DATA RESIDE ON FILE      : B:AIRMAY.DAT
TITLE FOR OUTPUT : AIR QUALITY MEASUREMENTS FOR NEW YORK
THE DATA WILL BE READ IN FREE FORMAT.
LARGE OUTPUT IS WANTED.
BOTH TYPES OF RESIDUAL PLOTS ARE WANTED.
THE EXTENSIVE SEARCH ALGORITHM WILL BE USED.
TREATMENT OF MISSING VALUES IN OPTION 1: THIS MEANS THAT A CASE WITH A
MISSING VALUE FOR AT LEAST ONE VARIABLE WILL BE DELETED.
YOUR OUTPUT WILL BE WRITTEN ON : B:AIRMAY.RES

ARE ALL THESE OPTIONS OK ?  YES OR NO : YES

IS THERE A UNIQUE VALUE WHICH IS TO BE INTERPRETED
AS A MISSING MEASUREMENT FOR ANY VARIABLE?
ANSWER YES OR NO : NO
DOES VARIABLE SOLAR RADI CONTAIN MISSING VALUE(S)?
ANSWER YES OR NO : YES
ENTER THE VALUE OF THIS VARIABLE WHICH HAS TO BE INTERPRETED AS
THE MISSING VALUE CODE : 9999
DOES VARIABLE WINDSPEED  CONTAIN MISSING VALUE(S)?
ANSWER YES OR NO : NO
DOES VARIABLE TEMPERATUR CONTAIN MISSING VALUE(S)?
ANSWER YES OR NO : NO
DOES THE RESPONSE VARIABLE CONTAIN MISSING VALUE(S)?
ANSWER YES OR NO : YES
ENTER THE VALUE OF THIS VARIABLE WHICH HAS TO BE INTERPRETED AS
THE MISSING VALUE CODE : 999
```

The treatment of the missing values for this data set appears on the output of PROGRESS as follows:

```
TREATMENT OF MISSING VALUES IN OPTION 1: THIS MEANS THAT A CASE WITH A
MISSING VALUE FOR AT LEAST ONE VARIABLE WILL BE DELETED.

YOUR DATA RESIDE ON FILE        : B:AIRMAY.DAT
VARIABLE SOLAR RADI HAS A MISSING VALUE FOR    4 CASES.
VARIABLE OZONE ppb  HAS A MISSING VALUE FOR    5 CASES.

CASE HAS A MISSING VALUE FOR VARIABLES (VARIABLE NUMBER  5 IS THE RESPONSE)
----                         ---------
   5                            1   5
   6                            1
  10                            5
  11                            1
  25                            5
  26                            5
  27                            1   5


THERE ARE    24 CASES STAYING IN THE ANALYSIS.
THE OBSERVATIONS, AFTER TREATMENT OF MISSING VALUES :

       SOLAR RADI WINDSPEED  TEMPERATUR OZONE ppb
   1     190.0000    7.4000    67.0000    41.0000
   2     118.0000    8.0000    72.0000    36.0000
   3     149.0000   12.6000    74.0000    12.0000
   4     313.0000   11.5000    62.0000    18.0000
   7     299.0000    8.6000    65.0000    23.0000
   8      99.0000   13.8000    59.0000    19.0000
   9      19.0000   20.1000    61.0000     8.0000
  12     256.0000    9.7000    69.0000    16.0000
  13     290.0000    9.2000    66.0000    11.0000
  14     274.0000   10.9000    68.0000    14.0000
  15      65.0000   13.2000    58.0000    18.0000
  16     334.0000   11.5000    64.0000    14.0000
  17     307.0000   12.0000    66.0000    34.0000
  18      78.0000   18.4000    57.0000     6.0000
  19     322.0000   11.5000    68.0000    30.0000
  20      44.0000    9.7000    62.0000    11.0000
  21       8.0000    9.7000    59.0000     1.0000
  22     320.0000   16.6000    73.0000    11.0000
  23      25.0000    9.7000    61.0000     4.0000
  24      92.0000   12.0000    61.0000    32.0000
  28      13.0000   12.0000    67.0000    23.0000
  29     252.0000   14.9000    81.0000    45.0000
  30     223.0000    5.7000    79.0000   115.0000
  31     279.0000    7.4000    76.0000    37.0000
```

The least squares analysis of the reduced data set is printed in Table 7. Some of the fitted OZONE ppb values (cases 9 and 18) are negative, which is physically impossible. The strange behavior of this fit is easy to understand when comparing it to the robust analysis in Table 8. First of all, the equations of both fits differ substantially from each other. In the

Table 7. Air Quality Data: LS Fit with Estimated Response and Standardized Residuals

Variable	Coefficient	Standard Error	t-Value
SOLAR RADI	−0.01868	0.03628	−0.51502
WINDSPEED	−1.99577	1.14092	−1.74926
TEMPERATUR	1.96332	0.66368	2.95823
Constant	−79.99270	46.81654	−1.70864

Index	Estimated "OZONE ppb"	Standardized Residuals
1	33.231	0.43
2	43.195	−0.40
3	37.362	−1.41
4	12.933	0.28
7	24.873	−0.10
8	6.452	0.70
9	−0.700	0.48
12	31.334	−0.85
13	25.807	−0.82
14	26.639	−0.70
15	6.321	0.65
16	16.468	−0.14
17	19.901	0.78
18	−6.263	0.68
19	24.545	0.30
20	21.552	−0.59
21	16.335	−0.85
22	24.221	−0.73
23	19.944	−0.89
24	14.101	0.99
28	27.357	−0.24
29	44.591	0.02
30	59.567	<u>3.08</u>
31	49.238	−0.68

column comprising the standardized residuals in Table 8, case 30 emerges as an outlier. This single outlier is the cause of the bad LS fit because it has tilted the LS hyperplane in its direction. Because of this, the other points are not well fitted anymore by LS.

Of course, negative predicted values are to be expected whenever a linear equation is fitted. Indeed, when the regression surface is not horizontal, there always exist vectors **x** for which the corresponding

Table 8. Air Quality Data: RLS Fit Based on LMS

Variable	Coefficient	Standard Error	t-Value
SOLAR RADI	0.00559	0.02213	0.25255
WINDSPEED	−0.74884	0.71492	−1.04745
TEMPERATUR	0.99352	0.42928	2.31438
Constant	−37.51613	28.95417	−1.29571

Index	Estimated "OZONE ppb"	Standardized Residuals
1	24.571	1.52
2	28.686	0.68
3	27.402	−1.42
4	17.220	0.07
7	22.294	0.07
8	11.321	0.71
9	8.143	−0.01
12	25.204	−0.85
13	22.788	−1.09
14	23.413	−0.87
15	10.587	0.69
16	19.325	−0.49
17	20.786	1.22
18	5.772	0.02
19	23.232	0.63
20	17.065	−0.56
21	13.883	−1.19
22	24.369	−1.24
23	15.965	−1.11
24	14.617	1.61
28	20.137	0.26
29	33.210	1.09
30	37.950	<u>7.13</u>
31	34.010	0.28

predicted $\hat{y}$ is negative. This is still true for robust regression, where one can easily encounter a negative $\hat{y}_i$ in a leverage point $\mathbf{x}_i$. However, in the above example the LS predictions are negative in *good* observations!

3. EXAMPLES

In the preceding sections we have seen that the high-breakdown fits provide much to think about. In particular, the standardized residuals associated with a robust fit yield powerful diagnostic tools. For example, they can be displayed in residual plots. These graphics make it easy to detect outlying values and call attention to model failures. Also, the residuals can be used to determine a weight for each point. Such weights make it possible to bound the effect of the outliers by using them for RLS. The fit resulting from this reweighting describes the trend followed by the majority of the data. The statistics associated with this fit, like t- and F-values, are more trustworthy than those calculated from the ordinary LS regression.

Example 1:   Hawkins–Bradu–Kass Data

We shall use the data generated by Hawkins, Bradu, and Kass (1984) for illustrating some of the merits of a robust technique. Such artificial data offer the advantage that at least the position of the bad points is known exactly, which avoids some of the controversies that are inherent in the analysis of real data. In this way, the effectiveness of the technique can be measured. The data set is listed in Table 9 and consists of 75 observations in four dimensions (one response and three explanatory variables). The first 10 observations are bad leverage points, and the next four points are good leverage points (i.e., their $\mathbf{x}_i$ are outlying, but the corresponding y_i fit the model quite well). We will compare the LS with our robust regression. In Hawkins et al. (1984), it is mentioned that M-estimators do not produce the expected results, because the outliers (the bad leverage points) are masked and the four good leverage points appear outlying because they possess large residuals from those fits. This should not surprise us, because M-estimators break down early in the presence of leverage points. A certain version of the "elemental sets" diagnostic of Hawkins et al. locates the outliers, but this technique would not have coped with a larger fraction of contamination. (More details about outlier diagnostics and their breakdown points will be provided in Chapter 6.)

Let us now restrict the discussion to LS and LMS. From the index plot associated with LS (see Figure 5), it appears that observations 11, 12, and 13 are outliers because they fall outside the ± 2.5 band. Unfortunately, from the generation one knows that these are good observations. The bad leverage points have tilted the LS fit totally in their direction. Therefore the first 10 points have small standardized LS residuals. (The index plots associated with M-estimators are very similar to that of LS.)

On the other hand, the index plot of the LMS (Figure 6) identifies the first 10 points as the influential observations. The four good leverage points fall in the neighborhood of the dashed line through 0. This means that these points are well accommodated by the LMS fit. Clearly, the

Table 9. Artificial Data Set of Hawkins, Bradu, and Kass (1984)

Index	x_1	x_2	x_3	y	Index	x_1	x_2	x_3	y
1	10.1	19.6	28.3	9.7	39	2.1	0.0	1.2	−0.7
2	9.5	20.5	28.9	10.1	40	0.5	2.0	1.2	−0.5
3	10.7	20.2	31.0	10.3	41	3.4	1.6	2.9	−0.1
4	9.9	21.5	31.7	9.5	42	0.3	1.0	2.7	−0.7
5	10.3	21.1	31.1	10.0	43	0.1	3.3	0.9	0.6
6	10.8	20.4	29.2	10.0	44	1.8	0.5	3.2	−0.7
7	10.5	20.9	29.1	10.8	45	1.9	0.1	0.6	−0.5
8	9.9	19.6	28.8	10.3	46	1.8	0.5	3.0	−0.4
9	9.7	20.7	31.0	9.6	47	3.0	0.1	0.8	−0.9
10	9.3	19.7	30.3	9.9	48	3.1	1.6	3.0	0.1
11	11.0	24.0	35.0	−0.2	49	3.1	2.5	1.9	0.9
12	12.0	23.0	37.0	−0.4	50	2.1	2.8	2.9	−0.4
13	12.0	26.0	34.0	0.7	51	2.3	1.5	0.4	0.7
14	11.0	34.0	34.0	0.1	52	3.3	0.6	1.2	−0.5
15	3.4	2.9	2.1	−0.4	53	0.3	0.4	3.3	0.7
16	3.1	2.2	0.3	0.6	54	1.1	3.0	0.3	0.7
17	0.0	1.6	0.2	−0.2	55	0.5	2.4	0.9	0.0
18	2.3	1.6	2.0	0.0	56	1.8	3.2	0.9	0.1
19	0.8	2.9	1.6	0.1	57	1.8	0.7	0.7	0.7
20	3.1	3.4	2.2	0.4	58	2.4	3.4	1.5	−0.1
21	2.6	2.2	1.9	0.9	59	1.6	2.1	3.0	−0.3
22	0.4	3.2	1.9	0.3	60	0.3	1.5	3.3	−0.9
23	2.0	2.3	0.8	−0.8	61	0.4	3.4	3.0	−0.3
24	1.3	2.3	0.5	0.7	62	0.9	0.1	0.3	0.6
25	1.0	0.0	0.4	−0.3	63	1.1	2.7	0.2	−0.3
26	0.9	3.3	2.5	−0.8	64	2.8	3.0	2.9	−0.5
27	3.3	2.5	2.9	−0.7	65	2.0	0.7	2.7	0.6
28	1.8	0.8	2.0	0.3	66	0.2	1.8	0.8	−0.9
29	1.2	0.9	0.8	0.3	67	1.6	2.0	1.2	−0.7
30	1.2	0.7	3.4	−0.3	68	0.1	0.0	1.1	0.6
31	3.1	1.4	1.0	0.0	69	2.0	0.6	0.3	0.2
32	0.5	2.4	0.3	−0.4	70	1.0	2.2	2.9	0.7
33	1.5	3.1	1.5	−0.6	71	2.2	2.5	2.3	0.2
34	0.4	0.0	0.7	−0.7	72	0.6	2.0	1.5	−0.2
35	3.1	2.4	3.0	0.3	73	0.3	1.7	2.2	0.4
36	1.1	2.2	2.7	−1.0	74	0.0	2.2	1.6	−0.9
37	0.1	3.0	2.6	−0.6	75	0.3	0.4	2.6	0.2
38	1.5	1.2	0.2	0.9					

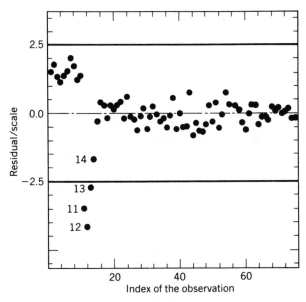

Figure 5. Hawkins–Bradu–Kass data: Index plot associated with LS regression.

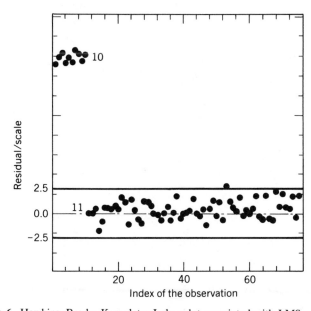

Figure 6. Hawkins–Bradu–Kass data: Index plot associated with LMS regression.

conclusions drawn from the LMS index plot agree with the construction of the data.

The following example illustrates the use of residual plots for model specification.

Example 2: *Cloud Point Data*

Table 10 shows a set of measurements concerning the cloud point of a liquid (Draper and Smith 1966, p. 162). The cloud point is a measure of the degree of crystallization in a stock and can be measured by the refractive index. The purpose is to construct a model where the percentage of I-8 in the base stock can be used as a predictor for the cloud point. Because the data contain only two variables, it is possible to explore the relation between these variables in a scatterplot. The scatterplot associated with Table 10 can be found in Figure 7.

The curved pattern in Figure 7 indicates that a simple linear model is not adequate. We will now examine whether the residual plot associated

Table 10. Cloud Point of a Liquid

Index (i)	Percentage of I-8 (x)	Cloud Point (y)
1	0	22.1
2	1	24.5
3	2	26.0
4	3	26.8
5	4	28.2
6	5	28.9
7	6	30.0
8	7	30.4
9	8	31.4
10	0	21.9
11	2	26.1
12	4	28.5
13	6	30.3
14	8	31.5
15	10	33.1
16	0	22.8
17	3	27.3
18	6	29.8
19	9	31.8

Source: Draper and Smith (1969, p. 162).

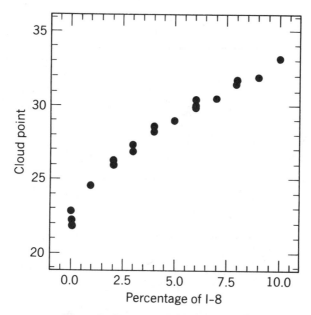

Figure 7. Cloud point data: Scatterplot.

with the linear fit would have suggested this. From the residual plot of the LS line in Figure 8, it is clear that the straight-line fit is imperfect because the residuals appear not to be randomly distributed about the zero line.

In order to be sure that this pattern is not caused by the presence of outliers, we will compare Figure 8 to the residual plot associated with the RLS line (Figure 9). The departure from linearity is magnified in this plot. The residuals inside the ±2.5 band tend to follow a parabolic curve. Those associated with cases 1, 10, and 16 fall outside the band, which indicates that they have large residuals from the RLS line. In spite of the fact that the slopes of both the LS and RLS lines are significantly different from zero (see the t-values in Table 11), the linear model is not appropriate.

Moreover, the value of R^2, which is a measure of model adequacy, is high. For LS, R^2 equals 0.955. The RLS value of R^2 is 0.977, showing that 97.7% of the total variation in the response is accounted for by the explanatory variable. Of course, R^2 is only a single number summary. It is not able to characterize an entire distribution or to indicate all possible defects in the functional form of the model. A large R^2 does not ensure that the data have been well fitted. On the other hand, the residual plots do embody the functional part of the model. Therefore, inspection of

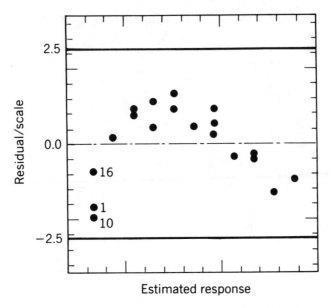

Figure 8. Cloud point data: Residual plot associated with LS.

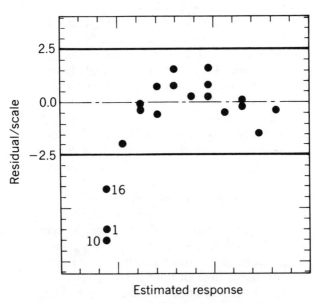

Figure 9. Cloud point data: Residual plot associated with LMS-based RLS.

Table 11. Cloud Point Data: Estimated Slope and Intercept by LS and RLS Regression, with Their t-Values

Variable	LS Results			RLS Results		
	$\hat{\theta}$	Standard Error	t-Value	$\hat{\theta}$	Standard Error	t-Value
Percentage of I-8	1.05	0.055	18.9	0.89	0.036	24.7
Constant	23.35	0.297	78.7	24.37	0.212	115.2
	$R^2 = 0.955$			$R^2 = 0.977$		

these plots is an indispensable part of regression analysis, even when R^2 is large or in the case of significant t-values. When this graphical display reveals an unexpected pattern, one has to remedy the defect by adapting the model. Depending on the anomalies in the pattern of the residuals, it may happen that an additional explanatory variable is necessary, or that some variables in the model have to be transformed. The possible improved models have to be restricted to those that are linear in the coefficients, or else we leave the domain of linear regression. However, several nonlinear functions are linearizable by using a suitable transformation. Daniel and Wood (1971) list various types of useful transformations.

A curved plot such as Figure 9 is one way to indicate nonlinearity. The usual approach to account for this apparent curvature is to consider the use of a quadratic model such as

$$y = \theta_1 x + \theta_2 x^2 + \theta_3 + e . \tag{3.1}$$

The LS and RLS estimates for this model are given in Table 12, along with some summary statistics.

The R^2 values for both the LS and the RLS have increased a little bit, whereas the t-values of the regression coefficients have hardly changed. For both fits, the coefficients are significantly different from zero at $\alpha = 5\%$. Let us now look at the distribution of the residuals associated with LS and RLS, in Figures 10 and 11, respectively.

Examining the LS residual plot, it would appear that the additional squared term has not been completely successful in restoring the distribution of the residuals. The pattern is not entirely neutral. Before deciding to change the model again, let us analyze the information provided by the RLS residual plot. From this plot, some outliers strike the eye. The observations 1, 10, and 15, which have obtained a zero weight from the LMS fit, fall outside the horizontal band. (Notwithstanding their zero

Table 12. Cloud Point Data: LS and RLS Estimates for the Quadratic Model along with Summary Values

Variable	$\hat{\theta}$	LS Results Standard Error	t-Value	$\hat{\theta}$	RLS Results Standard Error	t-Value
Percentage of I-8	1.67	0.099	16.9	1.57	0.084	18.8
(Percentage of I-8)^2	−0.07	0.010	−6.6	−0.07	0.009	−7.5
Constant	22.56	0.198	113.7	22.99	0.172	133.7
		$R^2 = 0.988$			$R^2 = 0.993$	

weights, the outliers are still indicated in the plot.) The presence of these points has affected the LS estimates because LS tries to make all the residuals small, even those associated with outliers, at the cost of an increased estimate of θ_1. On the other hand, the outlying observations do not act upon the RLS estimates, at which they obtain a large residual. The residuals inside the band in Figure 11, which correspond to the other observations, display no systematic pattern of variation. Summarizing the findings from Figures 10 and 11, one can conclude that the quadratic equation describes a more appropriate fit than the

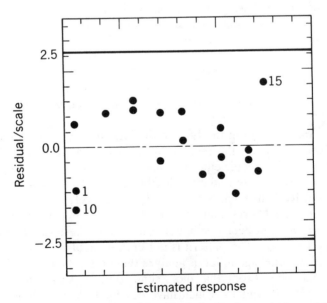

Figure 10. Cloud point data: LS residual plot for the quadratic model.

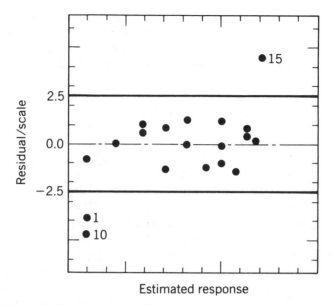

Figure 11. Cloud point data: RLS residual plot for the quadratic model.

simple line. Moreover, the residual plot of the RLS quadratic fit locates three observations that were responsible for the deformed pattern in the residual plot of the LS quadratic fit.

The presence of outliers or the choice of an appropriate model are not the only problems in regression analysis. A nearly linear dependence between two or more explanatory variables can also seriously disturb the estimated regression surface or make the regression coefficients uninterpretable. This phenomenon is called *multicollinearity*. (The terms *collinearity* and *ill-conditioning* are also employed in the literature.) The ideal situation would be that there is no relation among the variables in the factor space. In that case, it is easy to interpret a regression coefficient as the amount of change in the response when the corresponding explanatory variable grows with one unit, while the other explanatory variable(s) remain fixed. When collinearity is present in the data, the contribution of a single explanatory variable to the regression equation is hard to estimate.

The detection of multicollinearity may be very complicated. When there are only two explanatory variables, then collinearity leads to a high absolute value of the Pearson correlation or the alternative (more robust) Spearman rank correlation coefficient. For higher-dimensional factor spaces, the two-by-two correlation coefficients are not always sufficient for discovering collinearity, because the collinearity may involve several variables. Therefore, the squared multiple correlation coefficients R_j^2 of the regression of x_j on the remaining x-variables are possible diagnostics for measuring the degree to which any x_j is related to the other explanatory variables.

Chatterjee and Price (1977), Belsley et al. (1980), Weisberg (1980), and Hocking (1983), among others, report some tools for identifying collinearity. Most of these tools are based on the correlation matrix or its inverse. For example, the so-called Variance Inflation Factor (VIF) is based on the estimated variance of the ith regression coefficient (obtained from LS). There are some debates whether or not the data should be standardized first, because this may have a large effect on the resulting collinearity diagnostics (see the article by Belsley 1984 and its discussion). Another approach to deal with collinearity is ridge regression (Hoerl and Kennard 1970, 1981), based on the principle of using a little bit of all the variables rather than all of some variables and none of the remaining ones (Marquardt and Snee 1975). We will not describe these techniques in further detail here. Unfortunately, most of them are not immune to the presence of contamination, as was also noted by Mason and Gunst (1985) and Ercil (1986). For instance, consider the nonrobustness of the Pearson correlation coefficient, which may be affected a great deal by outliers. This means that the correlation coefficient can be close to zero because of the presence of a single outlier disturbing an otherwise linear relationship, thereby hiding collinearity. On the other hand, the correlation coefficient can also be carried arbitrarily close to 1 by means of a far outlier, which appears to create collinearity. Therefore, the identification of linear dependencies in factor space, combined with the detection of outliers, is an important problem of regression analysis. Indeed, collinearity inflates the variance of the regression coefficients, may be responsible for a wrong sign of the coefficients, and may affect statistical inference in general. Therefore, in cases where the presence of collinearity is anticipated, we recommend the use of the classical collinearity diagnostics on both the original data set and on the reweighted one based on LMS, in which the outliers have been removed.

Example 3: Heart Catheterization Data

The "Heart Catheterization" data set of Table 13 (from Weisberg 1980, p. 218 and Chambers et al. 1983, p. 310) demonstrates some of the effects caused by collinearity.

A catheter is passed into a major vein or artery at the femoral region

Table 13. Heart Catheterization Data[a]

Index (i)	Height (x_1)	Weight (x_2)	Catheter Length (y)
1	42.8	40.0	37
2	63.5	93.5	50
3	37.5	35.5	34
4	39.5	30.0	36
5	45.5	52.0	43
6	38.5	17.0	28
7	43.0	38.5	37
8	22.5	8.5	20
9	37.0	33.0	34
10	23.5	9.5	30
11	33.0	21.0	38
12	58.0	79.0	47

[a]Patient's height is in inches, patient's weight in pounds, and catheter length is in centimeters.

Source: Weisberg (1980, p. 218).

and moved into the heart. The catheter can be maneuvered into specific regions to provide information concerning the heart function. This technique is sometimes applied to children with congenital heart defects. The proper length of the introduced catheter has to be guessed by the physician. For 12 children, the proper catheter length (y) was determined by checking with a fluoroscope that the catheter tip had reached the right position. The aim is to describe the relation between the catheter length and the patient's height (x_1) and weight (x_2). The LS computations, as well as those for RLS, for the model

$$y = \theta_1 x_1 + \theta_2 x_2 + \theta_3 + e$$

are given in Table 14.

For both regressions, the F-value is large enough to conclude that $\hat{\theta}_1$ and $\hat{\theta}_2$ together contribute to the prediction of the response. Looking at the t-test for the individual regression coefficients, it follows that the LS estimate for θ_1 is not significantly different from zero at $\alpha = 5\%$. The same can be said for θ_2. However, the F-statistic says that the two x-variables viewed en bloc are important. For the RLS fit, $\hat{\theta}_1$ is not yet significantly different from zero, which means that the corresponding explanatory variable contributes little to the model. Also, the sign of $\hat{\theta}_1$ makes no sense in this context. Such phenomena typically occur in

Table 14. Heart Catheterization Data: LS and RLS Results

	LS Results			RLS Results		
Variable	$\hat{\theta}$	t-Value	p-Value	$\hat{\theta}$	t-Value	p-Value
Height	0.211	0.6099	0.5570	−0.723	−2.0194	0.0900
Weight	0.191	1.2074	0.2581	0.514	3.7150	0.0099
Constant	20.38	2.4298	0.0380	48.02	4.7929	0.0030
	$F = 21.267$ ($p = 0.0004$)			$F = 32.086$ ($p = 0.0006$)		

situations where collinearity is present. Indeed, Table 15 shows the high correlations among the variables, as computed by PROGRESS.

The correlation between height and weight is so high that either variable almost completely determines the other. Moreover, the low t-value for the regression coefficients confirms that either of the explanatory variables may be left out of the model. For the heart catheterization data, we will drop the variable weight and look at the simple regression of catheter length on height. A scatterplot of this two-dimensional data set is given in Figure 12, along with the LS and RLS fits.

The LS yields the fit $\hat{y} = 0.612x + 11.48$ (dashed line in Figure 12). The RLS fit $\hat{y} = 0.614x + 11.11$ (which now has a positive $\hat{\theta}_1$) lies close to the LS. Note that cases 5, 6, 8, 10, and 11 lie relatively far from the RLS fit. Because they are nicely balanced above and below the RLS line, the LS and RLS fits do not differ visibly for this sample.

The alternative choice would be to use weight instead of height as our predictor. A plot of measurements of catheter length versus the patient's weight is shown in Figure 13. This scatterplot suggests that a linear model

Table 15. Heart Catheterization Data: Correlations Between the Variables

	Pearson Correlation Coefficients		
Height	1.00		
Weight	0.96	1.00	
Catheter length	0.89	0.90	1.00
	Height	Weight	Catheter length
	Spearman Rank Correlation Coefficients		
Height	1.00		
Weight	0.92	1.00	
Catheter length	0.80	0.86	1.00
	Height	Weight	Catheter length

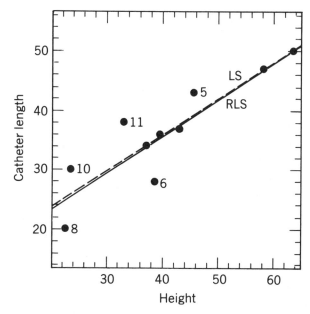

Figure 12. Heart catheterization data: Scatterplot of proper catheter length versus height for 12 children, with LS fit (dashed line) and RLS fit (solid line).

is not appropriate here. It is clear that a transformation is required. On physical grounds, one might try to use the cube root of the weight instead of the weight itself in order to obtain linearity.

Even when there is no evidence of multicollinearity, it may happen that the complete set of explanatory variables is too large for using them all in the model. In the first place, too many variables make it hard to understand the described mechanism. This should, however, be combined with the objective of explaining the variability of the response as much as possible, which leads to the consideration of more explanatory variables. But from a statistical point of view, one can sometimes say that the reduction of the number of variables improves the precision of the fit. Indeed, explanatory variables for which the associated regression coefficients are not significantly different from zero may increase the variance of the estimates. Subject-matter knowledge is sometimes sufficient to decide which of the possible equations is most appropriate. Otherwise one has to apply a statistical procedure for finding a suitable subset of the variables. This problem is called *variable selection*. It is closely linked to the problem of model specification, which we discussed above. Indeed, the problem of variable selection also includes the question, "In which form

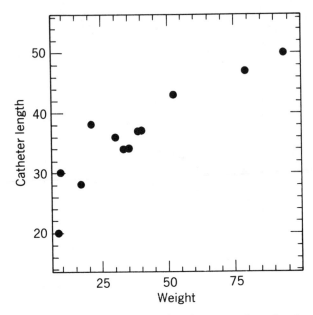

Figure 13. Heart catheterization data: Scatterplot of proper catheter length versus weight for 12 children.

should the explanatory variable enter the equation: as the original variable, or as a squared term, or as a logarithmic term, and so on?". For simplicity, one usually treats these problems separately in order to avoid the ideal, but intractable, approach consisting of a simultaneous treatment of both problems. The presence of outliers, either in y or in $\mathbf{x}$, complicates the situation even more. Therefore, we recommend starting the analysis with a high-breakdown regression on the full set of variables in order to determine the weights of the observations. Then, as a first approximation, one can use a classical technique for variable selection on the "cleaned" sample. Such variable selection techniques have been widely investigated in the literature, for instance by Aitkin (1974), Allen (1974), Diehr and Hoflin (1974), Hill and Hunter (1974), Narula and Wellington (1977, 1979), Snee (1977), Suich and Derringer (1977, 1980), Ellerton (1978), McKay (1979), Hintze (1980), Rencher and Pun (1980), Weisberg (1981), Wilkinson and Dallal (1981), Baskerville and Toogood (1982), Madsen (1982), and Young (1982). Hocking (1976) reviews the topic of variable selection. We will briefly describe the most widely used techniques.

When the set of candidate variables is not too large, one can consider all the possible models starting from this set. This is the so-called *all-subsets regression* procedure. It leads to

$$2^{\text{(number of available explanatory variables)}}$$

different models, a number that increases rapidly. Even when attention is focused on LS estimation (or our RLS, assuming that a very robust technique has been executed first), this number becomes computationally infeasible. This has motivated some people to develop more efficient algorithms, which appeal to numerical methods for calculating the LS estimates for the successive subsets. Usually, these procedures are based on either Gauss–Jordan reduction or a sweep operator (Beaton 1964, Seber 1977). Once the various fitted equations are at one's disposal, one needs a criterion for judging which subset of variables yields the "best" fit. Hocking (1976) describes different measures for this purpose, including the mean squared error, the coefficient of determination, and the C_p statistic (see also Daniel and Wood 1971 and Mallows 1973 for a thorough treatment of C_p). The Furnival and Wilson (1974) algorithm, which is available in the regression program P2R of BMDP, is a branch-and-bound type of procedure that cuts the computation time by searching only in certain promising directions.

Although it is felt that the investigation of all subsets produces the "best" set, it is not the most widely used method because of its computational cost. The so-called *stepwise procedures*, which consist of either adding or deleting one explanatory variable at a time, have been the favorite methods throughout. One distinguishes *forward selection* and *backward elimination* stepwise procedures, and a combination of both. Variations of these types have been implemented in several statistical packages, such as BMDP, SAS, and SPSS.

In *forward selection*, one starts with a simple regression model in which the explanatory variable is the variable that correlates best with the response. Then, at each subsequent step, one adds one variable to the model. At any step the selected variable x_j is the one producing the largest F_j-ratio among the candidates. This F_j-ratio is defined by

$$F_j = \frac{\text{SSE}_k - \text{SSE}_{k+(j)}}{\hat{\sigma}^2_{k+(j)}} ,$$

where SSE_k is the residual error sum of squares corresponding to the model with k terms, and $\text{SSE}_{k+(j)}$ is the one corresponding to the model where x_j is added. Variable x_j will be included in the equation if F_j is larger than a prespecified value. This prespecified value is often referred to as the *stopping rule*. One can choose this value such that the procedure will run the full course. That way one obtains one subset of each size. Moreover, one can then use a criterion for selecting the "best" of these subsets.

In the *backward elimination* methods, one works in the inverse way, starting from an equation containing all the variables. At each step, one eliminates the "worst" variable. For instance, the variable x_j will be a candidate for elimination from the current model (consisting of k terms) if it produces the smallest F_j-ratio, where

$$F_j = \frac{SSE_{k-(j)} - SSE_k}{\hat{\sigma}_k^2} .$$

Again several stopping rules similar to those for forward selection have been suggested (see Hocking 1976). Efroymson (1960) combined both ideas: His method is basically of the forward selection type, but at each step the elimination of a variable is also possible.

Faced with a problem of variable selection, one has to be aware of the weak points of the available techniques. For instance, the stepwise procedures do not necessarily yield the "best" subset of a given size. Moreover, these techniques induce a ranking on the explanatory variables which is often misused in practice. The order of deletion or inclusion is very deceptive, because the first variable deleted in backward elimination (or similarly the first one added in forward selection) is not necessarily the worst (or the best) in an absolute sense. It may, for example, happen that the first variable entered in forward selection becomes unnecessary in the presence of other variables. Also, forward and backward stepwise techniques may lead to totally different "best" subsets of variables. Berk (1978) compares the stepwise procedures with all-subsets regression. He shows that if forward selection agrees with all-subsets regression for every subset size, then backward elimination will also agree with all-subsets regression for every subset size, and inversely. All-subsets regression, however, is not the ideal way to avoid the disadvantages of the stepwise techniques. Indeed, the evaluation of all the possible subsets also largely depends on the employed criterion. Moreover, although the all-subsets procedure produces the "best" set for each subset size, this will not necessarily be the case in the whole population. It is only the "best" in the sample. The observation made by Gorman and Toman (1966) is perhaps suitable to conclude the topic of variable selection: "It is unlikely that there is a single best subset, but rather several equally good ones."

Example 4: Education Expenditure Data

The problem of *heteroscedasticity* has already been mentioned in Section 4 of Chapter 2, in the discussion on the diagnostic power of residual plots. The following data set, described by Chatterjee and Price (1977, p. 108) provides an interesting example of heteroscedasticity. It deals with education expenditure variables for 50 U.S. states. The data are reproduced in Table 16. The y-variable in Table 16 is the per capita expenditure on public education in a state, projected for 1975. The aim is to explain y by means of the explanatory variables x_1 (number of residents per thousand residing in urban areas in 1970), x_2 (per capita personal income in 1973), and x_3 (number of residents per thousand under 18 years of age in 1974).

Often the index i of an observation is time-related. In that case, the pattern of an index plot may point to nonconstancy of the spread of the residuals with respect to time. However, the magnitude of the residuals may also appear to vary systematically with $\hat{y}_i$ or an explanatory variable, or with another ordering of the cases besides the time ordering. The objective for the present data set is to analyze the constancy of the relationships with regard to a spatial ordering of the cases. The data in Table 16 are grouped by geographic region. One can distinguish four groups: the northeastern states (indices 1–9), the north central states (indices 10–21), the southern states (indices 22–37), and the western states (indices 38–50).

The routine application of least squares to these data yields the coefficients in Table 17, which also contains the results of reweighted least squares based on the LMS. Let us now compare the LS index plot (Figure 14a) with that of RLS (Figure 14b). Their main difference is that the fiftieth case (Alaska) can immediately by recognized as an outlier in the RLS plot, whereas it does not stand out clearly in the LS plot. (It seems that the education expenditure in Alaska is much higher than could be expected on the basis of its population characteristics x_1, x_2, and x_3 alone.) On the other hand, both plots appear to indicate that the dispersion of the residuals changes for the different geographical regions. This phenomenon is typical for heteroscedasticity. The defect can be remedied by handling the four clusters separately, but then the number of cases becomes very limited (in this example). Chatterjee and Price (1977) analyze these data by using another type of weighted LS regression. They assign weights to each of the four regions in order to compute a weighted sum of squared residuals. These weights are estimated in a first stage by using the mean (in a certain region) of the squared residuals resulting from ordinary LS. Chatterjee and Price also considered Alaska as an outlier and decided to omit it.

Table 16. Education Expenditure Data

Index	State		x_1	x_2	x_3	y
1	ME		508	3944	325	235
2	NH		564	4578	323	231
3	VT		322	4011	328	270
4	MA		846	5233	305	261
5	RI	Northeastern	871	4780	303	300
6	CT		774	5889	307	317
7	NY		856	5663	301	387
8	NJ		889	5759	310	285
9	PA		715	4894	300	300
10	OH		753	5012	324	221
11	IN		649	4908	329	264
12	IL		830	5753	320	308
13	MI		738	5439	337	379
14	WI		659	4634	328	342
15	MN	North central	664	4921	330	378
16	IA		572	4869	318	232
17	MO		701	4672	309	231
18	ND		443	4782	333	246
19	SD		446	4296	330	230
20	NB		615	4827	318	268
21	KS		661	5057	304	337
22	DE		722	5540	328	344
23	MD		766	5331	323	330
24	VA		631	4715	317	261
25	WV		390	3828	310	214
26	NC		450	4120	321	245
27	SC		476	3817	342	233
28	GA		603	4243	339	250
29	FL		805	4647	287	243
30	DY	Southern	523	3967	325	216
31	TN		588	3946	315	212
32	AL		584	3724	332	208
33	MS		445	3448	358	215
34	AR		500	3680	320	221
35	LA		661	3825	355	244
36	OK		680	4189	306	234
37	TX		797	4336	335	269
38	MT		534	4418	335	302
39	ID		541	4323	344	268
40	WY		605	4813	331	323
41	CO		785	5046	324	304
42	NM		698	3764	366	317
43	AZ		796	4504	340	332
44	UT	Western	804	4005	378	315
45	NV		809	5560	330	291
46	WA		726	4989	313	312
47	OR		671	4697	305	316
48	CA		909	5438	307	332
49	AK		831	5309	333	311
50	HI		484	5613	386	546

Source: Chatterjee and Price (1977).

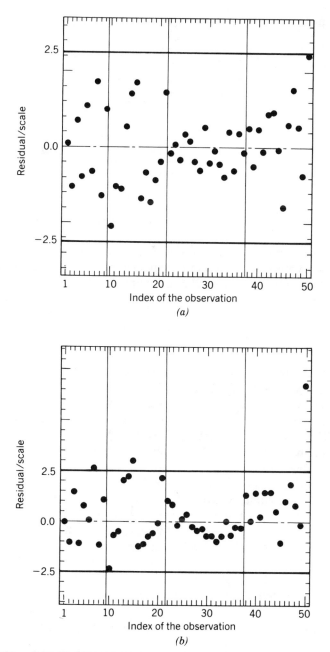

Figure 14. Education expenditure data: (*a*) Index plot associated with LS. (*b*) Index plot associated with LMS-based RLS.

111

Table 17. LS and RLS Results on Education Expenditure Data: Coefficients, Standard Errors, and t-Values

Variable	LS Results			RLS Results		
	$\hat{\theta}$	Standard Error	t-Value	$\hat{\theta}$	Standard Error	t-Value
x_1	−0.004	0.051	−0.08	0.075	0.042	1.76
x_2	0.072	0.012	6.24	0.038	0.011	3.49
x_3	1.552	0.315	4.93	0.756	0.292	2.59
Constant	−556.6	123.2	−4.52	−197.3	117.5	−1.68

*4. PROPERTIES OF THE LMS, THE LTS, AND S-ESTIMATORS

This section contains some theoretical results and may be skipped by those who are only interested in the application of robust regression and not its mathematical aspects. The first part is about the LMS estimator, given by

$$\underset{\hat{\theta}}{\text{Minimize}} \, \underset{i}{\text{med}} \, r_i^2 . \tag{4.1}$$

The existence of this estimator will be proven, and its breakdown and exact fit properties are stated. It is also shown that it attains the maximal breakdown point among all regression equivariant estimators. Then some results on one-step M-estimators are presented. Finally, least trimmed squares (LTS) and S-estimators are covered. Most of the material in this section follows Rousseeuw (1984) and Rousseeuw and Yohai (1984).

The n observations $(\mathbf{x}_i, y_i) = (x_{i1}, \ldots, x_{ip}, y_i)$ belong to the linear space of row vectors of dimension $p + 1$. The unknown parameter $\boldsymbol{\theta}$ is a p-dimensional column vector $(\theta_1, \ldots, \theta_p)^t$. The (unperturbed) linear model states that $y_i = \mathbf{x}_i \boldsymbol{\theta} + e_i$ where e_i is distributed according to $N(0, \sigma^2)$. Throughout this section it is assumed that all observations with $\mathbf{x}_i = \mathbf{0}$ have been deleted, because they give no information on $\boldsymbol{\theta}$. This condition is automatically satisfied if the model has an intercept because then the last coordinate of each $\mathbf{x}_i$ equals 1. Moreover, it is assumed that in the $(p + 1)$-dimensional space of the $(\mathbf{x}_i, y_i)$, there is no vertical hyperplane through zero containing more than $[n/2]$ observations. (Such a vertical hyperplane is a p-dimensional subspace that contains $(0, \ldots, 0)$ and $(0, \ldots, 0, 1)$. We call this subspace a hyperplane because its dimension is p, which is one less than the dimension of the total space. The notation $[q]$ stands for the largest integer less than or equal to q.)

The first theorem guarantees that the minimization in (4.1) always leads to a solution.

Theorem 1. There always exists a solution to (4.1).

Proof. We work in the $(p+1)$-dimensional space E of the observations $(\mathbf{x}_i, y_i)$. The space of the $\mathbf{x}_i$ is the horizontal hyperplane through the origin, which is denoted by $(y = 0)$ because the y-coordinates of all points in this plane are zero. Two cases have to be considered:

CASE A. This is really a special case, in which there exists a $(p-1)$-dimensional subspace V of $(y = 0)$ going through zero and containing at least $[n/2] + 1$ of the $\mathbf{x}_i$. The observations $(\mathbf{x}_i, y_i)$ corresponding to these $\mathbf{x}_i$ now generate a subspace S of E (in the sense of linear algebra), which is at most p-dimensional. Because it was assumed that E has no vertical hyperplane containing $[n/2] + 1$ observations, it follows that S does not contain $(0, \ldots, 0, 1)$; hence the dimension of S is at most $p - 1$. This means that there exists a nonvertical hyperplane H given by some equation $y = \mathbf{x}\boldsymbol{\theta}$ which includes S. For this value of $\boldsymbol{\theta}$, clearly $\operatorname{med}_i r_i^2 = 0$, which is the minimal value. This reasoning can be illustrated by taking the value of p equal to 2 and considering a linear model without intercept term. Figure 15 illustrates the positions of the subspaces S and V of E.

CASE B. Let us now assume that we are in the general situation in which case A does not hold. The rest of the proof will be devoted to showing that there exists a ball around the origin in the space of all $\boldsymbol{\theta}$, to which attention can be restricted for finding a minimum of $\operatorname{med}_i r_i^2(\boldsymbol{\theta})$. Because the objective function $\operatorname{med}_i r_i^2(\boldsymbol{\theta})$ is continuous in $\boldsymbol{\theta}$, this is sufficient for the existence of a minimum. Put

$$\delta = \tfrac{1}{2} \inf \{\tau > 0; \text{ there exists a } (p-1)\text{-dimensional subspace } V \text{ of } (y = 0)$$
$$\text{such that } V^\tau \text{ covers at least } [n/2] + 1 \text{ of the } \mathbf{x}_i\},$$

where V^τ is the set of all $\mathbf{x}$ with distance to V not larger than τ. Case A corresponds to $\delta = 0$, but now $\delta > 0$. Denote $M := \max_i |y_i|$. Now attention may be restricted to the closed ball around the origin with radius $(\sqrt{2} + 1)M/\delta$. Indeed, for any $\boldsymbol{\theta}$ with $\|\boldsymbol{\theta}\| > (\sqrt{2} + 1)M/\delta$, it will be shown that

$$\operatorname*{med}_i r_i^2(\boldsymbol{\theta}) > \operatorname*{med}_i y_i^2 = \operatorname*{med}_i r_i^2(\mathbf{0}),$$

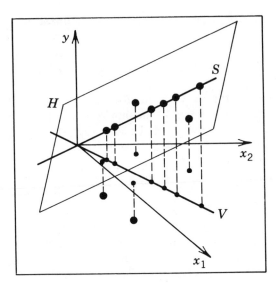

Figure 15. Illustration of the position of the subspaces S and V in the space E, as defined in the proof of Theorem 1 (Case A). There are 10 observations and $p = 2$.

so smaller objective functions cannot be found outside the ball. A geometrical construction (illustrated in Figure 16) is needed to prove this. Such a $\boldsymbol{\theta}$ determines a nonvertical hyperplane H given by $y = \mathbf{x}\boldsymbol{\theta}$. By the dimension theorem of linear algebra,

$$\dim\,(H \cap (y = 0)) = \dim\,(H) + \dim\,(y = 0) - \dim\,(H + (y = 0))$$

$$= p + p - (p + 1)$$

$$= p - 1$$

because $\|\boldsymbol{\theta}\| > (\sqrt{2} + 1)M/\delta$ implies that $\boldsymbol{\theta} \neq \mathbf{0}$ and hence $H \neq (y = 0)$. Therefore, $(H \cap (y = 0))^{\delta}$ contains at most $[n/2]$ of the $\mathbf{x}_i$. For each of the remaining observations $(\mathbf{x}_i, y_i)$, we construct the vertical two-dimensional plane P_i through $(\mathbf{x}_i, y_i)$, which is orthogonal to $(H \cap (y = 0))$. (This plane does not pass through zero, so to be called vertical, it has to go through both $(\mathbf{x}_i, y_i)$ and $(\mathbf{x}_i, y_i + 1)$.) We see that

$$|r_i| = |\mathbf{x}_i\boldsymbol{\theta} - y_i| \geq |\,|\mathbf{x}_i\boldsymbol{\theta}| - |y_i|\,|$$

with $|\mathbf{x}_i\boldsymbol{\theta}| > \delta\,|\tan\,(\alpha)|$, where α is the angle in $(-\pi/2, \pi/2)$ formed by H and the horizontal line in P_i. Therefore $|\alpha|$ is the angle between the line

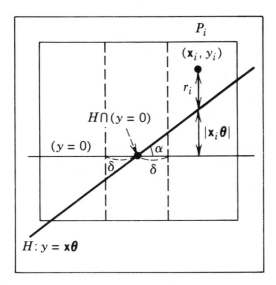

Figure 16. Illustration of a geometrical construction in the proof of Theorem 1 (Case B).

orthogonal to H and $(0, 1)$, hence

$$|\alpha| = \arccos\left\{ \frac{|(-\boldsymbol{\theta}, 1)(0, 1)'|}{\|(-\boldsymbol{\theta}, 1)\|\|(0, 1)\|} \right\} = \arccos\left\{ \frac{1}{\sqrt{1 + \|\boldsymbol{\theta}\|^2}} \right\}$$

and finally $|\tan(\alpha)| = \|\boldsymbol{\theta}\|$.

Because $\|\boldsymbol{\theta}\| > (\sqrt{2} + 1)M/\delta$, it follows that

$$|\mathbf{x}_i\boldsymbol{\theta}| > \delta\|\boldsymbol{\theta}\| > M \geq |y_i|$$

so

$$|r_i(\boldsymbol{\theta})| > (\delta\|\boldsymbol{\theta}\| - |y_i|) .$$

But then

$$r_i^2(\boldsymbol{\theta}) > ((\sqrt{2} + 1)M - |y_i|)^2$$
$$> ((\sqrt{2} + 1)M - M)^2$$
$$> 2M^2$$

for at least $n - [n/2]$ observations. Hence

$$\operatorname*{med}_i r_i^2(\boldsymbol{\theta}) > M^2 \geq \operatorname*{med}_i (y_i^2) .$$

So the objective function associated with such a $\boldsymbol{\theta}$ is larger than the one for $\boldsymbol{\theta} = \mathbf{0}$. Therefore, we only have to search for a solution $\boldsymbol{\theta}$ in the closed ball $B(\mathbf{0}, (\sqrt{2} + 1)M/\delta)$. Because this set is compact and $\text{med}_i\, r_i^2(\boldsymbol{\theta})$ is continuous in $\boldsymbol{\theta}$, the infimum is a minimum. $\square$

REMARK. This proof is not constructive. To actually find a solution to (4.1) we use the algorithm described in Chapter 5.

Let us now discuss some equivariance properties of the LMS. For regression estimators, one can consider three types of equivariance. Ranked from higher to lower priority, there exists regression, scale, and affine equivariance.

An estimator T is called *regression equivariant* if

$$T(\{(\mathbf{x}_i, y_i + \mathbf{x}_i \mathbf{v}); i = 1, \ldots, n\}) = T(\{(\mathbf{x}_i, y_i); i = 1, \ldots, n\}) + \mathbf{v}, \quad (4.2)$$

where $\mathbf{v}$ is any column vector. Regression equivariance is just as crucial as translation equivariance for a multivariate location estimator, but not as often formulated. It is implicit in the notion of a regression estimator. For instance, many proofs of asymptotic properties or descriptions of Monte Carlo studies begin with the phrase "without loss of generality, let $\boldsymbol{\theta} = \mathbf{0}$", which assumes that the results are valid at any parameter vector through application of (4.2). On the other hand, note that the coefficient of determination (R^2) is *not* regression invariant, because it depends on the inclination of the regression surface (Barrett 1974).

An estimator T is said to be *scale equivariant* if

$$T(\{(\mathbf{x}_i, cy_i); i = 1, \ldots, n\}) = cT(\{(\mathbf{x}_i, y_i); i = 1, \ldots, n\}) \quad (4.3)$$

for any constant c. It implies that the fit is essentially independent of the choice of measurement unit for the response variable y.

One says that T is *affine equivariant* if

$$T(\{(\mathbf{x}_i \mathbf{A}, y_i); i = 1, \ldots, n\}) = \mathbf{A}^{-1} T(\{(\mathbf{x}_i, y_i); i = 1, \ldots, n\}) \quad (4.4)$$

for any nonsingular square matrix $\mathbf{A}$. In words, affine equivariance means that a linear transformation of the $\mathbf{x}_i$ should transform the estimator T accordingly, because $\hat{y}_i = \mathbf{x}_i T = (\mathbf{x}_i \mathbf{A})(\mathbf{A}^{-1} T)$. This allows us to use another coordinate system for the explanatory variables, without affecting the estimated $\hat{y}_i$.

The LMS satisfies all three equivariance properties:

Lemma 1. The LMS estimator is regression equivariant, scale equivariant, and affine equivariant.

Proof. This follows from

$$\operatorname*{med}_{i} (\{y_i + \mathbf{x}_i \mathbf{v}\} - \mathbf{x}_i \{\boldsymbol{\theta} + \mathbf{v}\})^2 = \operatorname*{med}_{i} (y_i - \mathbf{x}_i \boldsymbol{\theta})^2 ,$$

$$\operatorname*{med}_{i} (c y_i - \mathbf{x}_i \{c\boldsymbol{\theta}\})^2 = c^2 \operatorname*{med}_{i} (y_i - \mathbf{x}_i \boldsymbol{\theta})^2 ,$$

and

$$\operatorname*{med}_{i} (y_i - \{\mathbf{x}_i \mathbf{A}\} \{\mathbf{A}^{-1}\boldsymbol{\theta}\})^2 = \operatorname*{med}_{i} (y_i - \mathbf{x}_i \boldsymbol{\theta})^2 ,$$

respectively. □

On the other hand, it may be noted that the repeated median, defined in (2.14) of Chapter 1, is regression and scale equivariant but not affine equivariant.

In what follows we shall say the observations are in *general position* when any p of them give a unique determination of $\boldsymbol{\theta}$. For example, in case $p = 2$ this means that any pair of observations (x_{i1}, x_{i2}, y_i) and (x_{j1}, x_{j2}, y_j) determines a unique nonvertical plane through zero, which implies that $(0, 0, 0)$, (x_{i1}, x_{i2}, y_i), and (x_{j1}, x_{j2}, y_j) may not be collinear. When the observations come from continuous distributions, this event has probability one.

As promised in Chapter 1 we will now investigate the breakdown properties of the LMS, using the finite-sample version of the breakdown point introduced by Donoho and Huber (1983). Take any sample Z of n data points $(\mathbf{x}_1, y_1), \ldots, (\mathbf{x}_n, y_n)$ and a regression estimator T. This means that applying T to Z yields a regression estimate $\hat{\boldsymbol{\theta}}$. Let bias $(m; T, Z)$ be the supremum of $\|T(Z') - T(Z)\|$ for all corrupted samples Z', where any m of the original data points are replaced by arbitrary values. Then the *breakdown point* of T at Z is

$$\varepsilon_n^*(T, Z) = \min \{m/n; \text{bias} (m; T, Z) \text{ is infinite}\} . \qquad (4.5)$$

We prefer replacing observations to adding observations, which some authors do, because replacement contamination is simple, realistic, and generally applicable. Indeed, from an intuitive point of view, outliers are not some faulty observations that are added at the end of the sample, but they treacherously hide themselves by replacing some of the data points that should have been observed. Replacement contamination does not cause any formal problems because the contaminated sample has the same size as the original one, so we only have to consider one estimator

T_n and not several T_{n+m}. This means that replacement still applies to many situations where adding observations does not make sense (for instance, one cannot just add cells to a two-way table). Therefore, we would like to defend the standard use of the above definition.

Theorem 2. If $p > 1$ and the observations are in general position, then the breakdown point of the LMS method is

$$([n/2] - p + 2)/n .$$

Proof. 1. We first show that $\varepsilon_n^*(T, Z) \geq ([n/2] - p + 2)/n$ for any sample $Z = \{(\mathbf{x}_i, y_i); i = 1, \ldots, n\}$ consisting of n observations in general position. By the first theorem, the sample Z yields a solution $\boldsymbol{\theta}$ of (4.1). We now have to show that the LMS remains bounded when $n - ([n/2] - p + 2) + 1$ points are unchanged. For this purpose, construct any corrupted sample $Z' = \{(\mathbf{x}_i', y_i'); i = 1, \ldots, n\}$ by retaining $n - [n/2] + p - 1$ observations of Z, which will be called the "good" observations, and by replacing the others by arbitrary values. It suffices to prove that $\|\boldsymbol{\theta} - \boldsymbol{\theta}'\|$ is bounded, where $\boldsymbol{\theta}'$ corresponds to Z'. For this purpose, some geometry is needed. We again work in the $(p + 1)$-dimensional space E of the observations $(\mathbf{x}_i, y_i)$ and in its horizontal hyperplane through the origin, denoted by $(y = 0)$. Put

$$\rho = \tfrac{1}{2} \inf \{\tau > 0; \text{ there exists a } (p - 1)\text{-dimensional subspace } V \text{ of}$$
$$(y = 0) \text{ through the origin such that } V^\tau \text{ covers}$$
$$\text{at least } p \text{ of the } \mathbf{x}_i\} ,$$

where V^τ is the set of all $\mathbf{x}$ with distance to V not larger than τ. Because Z is in general position, it holds that $\rho > 0$. Also, put $M := \max_i |r_i|$, where r_i are the residuals $y_i - \mathbf{x}_i \boldsymbol{\theta}$. The rest of the proof of part 1 will be devoted to showing that

$$\|\boldsymbol{\theta} - \boldsymbol{\theta}'\| < 2(\|\boldsymbol{\theta}\| + M/\rho) ,$$

which is sufficient because the right member is a finite constant. Denote by H the nonvertical hyperplane given by the equation $y = \mathbf{x}\boldsymbol{\theta}$, and let H' correspond in the same way to $\boldsymbol{\theta}'$. Without loss of generality assume that $\boldsymbol{\theta}' \neq \boldsymbol{\theta}$, hence $H' \neq H$. By the dimension theorem of linear algebra, the intersection $H \cap H'$ has dimension $p - 1$. If $\mathrm{pr}\,(H \cap H')$ denotes the vertical projection of $H \cap H'$ on $(y = 0)$, it follows that at most $p - 1$ of the good $\mathbf{x}_i$ can lie on $(\mathrm{pr}\,(H \cap H'))^p$. Define A as the set of remaining good observations, containing at least $n - [n/2] + p - 1 - (p - 1) = n -$

$[n/2]$ points. Now consider any $(\mathbf{x}_a, y_a)$ belonging to A, and put $r_a = y_a - \mathbf{x}_a\boldsymbol{\theta}$ and $r'_a = y_a - \mathbf{x}_a\boldsymbol{\theta}'$. Construct the vertical two-dimensional plane P_a through $(\mathbf{x}_a, y_a)$ and orthogonal to pr $(H \cap H')$. It follows, as in the proof of Theorem 1, that

$$|r'_a - r_a| = |\mathbf{x}_a\boldsymbol{\theta}' - \mathbf{x}_a\boldsymbol{\theta}| > \rho|\tan(\alpha') - \tan(\alpha)|$$

$$\geq \rho|\,|\tan(\alpha')| - |\tan(\alpha)|\,|$$

$$= \rho|\,\|\boldsymbol{\theta}'\| - \|\boldsymbol{\theta}\|\,|,$$

where α is the angle formed by H and some horizontal line in P_a and α' corresponds in the same way to H'. Since

$$\|\boldsymbol{\theta}' - \boldsymbol{\theta}\| \leq \|\boldsymbol{\theta}\| + \|\boldsymbol{\theta}'\| = 2\|\boldsymbol{\theta}\| + (\|\boldsymbol{\theta}'\| - \|\boldsymbol{\theta}\|) \leq |\,\|\boldsymbol{\theta}'\| - \|\boldsymbol{\theta}\|\,| + 2\|\boldsymbol{\theta}\|,$$

it follows that

$$|r'_a - r_a| > \rho(\|\boldsymbol{\theta}' - \boldsymbol{\theta}\| - 2\|\boldsymbol{\theta}\|).$$

Now the median of the squared residuals of the new sample Z' with respect to the old $\boldsymbol{\theta}$, with at least $n - [n/2] + p - 1$ of these residuals being the same as before, is less than or equal to M^2. Because $\boldsymbol{\theta}'$ is a solution of (4.1) for Z', it follows that also

$$\underset{i}{\text{med}}\,(y'_i - \mathbf{x}'_i\boldsymbol{\theta}')^2 \leq M^2.$$

If we now assume that $\|\boldsymbol{\theta}' - \boldsymbol{\theta}\| \geq 2(\|\boldsymbol{\theta}\| + M/\rho)$, then for all a in A it holds that

$$|r'_a - r_a| > \rho(\|\boldsymbol{\theta}' - \boldsymbol{\theta}\| - 2\|\boldsymbol{\theta}\|)$$

$$> \rho(2\|\boldsymbol{\theta}\| + 2M/\rho - 2\|\boldsymbol{\theta}\|) = 2M,$$

so

$$|r'_a| \geq |r'_a - r_a| - |r_a| > 2M - M = M$$

and finally

$$\underset{i}{\text{med}}\,(y'_i - \mathbf{x}'_i\boldsymbol{\theta}')^2 > M^2,$$

a contradiction. Therefore,

$$\|\boldsymbol{\theta}' - \boldsymbol{\theta}\| < 2(\|\boldsymbol{\theta}\| + M/\rho)$$

for any Z'.

2. Let us now show that the breakdown point can be no larger than the announced value. For this purpose, consider corrupted samples in which only $n - [n/2] + p - 2$ of the good observations are retained. Start by taking $p - 1$ of the good observations, which determine a $(p - 1)$-dimensional subspace L through zero. Now construct any nonvertical hyperplane H' through L, which determines some $\boldsymbol{\theta}'$ by means of the equation $y = \mathbf{x}\boldsymbol{\theta}'$. If all of the "bad" observations are put on H', then Z' has a total of

$$([n/2] - p + 2) + (p - 1) = [n/2] + 1$$

points that satisfy $y_i' = \mathbf{x}_i'\boldsymbol{\theta}'$ exactly; so the median squared residual of Z' with respect to $\boldsymbol{\theta}'$ is zero, hence $\boldsymbol{\theta}'$ satisfies (4.1) for Z'. By choosing H' steeper and steeper, one can make $\|\boldsymbol{\theta}' - \boldsymbol{\theta}\|$ as large as one wants.   □

Note that the breakdown point depends only slightly on n. In order to obtain a single value, one often considers the limit for $n \to \infty$ (with p fixed), so it can be said that the classical LS has a breakdown point of 0%, whereas the breakdown point of the LMS technique is as high as 50%, the best that can be expected. Indeed, 50% is the highest possible value for the breakdown point, since for larger amounts of contamination it becomes impossible to distinguish between the good and the bad parts of the sample, as will be proved in Theorem 4 below.

Once we know that an estimator does not break down for a given fraction m/n of contamination, it is of interest just how large the bias can be. Naturally, it is hoped that bias $(m; T, Z) = \sup \| T(Z') - T(Z) \|$ does not become too big. For this purpose, Martin et al. (1987) compute the maximal asymptotic bias of several regression methods and show that the LMS minimizes this quantity among a certain class of estimators. [The same property is much more generally true for the sample median in univariate location, as proved by Huber (1981, p. 74).]

We will now investigate another aspect of robust regression, namely the *exact fit property*. If the majority of the data follow a linear relationship *exactly*, then a robust regression method should yield this equation. If it does, the regression technique is said to possess the exact fit property. (In Section 5 of Chapter 2, this property was illustated for the case of a straight line fit.) The following example provides an illustration for the multivariate case. We created a data set of 25 observations, which are listed in Table 18. The first 20 observations satisfy the equation

$$y = x_1 + 2x_2 + 3x_3 + 4x_4 , \tag{4.6}$$

and five observations fall outside this hyperplane. Applying LS regression

Table 18. Artificial Data Set Illustrating the Exact Fit Property[a]

Index	x_1	x_2	x_3	x_4	y
1	1	0	0	0	1
2	0	1	0	0	2
3	0	1	1	0	5
4	0	0	1	1	7
5	1	1	0	1	7
6	1	1	1	0	6
7	0	1	1	1	9
8	1	0	0	1	5
9	1	1	1	1	10
10	0	1	0	1	6
11	1	0	1	0	4
12	1	0	1	1	8
13	1	0	2	3	19
14	2	0	1	3	17
15	1	2	3	0	14
16	2	3	1	0	11
17	2	0	3	1	15
18	2	1	1	3	19
19	1	0	2	1	11
20	1	1	2	2	17
21	1	2	0	1	11
22	2	1	0	1	10
23	2	2	1	0	15
24	1	1	2	2	20
25	1	2	3	4	40

[a] The first 20 points lie on the hyperplane $y = x_1 + 2x_2 + 3x_3 + 4x_4$

without intercept to this data set leads to the fit

$$\hat{y} = 0.508x_1 + 3.02x_2 + 3.08x_3 + 4.65x_4 .$$

Although a large proportion of the points lie on the same hyperplane, the LS does not manage to find it. The outlying points even produce small residuals, some of them smaller than the residuals of certain good points. This is visualized in the LS index plot in Figure 17.

On the other hand, the LMS looks for the pattern followed by the majority of the data, and it yields exactly equation (4.6) as its solution. The 20 points lying on the same hyperplane now have a zero residual (see

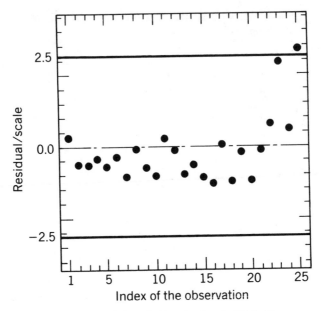

Figure 17. LS index plot for the data in Table 18.

the LMS index plot in Figure 18). Consequently, the scale estimate associated with LMS equals zero. In the index plot, the five outlying points fall far from the line through zero. The magnitude of their residual corresponds to their vertical distance from the hyperplane. In the exact fit case, reweighting on the basis of the LMS residuals is unnecessary, because the reduced data set would contain only the points that lie on the hyperplane.

The following theorem shows that the LMS satisfies the exact fit property.

Theorem 3. If $p > 1$ and there exists a θ such that at least $n - [n/2] + p - 1$ of the observations satisfy $y_i = \mathbf{x}_i\theta$ exactly and are in general position, then the LMS solution equals θ whatever the other observations are.

Proof. There exists some θ such that at least $n - [n/2] + p - 1$ of the observations lie on the hyperplane H given by the equation $y = \mathbf{x}\theta$. Then θ is a solution of (4.1), because $\text{med}_i\, r_i^2(\theta) = 0$. Suppose that there is another solution $\theta' \neq \theta$, corresponding to a hyperplane $H' \neq H$ and yielding residuals $r_i(\theta')$. As in the proof of Theorem 2, $(H \cap H')$ has

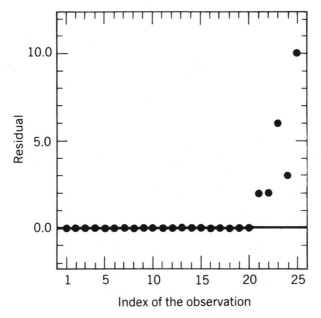

Figure 18. LMS index plot for the data in Table 18.

dimension $p - 1$ and thus contains, at most, $p - 1$ observations. For all remaining observations in H it holds that $r_i^2(\boldsymbol{\theta}') > 0$, and there are at least $n - [n/2]$ of them. Therefore $\text{med}_i \, r_i^2(\boldsymbol{\theta}') > 0$, so $\boldsymbol{\theta}'$ cannot be a solution.
□

REMARK 1. It appears that Theorem 3 is a special case of a more general relation between breakdown and exact fit. (This remark was derived from joint work with D. Donoho in 1983.) Let us consider a possible formalization of this connection, by defining the *exact fit point* as

$$\delta_n^*(T, Z) = \min \{m/n; \text{ there exists } Z' \text{ such that } T(Z') \neq \boldsymbol{\theta}\}, \quad (4.7)$$

where Z is a sample $\{(\mathbf{x}_1, y_1), \dots, (\mathbf{x}_n, y_n)\}$ such that $y_i = \mathbf{x}_i\boldsymbol{\theta}$ for all i, and Z' ranges over all corrupted samples where any m points of Z are replaced by arbitrary values. The smallest fraction of contamination capable of pulling T away from $\boldsymbol{\theta}$ is the exact fit point. If T is regression and scale equivariant, then

$$\delta_n^*(T, Z) \geq \varepsilon_n^*(T, Z). \quad (4.8)$$

Indeed, by regression equivariance we may assume that $\theta = 0$, so all $y_i = 0$. Take any $m \geq n\delta_n^*(T, Z)$. Then there exists $Z' = \{(\mathbf{x}_1', y_1'), (\mathbf{x}_2', y_2'), \ldots, (\mathbf{x}_n', y_n')\}$ obtained by replacing m points of Z, such that $T(Z') \neq 0$. Now construct the sample $Z'' = \{(\mathbf{x}_1', cy_1'), (\mathbf{x}_2', cy_2'), \ldots, (\mathbf{x}_n', cy_n')\}$, where c is a positive constant. But then

$$\| T(Z'') - T(Z) \| = c \| T(Z') \| \neq 0$$

and Z'' differs from Z in at most m points (because at most m of the cy_i' can be different from zero). By choosing c large we see that T breaks down, so $\varepsilon_n^*(T, Z) \leq m/n$, which proves (4.8). This result becomes particularly useful if T is well behaved so that $\varepsilon_n^*(T, Z)$ is the same at most Z (say, at any Z in general position), as is the case for LMS.

Unfortunately, the reverse inequality is not generally true, because one can construct counterexamples for which ε_n^* is *strictly* smaller than δ_n^*. Consider, for instance, an estimator (for $n = 20$ and $p = 2$) that gives the right answer whenever at least 15 observations are in an exact fit situation, but which is put equal to LS in all other cases. Nevertheless, this example is clearly pathological, and it should be possible to prove $\varepsilon_n^* = \delta_n^*$ under some "reasonable" conditions.

REMARK 2. The breakdown point in Theorem 2 is slightly smaller than that of the repeated median, although they are both 50% breakdown estimators. We are indebted to A. Siegel (personal communication) for a way to overcome this. Instead of taking the median of the ordered squared residuals, consider the hth order statistic $(r^2)_{h:n}$ and

$$\underset{\hat{\theta}}{\text{Minimize}} \ (r^2)_{h:n}, \quad \text{where } h = [n/2] + [(p+1)/2]. \qquad (4.9)$$

It turns out (analogous to the proof of Theorem 2) that this variant of the LMS has breakdown point equal to $([(n-p)/2]+1)/n$, which is exactly the same value as for Siegel's repeated median. In Theorem 4, we shall show that this is the best possible result. Therefore, we actually use this version of the LMS in PROGRESS. For this variant of the LMS, Theorem 3 holds whenever strictly more than $\frac{1}{2}(n + p - 1)$ of the observations are in an exact fit situation. (This can be proven as in Theorem 3, or by making use of Remark 1 above.)

REMARK 3. The LMS estimator can be viewed as a special case of a larger family of estimators, namely the *least quantile of squares* estimators (LQS), which are defined by

$$\underset{\hat{\theta}}{\text{Minimize}} \ (r^2)_{([(1-\alpha)n]+[\alpha(p+1)]):n}, \qquad (4.10)$$

where $0 \le \alpha \le 50\%$. For α tending to 50%, (4.10) is asymptotically equivalent to the LMS. The breakdown point of the LQS is equal to α for $n \to \infty$. Putting $\alpha = 0\%$ in (4.10), one finds the L_∞ estimator

$$\text{Minimize} \max_i r_i^2(\boldsymbol{\theta}) ,$$

which is also referred to as *minimax regression* because the largest (absolute) residual is minimized. This method was already considered by Euler, Lambert, and Laplace (see Sheynin 1966 and Plackett 1972). Unfortunately, it is even *less* robust than least squares (see exercise 10).

Theorem 4. Any regression equivariant estimator T satisfies

$$\varepsilon_n^*(T, Z) \le ([(n - p)/2] + 1)/n$$

at all samples Z.

Proof. Suppose that the breakdown point is strictly larger than $([(n - p)/2] + 1)/n$. This would mean that there exists a finite constant b such that $T(Z')$ lies in the ball $B(T(Z), b)$ for all samples Z' containing at least $n - [(n - p)/2] - 1$ points of Z. Set $q = n - [(n - p)/2] - 1$, which also equals $[(n + p + 1)/2] - 1$. Here $B(T(Z), b)$ is defined as the set of all $\boldsymbol{\theta}$ for which $\|T(Z) - \boldsymbol{\theta}\| \le b$. Now construct a p-dimensional column vector $\mathbf{v} \ne \mathbf{0}$ such that $\mathbf{x}_1 \mathbf{v} = 0, \ldots, \mathbf{x}_{p-1} \mathbf{v} = 0$. If $n + p + 1$ is even, then $2q - (p - 1) = n$; otherwise $2q - (p - 1) = n - 1$. In general one can say that $2q - (p - 1) \le n$. Therefore, the first $2q - (p - 1)$ points of Z can be replaced by

$$(\mathbf{x}_1, y_1), \ldots, (\mathbf{x}_{p-1}, y_{p-1}), (\mathbf{x}_p, y_p), \ldots, (\mathbf{x}_q, y_q) ,$$

$$(\mathbf{x}_p, y_p + \mathbf{x}_p \tau \mathbf{v}), \ldots, (\mathbf{x}_q, y_q + \mathbf{x}_q \tau \mathbf{v})$$

for any $\tau > 0$. For this new sample Z', the estimate $T(Z')$ belongs to $B(T(Z), b)$, since Z' contains q points of Z. But looking at Z' in another way reveals that $T(Z')$ can also be written as $T(Z'') + \tau \mathbf{v}$, where $T(Z'')$ is in $B(T(Z), b)$. Therefore, $T(Z')$ belongs to $B(T(Z) + \tau \mathbf{v}, b)$. This is a contradiction, however, because the intersection of $B(T(Z), b)$ and $B(T(Z) + \tau \mathbf{v}, b)$ is empty for large enough values of τ. $\square$

Note that this maximal breakdown point is attained by the repeated median (Siegel 1982) and the version (4.9) of the LMS.

By putting $p = 1$ and $x_{i1} = 1$ for all i in (4.1), one obtains the special case of one-dimensional estimation of a location parameter θ of the sample $(y_i)_{i=1,\ldots,n}$. The LMS estimator then corresponds to

$$\text{Minimize} \underset{\theta}{\text{ med}} \underset{i}{\text{ }} (y_i - \theta)^2 . \qquad (4.11)$$

This estimator will be investigated more fully in Chapter 4, where it will be shown that the LMS estimate T for θ corresponds to the midpoint of the "shortest half" of the sample, because one can prove that

$$T - m_T \text{ and } T + m_T \text{ are both observations in the sample}, \quad (4.12)$$

where $m_T^2 = \text{med}_i (y_i - T)^2$ equals the minimum of (4.11). This property can also be used in the regression model with intercept term, obtained by putting $x_{ip} = 1$ for all i. From (4.12) it follows that for an LMS solution $\hat{\theta}$, both hyperplanes

$$y = x_1 \hat{\theta}_1 + \cdots + x_p \hat{\theta}_p - m_T$$

and

$$y = x_1 \hat{\theta}_1 + \cdots + x_p \hat{\theta}_p + m_T$$

contain at least one observation. Therefore, the LMS solution corresponds to finding the thinnest "hyperstrip" (i.e., the region between two parallel hyperplanes) covering half of the observations. To be exact, the thickness of the hyperstrip is measured in the vertical direction, and it must contain at least $[n/2] + 1$ points.

From previous experience with robustness, it seems natural to replace the square in (4.1) by the absolute value, yielding

$$\text{Minimize} \underset{\theta}{\text{ med}} \underset{i}{\text{ }} |r_i| . \qquad (4.13)$$

However, it turns out that (4.12) no longer holds for that estimator, because there may be a whole region of solutions with the same objective function (4.13). (This can only happen when n is even, because $\text{med}_i |r_i|$ is the average of two absolute residuals. An example will be given in Section 2 of Chapter 4.) We will show in Chapter 4 that every solution of (4.1) is also a solution of (4.13), but not vice versa. Things become much more simple when the hth ordered squared residual is to be minimized, as in (4.9) and (4.10), because this is always equivalent to

$$\text{Minimize} \underset{\theta}{\text{ }} |r|_{h:n} , \qquad (4.14)$$

where $|r|_{1:n} \leq |r|_{2:n} \leq \cdots \leq |r|_{n:n}$ are the ordered absolute residuals.

Steele and Steiger (1986) also investigated some properties of the estimator defined in (4.14), where they put $h = [n/2] + 1$. This reduces to the median for n odd and to the *high median*, or larger of the two middle values, if n is even. Their work is restricted to the case of simple regression $\hat{y} = \hat{\theta}_1 x + \hat{\theta}_2$. They propose necessary and sufficient conditions for a local minimum of the objective function, which yield necessary conditions for global minimizers. More details on these algorithms are given in Section 2 of Chapter 5.

In order to show that the objective functions of (4.13) and (4.14) satisfy a Lipschitz condition, we need the following lemma.

Lemma 2. Let $(a_1, a_2, \ldots, a_n)$ and $(b_1, b_2, \ldots, b_n)$ be any pair of samples with real elements. Then:

(i) For each integer $1 \le h \le n$, it holds that

$$|a_{h:n} - b_{h:n}| \le \max_k |a_k - b_k|.$$

(ii) Also $|\mathrm{med}_i\, a_i - \mathrm{med}_j\, b_j| \le \max_k |a_k - b_k|$.
(iii) The sharpest upper bound on (i) and (ii) is

$$\min_{f \in \mathscr{S}_n} \max_k |a_k - b_{f(k)}| = \max_h |a_{h:n} - b_{h:n}|,$$

(where $\mathscr{S}_n$ is the set of all permutations on $\{1, \ldots, n\}$), which is a metric on the set of all samples $\{a_1, \ldots, a_n\}$ in which the sequence of the observations is disregarded.

Proof. (i) Put $c := \max_k |a_k - b_k|$. We first show that

$$|a_{1:n} - b_{1:n}| \le c$$

(otherwise assume without loss of generality that $a_{1:n} + c < b_{1:n}$, but then there can be no element b_j such that $|a_{1:n} - b_j| \le c$). Analogously, we can show that $|a_{n:n} - b_{n:n}| \le c$. For the general case, assume that there exists some $1 < h < n$ such that $a_{h:n} + c < b_{h:n}$. From the definition of c, there exists a permutation f of $\{1, \ldots, n\}$ such that $|a_{j:n} - b_{f(j):n}| \le c$ for all j. However, $j \le h$ must imply $f(j) \le h$ because otherwise $b_{f(j):n} - a_{j:n} \ge b_{h:n} - a_{h:n} > c$. Therefore, $f(\{1, \ldots, h\}) = \{1, \ldots, h\}$, but this is a contradiction because h itself cannot be attained.
(ii) If n is odd ($n = 2h - 1$) then the median is simply the hth order

statistic. If n is even $(n = 2h)$, then

$$
\left| \operatorname*{med}_i a_i - \operatorname*{med}_j b_j \right| = \left| \tfrac{1}{2}(a_{h:n} + a_{h+1:n}) - \tfrac{1}{2}(b_{h:n} + b_{h+1:n}) \right|
$$

$$
\leq \tfrac{1}{2} |a_{h:n} - b_{h:n}| + \tfrac{1}{2} |a_{h+1:n} - b_{h+1:n}|
$$

$$
\leq c .
$$

(iii) The inequality $\leq$ is immediate because this combination corresponds to a particular choice of f, and $\geq$ follows from (i). To see why this is a metric, note that $\max_h |a_{h:n} - b_{h:n}| = 0$ if and only if $a_{h:n} = b_{h:n}$ for all h. The symmetry property and triangle inequality are also straightforward. $\square$

For continuous distribution functions F and G, this metric amounts to

$$
d(F, G) = \sup_t |F^{-1}(t) - G^{-1}(t)| ,
$$

which gives the right answer at translation families but can easily become infinite.

Theorem 5. (i) For each integer $1 \leq h \leq n$ it holds that

$$
\sup_{\theta \neq \theta'} \frac{\left| \, |r(\theta)|_{h:n} - |r(\theta')|_{h:n} \right|}{\|\theta - \theta'\|} \leq \max_i \|\mathbf{x}_i\| .
$$

(ii) $\displaystyle \sup_{\theta \neq \theta'} \frac{\left| \operatorname*{med}_i |r_i(\theta)| - \operatorname*{med}_j |r_j(\theta)| \right|}{\|\theta - \theta'\|} \leq \max_i \|\mathbf{x}_i\| .$

Proof. From (i) of Lemma 2 it follows that

$$
\left| \, |r(\theta)|_{h:n} - |r(\theta')|_{h:n} \right| \leq \max_i \left| \, |r_i(\theta)| - |r_i(\theta')| \right|
$$

$$
= \max_i \left| \, |y_i - x_i\theta| - |y_i - x_i\theta'| \right|
$$

$$
\leq \max_i |y_i - x_i\theta - y_i + x_i\theta'|
$$

$$
= \max_i |x_i(\theta - \theta')|
$$

$$
\leq \|\theta - \theta'\| \max_i \|\mathbf{x}_i\| .
$$

Part (ii) is completely analogous.

Note that Theorem 5 implies that $|r|_{h:n}$ and $\text{med}_i |r_i|$ are continuous in $\boldsymbol{\theta}$ (but they are not everywhere differentiable).

A disadvantage of the LMS method is its lack of efficiency (because of its $n^{-1/3}$ convergence, which is proved in Section 4 of Chapter 4) when the errors would really be normally distributed. Of course it is possible to take an extreme point of view, wanting to stay on the safe side, even if it costs a lot. After all, saying that the LS method is more efficient at the normal is merely a tautology, because Gauss actually introduced the normal distribution in order to suit that method (Huber 1972, p. 1042). However, it is not so difficult to improve the efficiency of the LMS estimator. One can use the LMS estimates as starting values for computing a one-step M-estimator (Bickel 1975) in the following way: suppose we have the LMS solution $(\theta_1^*, \ldots, \theta_p^*)^t$ and a corresponding scale estimate σ^*; then the *one-step M-estimator* (OSM) is defined as

$$\hat{\boldsymbol{\theta}}_{\text{OSM}} = \boldsymbol{\theta}^* + (\mathbf{X}'\mathbf{X})^{-1}\mathbf{X}'(\psi(r_1^*/\sigma^*), \ldots, \psi(r_n^*/\sigma^*))^t \frac{\sigma^*}{B(\psi, \Phi)}, \quad (4.15)$$

where

$$B(\psi, \Phi) = \int \psi'(u) \, d\Phi(u) \quad \text{and} \quad r_i^* = y_i - \mathbf{x}_i \boldsymbol{\theta}^*$$

and $\mathbf{X}$ is an n-by-p matrix, the rows of which are the vectors $\mathbf{x}_i$. Afterwards σ^* can be replaced by $\hat{\sigma}_{OSM}$, using a M-estimator for scale.

If one uses a *redescending* ψ-function in (4.15), that is, a function for which $\psi(x) = 0$ whenever $|x| \geq c$, the large outliers will not enter the computation. One possible choice is the hyperbolic tangent estimator (Hampel et al. 1981), which possesses maximal asymptotic efficiency subject to certain robustness requirements. It is given by

$$\psi(x) = \begin{cases} x, & 0 \leq |x| \leq d \\ (A(k-1))^{1/2} \tanh \{\tfrac{1}{2}((k-1)B^2/A)^{1/2}(c-|x|)\} \, \text{sgn}(x), & d \leq |x| \leq c \\ 0, & |x| \geq c, \end{cases} \quad (4.16)$$

where $0 < d < c$ satisfies $d = (A(k-1))^{1/2} \tanh \{\tfrac{1}{2}((k-1)B^2/A)^{1/2}(c-d)\}$, $A = \int \psi^2 \, d\Phi$, and $B = \int \psi' \, d\Phi$. (For instance, $c = 3.0$ and $k = 5.0$ yields $A = 0.680593$, $B = 0.769313$, and $d = 1.47$.) Another possibility is the biweight ψ-function (Beaton and Tukey, 1974) corresponding to

$$\psi(x) = \begin{cases} x(1 - (x/c)^2)^2, & |x| \leq c \\ 0, & |x| \geq c. \end{cases} \quad (4.17)$$

In either case, such a one-step M-estimator converges like $n^{-1/2}$ and possesses the same asymptotic efficiency (for normal errors) as a fully iterated M-estimator. This was proven by Bickel (1975) when the starting value is $n^{1/2}$ consistent, but in general it even holds when the starting value is better than $n^{1/4}$ consistent (Bickel, personal communication, 1983) as is the case for LMS. Formally,

$$\mathcal{L}(n^{1/2}(\hat{\boldsymbol{\theta}}_{OSM} - \boldsymbol{\theta})) \to N(\mathbf{0}, V(\psi_{\hat{\sigma}}, F)L^{-1}),$$

where $\boldsymbol{\theta}$ is the unknown parameter vector, $\hat{\sigma} = \hat{\sigma}_{OSM}$, $L = \lim_{n \to \infty} (\mathbf{X}'\mathbf{X}/n)$, and F is the true distribution of the errors. $V(\psi_{\hat{\sigma}}, F)$ is called the *asymptotic variance*. When the errors are really normally distributed, then $F(t) = \Phi(t/\hat{\sigma})$ and the asymptotic variance can be calculated. Indeed, in that case

$$V(\psi_{\hat{\sigma}}, F) = \frac{\int (\psi(t/\hat{\sigma}))^2 \, dF(t)}{\left[\int (\psi(t/\hat{\sigma}))' \, dF(t)\right]^2}$$

$$= \hat{\sigma}^2 \frac{\int \psi^2(t/\hat{\sigma}) \, d\Phi(t/\hat{\sigma})}{\left[\int \psi'(t/\hat{\sigma}) \, d\Phi(t/\hat{\sigma})\right]^2}$$

$$= \hat{\sigma}^2 \frac{\int \psi^2(u) \, d\Phi(u)}{\left[\int \psi'(u) \, d\Phi(u)\right]^2}$$

$$= \hat{\sigma}^2 V(\psi, \Phi). \qquad (4.18)$$

For ψ defined as in (4.16),

$$V(\psi, \Phi) = A/B^2. \qquad (4.19)$$

Consequently, one can say that the variance–covariance matrix of the estimated regression coefficients is (approximately) equal to the p-by-p matrix

$$\hat{\sigma}^2 V(\psi, \Phi)(\mathbf{X}'\mathbf{X})^{-1}. \qquad (4.20)$$

For the LS estimator, $\psi(r) = r$ and hence $V(\psi, \Phi)$ equals 1. Replacing this value in (4.20) one recovers the well-known formula. Furthermore, expression (4.20) can be used to calculate the asymptotic efficiency e for the combined procedure (LMS + one-step M) in a normal error model, namely

$$e = 1/V(\psi, \Phi).$$

Hampel et al. (1981, Table 2) give a list of values of e for different constants c and k, as well as the corresponding A, B, and d. For instance, for $c = 3$ and $k = 5$ in (4.16), they obtain $e = 86.96\%$.

The diagonal elements of the matrix in (4.20) are the variances of the estimated regression coefficients $\hat{\theta}_j$. Therefore, it is possible to construct an approximate $(1 - \alpha)$ confidence interval for each θ_j, namely

$$[\hat{\theta}_j - \hat{\sigma}\sqrt{V(\psi, \Phi)((\mathbf{X'X})^{-1})_{jj}}\, t_{n-p,1-\alpha/2},$$

$$\hat{\theta}_j + \hat{\sigma}\sqrt{V(\psi, \Phi)((\mathbf{X'X})^{-1})_{jj}}\, t_{n-p,1-\alpha/2}]. \quad (4.21)$$

Another possibility for improving the efficiency is to use reweighted least squares. To each observation $(\mathbf{x}_i, y_i)$, one assigns a weight w_i that is a function of the standardized LMS residuals r_i/σ^* (in absolute value). For this purpose, one can choose several types of functions. The first kind of weight function that we will consider here is of the form

$$w_i = \begin{cases} 1 & \text{if } |r_i/\sigma^*| \leq c_1 \\ 0 & \text{otherwise}. \end{cases} \quad (4.22)$$

This weight function, yielding only binary weights, produces a clear distinction between "accepted" and "rejected" points.

The second type of weight function is less radical. It consists of introducing a linear part that smoothes the transition from weight 1 to weight 0. In that way, far outliers (this means cases with large LMS residuals) disappear entirely and intermediate cases are gradually downweighted. In the general formula

$$w_i = \begin{cases} 1 & \text{if } |r_i/\sigma^*| \leq c_2 \\ (c_3 - |r_i/\sigma^*|)/(c_3 - c_2) & \text{if } c_2 \leq |r_i/\sigma^*| \leq c_3 \\ 0 & \text{otherwise} \end{cases} \quad (4.23)$$

the constants c_2 and c_3 have to be chosen.

A third weight function can be defined by means of the hyperbolic tangent function (4.16), namely

$$w_i = \begin{cases} 1 & \text{if } |r_i/\sigma^*| \leq d \\ \dfrac{(A(k-1))^{1/2} \tanh\{\frac{1}{2}((k-1)B^2/A)^{1/2}(c - |r_i/\sigma^*|)\}}{|r_i/\sigma^*|} & \text{if } d \leq |r_i/\sigma^*| \leq c \\ 0 & \text{otherwise}, \end{cases} \quad (4.24)$$

where the constants c, k, d, A, and B correspond to those already defined in (4.16).

Once a weight function is selected, one replaces all observations (x_i, y_i) by $(w_i^{1/2} x_i, w_i^{1/2} y_i)$. On these weighted observations, a standard least squares program may be used to obtain the final estimate. The RLS results in PROGRESS are obtained with the weight function (4.22) with $c_1 = 2.5$.

Another way to improve the slow rate of convergence of the LMS consists of using a different objective function. Instead of adding all the squared residuals as in LS, one can limit one's attention to a "trimmed" sum of squares. This quantity is defined as follows: first one orders the squared residuals from smallest to largest, denoted by

$$(r^2)_{1:n} \le (r^2)_{2:n} \le \cdots \le (r^2)_{n:n} .$$

Then one adds only the first h of these terms. In this way, Rousseeuw (1983) defined the *least trimmed squares* (LTS) estimator

$$\text{Minimize}_{\hat{\theta}} \sum_{i=1}^{h} (r^2)_{i:n} . \tag{4.25}$$

Putting $h = [n/2] + 1$, the LTS attains the same breakdown point as the LMS (see Theorem 2). Moreover, for $h = [n/2] + [(p+1)/2]$ the LTS reaches the maximal possible value for the breakdown point given in Theorem 4. (Note that the LTS has nothing to do with the trimmed least squares estimators described by Ruppert and Carroll 1980.) Before we investigate the robustness properties of the LTS, we will first verify its equivariance.

Lemma 3. The LTS estimator is regression, scale, and affine equivariant.

Proof. Regression equivariance follows from the identity

$$\sum_{i=1}^{h} ((y_i + x_i v - x_i \{v + \theta\})^2)_{i:n} = \sum_{i=1}^{h} ((y_i - x_i \theta)^2)_{i:n}$$

for any column vector v. Scale and affine equivariance are analogous.
□

Theorem 6. The breakdown point of the LTS method defined in (4.25) with $h = [n/2] + [(p+1)/2]$ equals

$$\varepsilon_n^*(T, Z) = ([(n-p)/2] + 1)/n .$$

Proof. In order to prove this theorem we again assume that all observations with $(x_{i1}, \ldots, x_{ip}) = \mathbf{0}$ have been deleted and that the observations are in general position.

1. We first show that $\varepsilon_n^*(T, Z) \geq ([(n - p)/2] + 1)/n$. Since the sample $Z = \{(\mathbf{x}_i, y_i); i = 1, \ldots, n\}$ consists of n points in general position, it holds that

$$\rho = \tfrac{1}{2} \inf \{\tau > 0; \text{ there exists a } (p - 1)\text{-dimensional subspace } V \text{ of } (y = 0) \text{ such that } V^\tau \text{ covers at least } p \text{ of the } \mathbf{x}_i\}$$

is strictly positive. Suppose $\boldsymbol{\theta}$ minimizes (4.25) for Z, and denote by H the corresponding hyperplane given by the equation $y = \mathbf{x}\boldsymbol{\theta}$. We put $M = \max_i |r_i|$, where $r_i = y_i - \mathbf{x}_i\boldsymbol{\theta}$. Now construct any contaminated sample $Z' = \{(\mathbf{x}_i', y_i'); i = 1, \ldots, n\}$ by retaining $n - [(n - p)/2] = [(n + p + 1)/2]$ observations of Z and by replacing the others by arbitrary values. It now suffices to prove that $\|\boldsymbol{\theta} - \boldsymbol{\theta}'\|$ is bounded, where $\boldsymbol{\theta}'$ corresponds to Z'. Without loss of generality assume $\boldsymbol{\theta}' \neq \boldsymbol{\theta}$, so the corresponding hyperplane H' is different from H. Repeating now the reasoning of the first part of the proof of Theorem 2, it follows that

$$|r_a' - r_a| > \rho(\|\boldsymbol{\theta}' - \boldsymbol{\theta}\| - 2\|\boldsymbol{\theta}\|) ,$$

where r_a and r_a' are the residuals associated with H and H' corresponding to the point $(\mathbf{x}_a, y_a)$. Now the sum of the first h squared residuals of the new sample Z' with respect to the old $\boldsymbol{\theta}$, with at least $[(n + p + 1)/2] \geq h$ of these residuals being the same as before, is less than or equal to hM^2. Because $\boldsymbol{\theta}'$ corresponds to Z' it follows that also

$$\sum_{i=1}^{h} ((y_i' - \mathbf{x}_i'\boldsymbol{\theta}')^2)_{i : n} \leq hM^2 .$$

If we now assume that

$$\|\boldsymbol{\theta}' - \boldsymbol{\theta}\| \geq 2\|\boldsymbol{\theta}\| + M(1 + \sqrt{h})/\rho ,$$

then for all a in A it holds that

$$|r_a' - r_a| > \rho(\|\boldsymbol{\theta}' - \boldsymbol{\theta}\| - 2\|\boldsymbol{\theta}\|) \geq M(1 + \sqrt{h}) ,$$

so

$$|r_a'| \geq |r_a' - r_a| - |r_a| > M(1 + \sqrt{h}) - M = M\sqrt{h} .$$

Now note that $n - |A| \leq h - 1$. Therefore any set of h of the $(\mathbf{x}_i', y_i')$ must contain at least one of the $(\mathbf{x}_a, y_a)$, so

$$\sum_{i=1}^{h} ((y_i' - \mathbf{x}_i'\boldsymbol{\theta}')^2)_{i:n} \geq (r_a')^2 > hM^2 \,,$$

a contradiction. This implies that

$$\|\boldsymbol{\theta}' - \boldsymbol{\theta}\| < 2\|\boldsymbol{\theta}\| + M(1 + \sqrt{h})/\rho < \infty$$

for all such samples Z'.

2. The opposite inequality $\varepsilon_n^*(T, Z) \leq ([(n - p)/2] + 1)/n$ immediately follows from Theorem 4 and Lemma 3.   □

REMARK 1.   Another way to interpret Theorem 6 is to say that T remains bounded whenever strictly more than $\frac{1}{2}(n + p - 1)$ observations are uncontaminated.

REMARK 2.   The value of h yielding the maximal value of the breakdown point can also be found by the following reasoning based on the proofs of Theorems 2 and 6. On the one hand, the number of bad observations $n - |A|$ must be strictly less than h; on the other hand, $|A| + p - 1$ must be at least h. The best value of h is then obtained by minimizing $|A|$ over h subject to $|A| - 1 \geq n - h$ and $|A| - 1 \geq h - p$, which yields $h = [n/2] + [(p + 1)/2]$.

REMARK 3.   In general, h may depend on some trimming proportion α, for instance by means of $h = [n(1 - \alpha)] + [\alpha(p + 1)]$ or $h = [n(1 - \alpha)] + 1$. Then the breakdown point ε_n^* is roughly equal to this proportion α. For α tending to 50%, one finds again the LTS estimator, whereas for α tending to 0%, the LS estimator is obtained.

The following corollary shows that also the LTS satisfies the exact fit property.

Corollary.   If there exists some $\boldsymbol{\theta}$ such that strictly more than $\frac{1}{2}(n + p - 1)$ of the observations satisfy $y_i = \mathbf{x}_i\boldsymbol{\theta}$ exactly and are in general position, then the LTS solution equals $\boldsymbol{\theta}$ whatever the other observations are.

For instance, in the case of simple regression it follows that whenever 11 out of 20 observations lie on one line, this line will be obtained.

Unlike the slow convergence rate of the LMS, the LTS converges like $n^{-1/2}$, with the same asymptotic efficiency at the normal distribution as the M-estimator defined by

$$\psi(x) = \begin{cases} x, & |x| \leq \Phi^{-1}(1 - \alpha/2) \\ 0, & \text{otherwise}, \end{cases} \qquad (4.26)$$

which is called a Huber-type skipped mean in the case of location (see Chapter 4 for details). The main disadvantage of the LTS is that its objective function requires sorting of the squared residuals, which takes $O(n \log n)$ operations compared with only $O(n)$ operations for the median.

REMARK. Until now we have considered the estimators obtained by substituting the sum in the definition of the LS estimator by a median, yielding LMS, and by a trimmed sum, leading to LTS. Another idea would be to replace the sum by a winsorized sum, yielding something that could be called *least winsorized squares* (LWS) regression, given by

$$\text{Minimize} \sum_{i=1}^{h} (r^2)_{i:n} + (n - h)(r^2)_{h:n}, \qquad (4.27)$$

where h may also depend on some fraction α. Like LMS and LTS, this estimator is regression, scale, and affine equivariant, and it possesses the same breakdown point for a given value of h. However, some preliminary simulations have revealed that the LWS is inferior to the LTS.

S-estimators (Rousseeuw and Yohai, 1984) form another class of high-breakdown affine equivariant estimators with convergence rate $n^{-1/2}$. They are defined by minimization of the dispersion of the residuals:

$$\text{Minimize } s(r_1(\boldsymbol{\theta}), \ldots, r_n(\boldsymbol{\theta})), \qquad (4.28)$$

with final scale estimate

$$\hat{\sigma} = s(r_1(\hat{\boldsymbol{\theta}}), \ldots, r_n(\hat{\boldsymbol{\theta}})). \qquad (4.29)$$

The dispersion $s(r_1(\boldsymbol{\theta}), \ldots, r_n(\boldsymbol{\theta}))$ is defined as the solution of

$$\frac{1}{n} \sum_{i=1}^{n} \rho\left(\frac{r_i}{s}\right) = K. \qquad (4.30)$$

K is often put equal to $E_\Phi[\rho]$, where Φ is the standard normal. The function ρ must satisfy the following conditions:

(S1) ρ is symmetric and continuously differentiable, and $\rho(0) = 0$.

(S2)   There exists $c > 0$ such that ρ is strictly increasing on $[0, c]$ and constant on $[c, \infty)$.

[If there happens to be more than one solution to (4.30), then put $s(r_1, \ldots, r_n)$ equal to the supremum of the set of solutions; this means $s(r_1, \ldots, r_n) = \sup \{s; (1/n) \sum \rho(r_i/s) = K\}$. If there exists no solution to (4.30), then put $s(r_1, \ldots, r_n) = 0$.]

The estimator in (4.28) is called an S-estimator because it is derived from a scale statistic in an implicit way. (Actually, s given by (4.30) is an M-estimator of scale.) Clearly S-estimators are regression, scale, and affine equivariant.

Because of condition (S2), $\psi(x) = \rho'(x)$ will always be zero from a certain value of x on, so ψ is redescending. An example is the ρ-function corresponding to

$$\rho(x) = \begin{cases} \dfrac{x^2}{2} - \dfrac{x^4}{2c^2} + \dfrac{x^6}{6c^4} & \text{for } |x| \le c \\ \dfrac{c^2}{6} & \text{for } |x| > c, \end{cases} \tag{4.31}$$

the derivative of which is Tukey's biweight function defined in (4.17). Another possibility is to take a ρ corresponding to the hyperbolic tangent estimator (4.16).

In order to show that the breakdown point of S-estimators is also 50% we need a preliminary lemma, in which an extra condition on the function ρ is needed:

(S3) $$\frac{K}{\rho(c)} = \frac{1}{2}$$

This condition is easy to fulfill. In the case of (4.31) with $K = E_\Phi[\rho]$, it is achieved by taking $c = 1.547$. Let us now look at the scale estimator $s(r_1, \ldots, r_n)$, which is defined by (4.30) for any sample $(r_1, \ldots, r_n)$.

Lemma 4.   For each ρ satisfying conditions (S1)–(S3) and for each n, there exist positive constants α and β such that the estimator s given by (4.30) satisfies

$$\alpha \operatorname*{med}_i |r_i| \le s(r_1, \ldots, r_n) \le \beta \operatorname*{med}_i |r_i|.$$

Here $\operatorname{med}_i |r_i|$ or $s(r_1, \ldots, r_n)$ may even be zero.

Proof. 1. We first consider the case n odd $(n = 2m + 1)$. Put $S = s(r_1, \ldots, r_n)$ for ease of notation. We will show that

$$\frac{\operatorname*{med}_i |r_i|}{c} \leq S \leq \frac{\operatorname*{med}_i |r_i|}{\rho^{-1}(\rho(c)/(n + 1))} .$$

Suppose $\operatorname{med}_i |r_i| > cS$. Because $\operatorname{med}_i |r_i| = |r|_{m+1:n}$ it holds that at least $m + 1$ of the $|r_i|/S$ are larger than c. Consequently,

$$\frac{1}{n} \sum_{i=1}^{n} \rho\left(\frac{|r_i|}{S}\right) \geq \frac{1}{n} (m + 1)\rho(c) > \rho(c)/2 = K ,$$

which is a contradiction. Therefore, $\operatorname{med}_i |r_i| \leq cS$.

Now suppose that $\operatorname{med}_i |r_i| < \rho^{-1}(\rho(c)/(n + 1))S$. This would imply that the first $m + 1$ of the $|r_i|/S$ are strictly smaller than $\rho^{-1}(\rho(c)/(n + 1))$. Introducing this in $(1/n) \sum_{i=1}^{n} \rho(|r_i|/S)$, we find that

$$\frac{1}{n} \sum_{i=1}^{n} \rho\left(\frac{|r_i|}{S}\right) < \frac{m + 1}{n} \rho\left(\rho^{-1}\left(\frac{\rho(c)}{n + 1}\right)\right) + \frac{m}{n} \rho(c)$$

$$\leq \frac{m + 1}{n} \left(\frac{\rho(c)}{n + 1}\right) + \frac{m}{n} \rho(c)$$

$$= \rho(c)\left\{\frac{1}{2n} + \frac{m}{n}\right\}$$

$$= \tfrac{1}{2}\rho(c) = K ,$$

which is again a contradiction, so $\operatorname{med}_i |r_i| \geq \rho^{-1}\{\rho(c)/(n + 1)\}S$.

2. Let us now treat the case where n is even $(n = 2m)$. We will prove that

$$\frac{\operatorname*{med}_i |r_i|}{c} \leq S \leq \frac{\operatorname*{med}_i |r_i|}{\tfrac{1}{2}\rho^{-1}(2\rho(c)/(n + 2))} .$$

Suppose first that $\operatorname{med}_i |r_i| > cS$. Since n is even, $\operatorname{med}_i |r_i| = \tfrac{1}{2}\{|r|_{m:n} + |r|_{m+1:n}\}$. Then, at least m of the $|r_i|/S$ are strictly larger than c, and

$$\frac{1}{n} \sum_{i=1}^{n} \rho\left(\frac{|r_i|}{S}\right) > \frac{m}{n} \rho(c) = \tfrac{1}{2}\rho(c) = K ,$$

except when all other $|r_i|$ are zero, but then the set of solutions of (4.30) is the interval $(0, 2 \operatorname{med}_i |r_i|/c]$, so $S = 2 \operatorname{med}_i |r_i|/c$. In either case, $\operatorname{med}_i |r_i| \leq cS$.

Suppose now that $\operatorname{med}_i |r_i| < \frac{1}{2} \rho^{-1} \{2\rho(c)/(n+2)\} S$. Then $|r|_{m+1:n} < \rho^{-1} \{2\rho(c)/(n+2)\} S$. Hence, the first $m+1$ of the $|r_i|/S$ are less than $\rho^{-1} \{2\rho(c)/(n+2)\}$. Therefore,

$$\frac{1}{n} \sum_{i=1}^{n} \rho\left(\frac{|r_i|}{S}\right) < \frac{m+1}{n} \frac{2\rho(c)}{n+2} + \frac{m-1}{n} \rho(c)$$

$$= \frac{\rho(c)}{n} + \frac{m-1}{n} \rho(c)$$

$$\leq (m/n)\rho(c) = \rho(c)/2 = K.$$

Finally, $\operatorname{med}_i |r_i| \geq \frac{1}{2} \rho^{-1} \{2\rho(c)/(n+2)\} S$.

3. We will now deal with special cases with zeroes.

Let us start with n odd $(n = 2m + 1)$. When $\operatorname{med}_i |r_i| = 0$, then the first $m+1$ of the $|r_i|$ are zero, and whatever the value of S, we always have $(1/n) \sum_{i=1}^{n} \rho(|r_i|/S) < \frac{1}{2}\rho(c)$. Therefore the set of solutions is empty, so $S = 0$ by definition.

On the other hand, when $\operatorname{med}_i |r_i| > 0$, then there are at least $m+1$ nonzero $|r_i|$, hence

$$\lim_{s \searrow 0} \left\{ \frac{1}{n} \sum_{i=1}^{n} \rho\left(\frac{|r_i|}{s}\right) \right\} \geq ((m+1)/n)\rho(c) > K$$

$$\lim_{s \nearrow \infty} \left\{ \frac{1}{n} \sum_{i=1}^{n} \rho\left(\frac{|r_i|}{s}\right) \right\} = 0 < K.$$

Therefore, there exists a strictly positive solution S.

Let us now consider n even $(n = 2m)$. When $\operatorname{med}_i |r_i| = 0$, then the first $m+1$ of the $|r_i|$ are zero, and whatever the value of S we have again that $(1/n) \sum_{i=1}^{n} \rho(|r_i|/S) < \frac{1}{2}\rho(c)$, so $S = 0$.

When $\operatorname{med}_i |r_i| > 0$, then we are certain that $|r|_{m+1:n} > 0$ too. We shall consider both the case $|r|_{m:n} > 0$ and $|r|_{m:n} = 0$. If $|r|_{m:n}$ is strictly positive, then there are at least $m+1$ nonzero $|r_i|$ and one can follow the same reasoning as in the case where n is odd. If on the other hand $|r|_{m:n} = 0$, then we are in the special case where the ordered $|r_i|$ can be written as a sequence of m zeroes and with $2 \operatorname{med}_i |r_i|$ in the $(m+1)$th position. By definition

$$S = \sup (0, 2 \operatorname{med}_i |r_i|/c] = 2 \operatorname{med}_i |r_i|/c > 0.$$

We may therefore conclude (for both odd and even n) that $\text{med}_i \, |r_i|$ is zero if and only if S is zero.   □

REMARK.   Note that Lemma 4 (as well as Theorems 7 and 8) do not rely on the assumption that $K = E_\Phi[\rho]$, which is only needed if one wants (4.30) to yield a consistent scale estimate for normally distributed residuals.

Theorem 7.   For any ρ satisfying (S1) to (S3), there always exists a solution to (4.28).

Proof.   Making use of the preceding lemma, this follows from the proof of Theorem 1, where the result was essentially given for the minimization of $\text{med}_i \, |r_i|$.   □

Theorem 8.   An S-estimator constructed from a function ρ satisfying (S1) to (S3) has breakdown point

$$\varepsilon_n^* = ([n/2] - p + 2)/n$$

at any sample $\{(\mathbf{x}_i, y_i); i = 1, \ldots, n\}$ in general position.

Proof.   This follows from Theorem 2 by making use of Lemma 4.   □

The breakdown point depends only slightly on n, and for $n \to \infty$ we obtain $\varepsilon^* = 50\%$, the best we can expect. The following result concerns the exact fit property for S-estimators, which again illustrates their high resistance.

Corollary.   If there exists some $\boldsymbol{\theta}$ such that at least $n - [n/2] + p - 1$ of the points satisfy $y_i = \mathbf{x}_i \boldsymbol{\theta}$ *exactly* and are in general position, then the S-estimate for the regression vector will be equal to $\boldsymbol{\theta}$ whatever the other observations are.

REMARK 1.   If condition (S3) is replaced by

$$\frac{K}{\rho(c)} = \alpha \, ,$$

where $0 < \alpha \le \frac{1}{2}$, then the corresponding S-estimators have a breakdown point tending to $\varepsilon^* = \alpha$ when $n \to \infty$. If it is assumed that $K = E_\Phi[\rho]$ in order to achieve a consistent scale estimate for normally distributed residuals, one can trade a higher asymptotic efficiency against a lower breakdown point.

REMARK 2. Note that S-estimators satisfy the same first-order necessary conditions as the M-estimators discussed in Section 2 of Chapter 1. Indeed, let $\boldsymbol{\theta}$ be any p-dimensional parameter vector. By definition, we know that

$$S(\boldsymbol{\theta}) = s(r_1(\boldsymbol{\theta}), \ldots, r_n(\boldsymbol{\theta})) \geq \hat{\sigma} = S(\hat{\boldsymbol{\theta}}) .$$

Keeping in mind that $S(\boldsymbol{\theta})$ satisfies

$$(1/n) \sum_{i=1}^{n} \rho(r_i(\boldsymbol{\theta})/S(\boldsymbol{\theta})) = K$$

and that $\rho(u)$ is nondecreasing in $|u|$, it follows that always

$$(1/n) \sum_{i=1}^{n} \rho(r_i(\boldsymbol{\theta})/\hat{\sigma}) \geq K .$$

At $\boldsymbol{\theta} = \hat{\boldsymbol{\theta}}$, this becomes an equality. Therefore, $\hat{\boldsymbol{\theta}}$ minimizes $(1/n) \sum_{i=1}^{n} \rho(r_i(\boldsymbol{\theta})/\hat{\sigma})$. (This fact cannot be used for determining $\hat{\boldsymbol{\theta}}$ in practice, because $\hat{\sigma}$ is fixed but unknown.) Differentiating with respect to $\boldsymbol{\theta}$, we find

$$(1/n) \sum_{i=1}^{n} \psi(r_i(\hat{\boldsymbol{\theta}})/\hat{\sigma}))\mathbf{x}_i = \mathbf{0} .$$

If we denote $\rho - K$ by χ, we conclude that $(\hat{\boldsymbol{\theta}}, \hat{\sigma})$ is a solution of the system of equations

$$\begin{cases} \dfrac{1}{n} \sum_{i=1}^{n} \psi(r_i(\hat{\boldsymbol{\theta}})/\hat{\sigma})\mathbf{x}_i = \mathbf{0} \\[2ex] \dfrac{1}{n} \sum_{i=1}^{n} \chi(r_i(\hat{\boldsymbol{\theta}})/\hat{\sigma}) = 0 \end{cases} \qquad (4.32)$$

(described in Section 2 of Chapter 1) for defining an M-estimator. Unfortunately, these equations cannot be used directly because there are infinitely many solutions (ψ is redescending) and the iteration procedures for the computation of M-estimators easily end in the wrong place if there are leverage points. [This means we still have to minimize (4.28) with brute force in order to actually compute the S-estimate in a practical situation.] Therefore it would be wrong to say that S-estimators *are* M-estimators, because their computation and breakdown are completely different, but they do satisfy similar first-order necessary conditions.

Besides their high resistance to contaminated data, S-estimators also behave well when the data are not contaminated. To show this, we will look at the asymptotic behavior of S-estimators at the central Gaussian model, where $(\mathbf{x}_i, y_i)$ are i.i.d. random variables satisfying

$$y_i = \mathbf{x}_i \boldsymbol{\theta}_0 + e_i , \tag{4.33}$$

$\mathbf{x}_i$ follows some distribution H, and e_i is independent of $\mathbf{x}_i$ and distributed like $\Phi(e/\sigma_0)$ for some $\sigma_0 > 0$.

Theorem 9.   Let ρ be a function satisfying (S1) and (S2), with derivative $\rho' = \psi$. Assume that:

(1) $\psi(u)/u$ is nonincreasing for $u > 0$;
(2) $E_H[\|\mathbf{x}\|] < \infty$, and H has a density.

Let $(\mathbf{x}_i, y_i)$ be i.i.d. according to the model in (4.33), and let $\hat{\boldsymbol{\theta}}_n$ be a solution of (4.28) for the first n points, and $\hat{\sigma}_n = s(r_1(\hat{\boldsymbol{\theta}}_n), \ldots, r_n(\hat{\boldsymbol{\theta}}_n))$. If $n \to \infty$ then

$$\hat{\boldsymbol{\theta}}_n \to \boldsymbol{\theta}_0 \quad \text{a.s.}$$

and

$$\hat{\sigma}_n \to \sigma_0 \quad \text{a.s.}$$

Proof. This follows from Theorem 2.2 and 3.1 of Maronna and Yohai (1981), because S-estimators satisfy the same first-order necessary conditions as M-estimators (according to Remark 2 above).   □

Let us now show the asymptotic normality of S-estimators.

Theorem 10.   Without loss of generality let $\boldsymbol{\theta}_0 = 0$ and $\sigma_0 = 1$. If the conditions of Theorem 9 hold and

(3) ψ is differentiable in all but a finite number of points, $|\psi'|$ is bounded, and $\int \psi' \, d\Phi > 0$;
(4) $E_H[\mathbf{x}'\mathbf{x}]$ is nonsingular and $E_H[\|\mathbf{x}\|^3] < \infty$, then

$$\mathscr{L}(n^{1/2}(\hat{\boldsymbol{\theta}}_n - \boldsymbol{\theta}_0)) \to N\left(\boldsymbol{0}, E_H[\mathbf{x}'\mathbf{x}]^{-1}\left\{\int \psi^2 \, d\Phi\right\} \Big/ \left\{\int \psi' \, d\Phi\right\}^2\right)$$

and

$$\mathscr{L}(n^{1/2}(\hat{\sigma}_n - \sigma_0)) \to N\left(0, \frac{\int (\rho(y) - E_\Phi[\rho])^2 \, d\Phi(y)}{\left\{\int y\psi(y) \, d\Phi(y)\right\}^2}\right).$$

Proof. This follows from Theorem 4.1 of Maronna and Yohai (1981) using Remark 2. □

As a consequence of Theorem 10, we can compute the asymptotic efficiency e of an S-estimator at the Gaussian model as

$$e = \frac{\left(\int \psi' \, d\Phi\right)^2}{\left(\int \psi^2 \, d\Phi\right)}.$$

Table 19 gives the asymptotic efficiency of the S-estimators corresponding to the function ρ defined in (4.31) for different values of the breakdown point ε^*. From this table it is apparent that values of c larger than 1.547 yield better asymptotic efficiencies at the Gaussian central model, but yield smaller breakdown points. Furthermore we note that taking $c = 2.560$ yields a value of e which is larger than that of L_1 (for

Table 19. Asymptotic Efficiency of S-Estimators for Different Values of ε^*, Making use of Tukey's Biweight Function

ε^*	e	c	K
50%	28.7%	1.547	0.1995
45%	37.0%	1.756	0.2312
40%	46.2%	1.988	0.2634
35%	56.0%	2.251	0.2957
30%	66.1%	2.560	0.3278
25%	75.9%	2.937	0.3593
20%	84.7%	3.420	0.3899
15%	91.7%	4.096	0.4194
10%	96.6%	5.182	0.4475

which e is about 64%), and gains us a breakdown point of 30%. In practice, we do not recommend the estimators in the table with a breakdown point smaller than 25%. It appears to be better to apply the $c = 1.547$ estimator because of its high breakdown point. From this first solution, one can then compute a one-step M-estimator or a one-step reweighted least squares in order to make up for the initial low efficiency. Such a two-stage procedure inherits the 50% breakdown point from the first stage and inherits the high asymptotic efficiency from the second. An algorithm for computing S-estimators will be described in Chapter 5.

5. RELATION WITH PROJECTION PURSUIT

The goal of projection pursuit (PP) procedures is to discover structure in a multivariate data set by projecting these data in a lower-dimensional space. Such techniques were originally proposed by Roy (1953), Kruskal (1969), and Switzer (1970). The name "projection pursuit" itself was coined by Friedman and Tukey (1974), who developed a successful algorithm. The main problem is to find "good" projections, because arbitrary projections are typically not very informative. Friedman and Stuetzle (1982) give some examples of point configurations with strong structure, which possess projections in which no structure is apparent. Therefore, PP tries out many low-dimensional projections of a high-dimensional point cloud in search for a "most interesting" one, by numerically optimizing a certain objective function (which is also called a "projection index"). Some important applications are PP classification (Friedman and Stuetzle 1980), PP regression (Friedman and Stuetzle 1981), robust principal components (Ruymgaart 1981), and PP density estimation (Friedman et al. 1984). A recent survey of the field has been given by Huber (1985). The program MACSPIN (D^2 Software 1986) enables us to look at two-dimensional projections in a dynamic way.

Let us now show that there is a relation between robust regression and PP. To see this, consider the $(p + 1)$-dimensional space of the $(\mathbf{x}_i, y_i)$, where the last component of $\mathbf{x}_i$ equals 1 in the case of regression with a constant. In this space, linear models are defined by

$$(\mathbf{x}, y)\begin{pmatrix} \boldsymbol{\theta} \\ -1 \end{pmatrix} = 0 \tag{5.1}$$

for some p-dimensional column vector $\boldsymbol{\theta}$. In order to find a "good" $\hat{\boldsymbol{\theta}}$, we start by projecting the point cloud on the y-axis in the direction orthogonal to $(\boldsymbol{\theta}, -1)'$ for any vector $\boldsymbol{\theta}$. This means that each $(\mathbf{x}_i, y_i)$ is projected onto $(0, r_i(\boldsymbol{\theta}))$, where $r_i(\boldsymbol{\theta}) = y_i - \mathbf{x}_i\boldsymbol{\theta}$. Following Rousseeuw (1984, p.

874) and Donoho et al. (1985), we measure the "interestingness" of any such projection by its dispersion

$$s(r_1(\boldsymbol{\theta}), \ldots, r_n(\boldsymbol{\theta})) , \qquad (5.2)$$

where the objective s is scale equivariant

$$s(\tau r_1, \ldots, \tau r_n) = |\tau| s(r_1, \ldots, r_n) \quad \text{for all } \tau \qquad (5.3)$$

but not translation invariant. The PP estimate $\hat{\boldsymbol{\theta}}$ is then obtained by minimization of the projection index (5.2). If $s(r_1, \ldots, r_n) = (\Sigma_{i=1}^n r_i^2/n)^{1/2}$, then this "most interesting" $\hat{\boldsymbol{\theta}}$ is simply the vector of least squares coefficients. Analogously, $s(r_1, \ldots, r_n) = \Sigma_{i=1}^n |r_i|/n$ yields the L_1 estimator, and minimizing $(\Sigma_{i=1}^n |r_i|^q/n)^{1/q}$ gives the L_q-estimators (Gentleman 1965, Sposito et al. 1977). Using a very robust s brings us back to our high-breakdown regression estimators: $s = (\text{med}_i\, r_i^2)^{1/2}$ yields the LMS, $s = (\Sigma_{i=1}^h (r^2)_{i:n}/n)^{1/2}$ yields the LTS, and by putting s equal to a robust M-estimator of scale we obtain S-estimators. Note that the minimization of any s satisfying (5.3) will yield a regression estimator that is regression, scale, and affine equivariant (as discussed in Section 4). So any nice s defines a type of regression estimator; by varying s one obtains an entire class of regression estimators belonging to the PP family. Thus the PP principle extends to cover the linear regression problem and to encompass both classical procedures and high-breakdown methods. This notion is best thought of as "PP Linear Regression," to distinguish it from Friedman and Stuetzle's (1981) nonlinear and nonrobust "PP Regression."

It appears that the only affine equivariant high-breakdown regression estimators known so far (the LMS, the LTS, and S-estimators) are related to PP. (GM-estimators do not have high breakdown, and the repeated median is not affine equivariant.) There is a reason for this apparent relation between high breakdown and PP. Indeed, Donoho, Rousseeuw, and Stahel have found that breakdown properties are determined by behavior near situations of *exact fit*: These are situations where most of the data lie exactly in a regression hyperplane (see Remark 1 following Theorem 3 of Section 4). Such configurations are precisely those having a projection in which most of the data collapse to a point. In other words, high breakdown appears to depend on an estimator's behavior in those situations where certain special kinds of projections occur. Since PP can, in principle, be used to search for such projections, the usefulness of PP in synthesizing high-breakdown procedures is not surprising.

Table 20. Schematic Overview of Some Affine Equivariant Regression Techniques

Criterion	Method	Computation	ε^*
Best linear unbiased	LS	Explicit	0%
Minimax variance	M	Iterative	0%
Bounded influence	GM	Iterative, with weights on $\mathbf{x}_i$ (harder)	Down to 0% if p increases
High breakdown	LMS, LTS, S	Projection pursuit techniques	Constant, up to 50%

Note, however, that PP is not necessarily the only way to obtain high-breakdown equivariant estimators, at least not in multivariate location where Rousseeuw (1983) gives the example of the minimal volume ellipsoid containing at least half the data. Also, not every PP-based affine equivariant estimator is going to have high breakdown (Fill and Johnstone 1984).

The relation of our robust regression estimators with PP also gives a clue to their computational complexity. In principle, all possible projections must be tried out (although the actual algorithm in Chapter 5 can exploit some properties to speed things up). This means that the LMS, the LTS, and S-estimators belong to the highly computer-intensive part of statistics, just like PP and the bootstrap, to which the algorithm is also related. In Table 20 we have a schematic overview of criteria in affine equivariant regression.

*6. OTHER APPROACHES TO ROBUST MULTIPLE REGRESSION

We have seen that the conditions under which the LS criterion is optimal are rarely fulfilled in realistic situations. In order to define more robust regression alternatives, many statisticians have exploited the resistance of the sample median to extreme values. For instance, the L_1 criterion can be seen as a generalization of the univariate median, because the minimization of $\sum_{i=1}^{n} |y_i - \hat{\theta}|$ defines the median of n observations y_i. In the regression problem, the L_1 estimator is given by

$$\underset{\hat{\boldsymbol{\theta}}}{\text{Minimize}} \sum_{i=1}^{n} |r_i| . \tag{6.1}$$

The substitution of the square by the absolute value leads to a considerable gain in robustness. However, in terms of the breakdown point, L_1 is not really better than LS, because the L_1 criterion is robust with respect to outliers in the y_i but is still vulnerable to leverage points. Moreover, Wilson (1978) showed that the efficiency of the L_1 estimator decreases when n increases. From a numerical point of view, the minimization in (6.1) amounts to the solution of a linear program:

$$\underset{\hat{\theta}}{\text{Minimize}} \sum_{i=1}^{n} (u_i + v_i)$$

under the constraints

$$y_i = \sum_{k=1}^{p} \theta_k x_{ik} + u_i - v_i , \qquad u_i \geq 0, v_i \geq 0 .$$

Barrodale and Roberts (1974) and Sadovski (1977) described algorithms and presented FORTRAN routines for calculating L_1 regression coefficients.

The minimization of an L_q norm (for $1 \leq q \leq 2$) of the residuals has been considered by Gentleman (1965), Forsythe (1972), and Sposito et al. (1977), who presented an algorithm (with FORTRAN code) for the L_q fit of a straight line. Dodge (1984) suggested a regression estimator based on a convex combination of the L_1 and L_2 norms, resulting in

$$\underset{\hat{\theta}}{\text{Minimize}} \sum_{i=1}^{n} \left((1 - \delta) \frac{r_i^2}{2} + \delta |r_i| \right) \quad \text{with } 0 \leq \delta \leq 1 .$$

Unfortunately, all these proposals possess a zero breakdown point.

In Section 7 of Chapter 2, we listed some estimators for simple regression which are also based on the median. Some of them have been generalized to multiple regression by means of a "sweep" operator (see Andrews 1974).

The idea behind Theil's (1950) estimator, which consists of looking at the median of all pairwise slopes (see Section 7 of Chapter 2), has also been the source of extensions and modifications. A recent proposal comes from Oja and Niinimaa (1984). For each subset $J = \{i_1, i_2, \ldots, i_p\}$ of $\{1, 2, \ldots, n\}$ containing p indices, they define

$$\boldsymbol{\theta}_J = \boldsymbol{\theta}(i_1, i_2, \ldots, i_p) \tag{6.2}$$

as the parameter vector corresponding to the hyperplane going exactly through the p points $(\mathbf{x}_{i_1}, y_{i_1}), \ldots, (\mathbf{x}_{i_p}, y_{i_p})$. They call these $\boldsymbol{\theta}_J$ *pseudo-*

observations, and there are C_n^p of them. Their idea is now to compute a multivariate location estimator (in p-dimensional space) of all these $\boldsymbol{\theta}_J$. A certain weighted average of the $\boldsymbol{\theta}_J$ yields LS, but of course they want to insert a robust multivariate estimator. If one computes

$$\hat{\theta}_j = \underset{J}{\mathrm{med}}\,(\boldsymbol{\theta}_J)_j \quad \text{for all } j = 1, \ldots, p \tag{6.3}$$

(coordinatewise median over all subsets J), then one obtains a regression estimator that fails to be affine equivariant. For simple regression, (6.3) indeed yields the Theil–Sen estimator described in Section 7 of Chapter 2. In order to obtain an affine equivariant regression estimator, one has to apply a multivariate location estimator T which is itself affine equivariant, meaning that

$$T(\mathbf{z}_1\mathbf{A} + \mathbf{b}, \ldots, \mathbf{z}_k\mathbf{A} + \mathbf{b}) = T(\mathbf{z}_1, \ldots, \mathbf{z}_k)\mathbf{A} + \mathbf{b} \tag{6.4}$$

for any sample $\{\mathbf{z}_1, \ldots, \mathbf{z}_k\}$ of p-dimensional row vectors, for any nonsingular square matrix $\mathbf{A}$, and for any p-dimensional vector $\mathbf{b}$. For this purpose they propose to apply the *generalized median*, an ingenious construction of Oja (1983) which is indeed affine equivariant and will be discussed in Section 1 of Chapter 7. Unfortunately, the computation complexity of this generalized median is enormous (and it would have to be applied to a very large set of $\boldsymbol{\theta}_J$ vectors!). Even when the coordinatewise median (6.3) is applied, the Oja–Niinimaa regression estimator needs considerable computation time because of the C_n^p pseudo-observations. In either case the consideration of all $\boldsymbol{\theta}_J$ is impossible, so it might be useful to consider a random subpopulation of pseudo-observations.

Let us now consider the breakdown point of this technique. We can only be sure that a pseudo-observation $\boldsymbol{\theta}_J = \boldsymbol{\theta}(i_1, \ldots, i_p)$ is "good" when the p points $(\mathbf{x}_{i_1}, y_{i_1}), \ldots, (\mathbf{x}_{i_p}, y_{i_p})$ are all good. If there is a fraction ε of outliers in the original data, then we can only be certain of a proportion $(1 - \varepsilon)^p$ of "good" pseudo-observations. Therefore, we must have that

$$(1 - \varepsilon)^p \geq \tfrac{1}{2} \tag{6.5}$$

because the best possible breakdown point of a multivariate location estimator is 50%, which means that still 50% of "good" pseudo-observations are needed. Formula (6.5) yields an upper bound on the amount of contamination that is allowed in the original data, namely

$$\varepsilon^* = 1 - (\tfrac{1}{2})^{1/p}. \tag{6.6}$$

For $p = 2$, one finds again the breakdown point of Theil's estimator. The value of ε^* in (6.6) decreases very fast with p, as shown in Table 21.

M-estimators (Huber 1973) marked an important step forward in robust estimation. Much research has been concentrated on constructing functions ρ and ψ (see Chapter 1) such that the associated M-estimators were as robust as possible on the one hand, but still fairly efficient (in the case of a normal error distribution) on the other hand. Note that LS is also an M-estimator with $\psi(t) = t$ and that L_1 regression corresponds to $\psi(t) = \text{sgn}(t)$. Huber (1964) proposed the following ψ function:

$$\psi(t) = \begin{cases} t & \text{if } |t| < b \\ b \text{ sgn}(t) & \text{if } |t| \geq b , \end{cases} \tag{6.7}$$

where b is a constant. Actually, in a univariate location setting, this estimator was already constructed by the Dutch astronomer Van de Hulst in 1942 (see van Zwet 1985). Asymptotic properties of this estimator are discussed in Huber (1973). Hampel (1974) defined a function that protects the fit even more against strongly outlying observations, by means of

$$\psi(t) = \begin{cases} t & \text{if } |t| < a \\ a \text{ sgn}(t) & \text{if } a \leq |t| < b \\ \{(c - |t|)/(c - b)\} a \text{ sgn}(t) & \text{if } b \leq |t| \leq c \\ 0 & \text{otherwise} , \end{cases} \tag{6.8}$$

which is called a three-part redescending M-estimator. Figure 19 shows ψ-functions of both types. In the literature one can find many more ψ-functions (see Hampel et al. 1986 for a detailed description).

Of course, it is not sufficient to define new estimators and to study their asymptotic properties. Besides that, one has to develop a method for calculating the estimates. The solution of the system of equations (4.32) corresponding to M-estimates is usually performed by an iterative

Table 21. Value of $\varepsilon^* = 1 - (\frac{1}{2})^{1/p}$ for Some Values of p

p	ε^*	p	ε^*
1	50%	6	11%
2	29%	7	9%
3	21%	8	8%
4	16%	9	7%
5	13%	10	7%

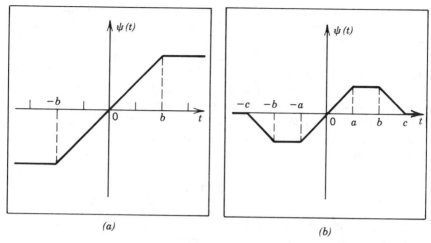

Figure 19. (a) Huber-type ψ-function. (b) Hampel-type ψ-function.

procedure. In each step, one has to estimate the coefficients and the scale simultaneously. However, it is very important to start the iteration with a "good" starting value, that is, an estimate which is already sufficiently robust. Without this precaution, one can easily end up in a local minimum that does not correspond at all to the expected robust solution. Dutter (1977) proposed some algorithms for solving the numerical problems associated with M-estimators. The calculation of GM-estimators or bounded-influence estimators [Chapter 1, equation (2.12)] presents similar problems. Dutter (1983a) described a user-oriented and portable computer program (BLINWDR) for calculating these estimates. Marazzi (1986) developed the subroutine library ROBETH, which computes M-estimators and bounded influence estimators. ROBETH also contains robust tests for linear models (Ronchetti 1982) and a variable selection procedure (see also Ronchetti 1985). The corresponding system control program is called ROBSYS (Marazzi 1987). TROLL (Samarov and Welsch 1982, Peters et al. 1982) is a large interactive system for statistical analysis which also includes the computation of bounded influence regression estimators.

Note that the breakdown point of M- and GM-estimators is quite different. Where ε^* is again 0% for M-estimators because of their vulnerability to leverage points, it becomes nonzero for GM-estimators (Maronna et al. 1979, Donoho and Huber 1983). For p tending to ∞, the breakdown point of GM-estimators drops to 0% (a small numerical study was performed by Kamber 1985). Yohai (1985) observes that some

GM-estimators have very low efficiency in the presence of good leverage points. Exact breakdown points of GM-estimators are computed by Martin et al. (1987).

Another approach to robust regression is based on the ranks of the residuals. In the framework of univariate location, these so-called *R-estimators* are due to Hodges and Lehmann (1963). The idea of using rank statistics has been extended to the domain of multiple regression by Adichie (1967), Jurecková (1971), and Jaeckel (1972). The proposal of Jaeckel leads to the following definition: If R_i is the rank of $r_i = y_i - \mathbf{x}_i \boldsymbol{\theta}$, then the objective is to

$$\underset{\hat{\boldsymbol{\theta}}}{\text{Minimize}} \sum_{i=1}^{n} a_n(R_i) r_i , \qquad (6.9)$$

where the scores function $a_n(i)$ is monotone and satisfies $\sum_{i=1}^{n} a_n(i) = 0$. Some possibilities for the scores $a_n(i)$ are:

Wilcoxon scores: $a_n(i) = i - (n+1)/2$

Van der Waerden scores: $a_n(i) = \Phi^{-1}(i/(n+1))$

median scores: $a_n(i) = \text{sgn}\,(i - (n+1)/2)$

bounded normal scores: $a_n(i) = \min(c, \max\{\Phi^{-1}(i/(n+1)), -c\})$.

(The latter scores were proposed by Rousseeuw 1979 and Ronchetti 1979.) In the case of regression with intercept, one has to estimate the constant term separately, since the objective function is invariant with respect to the intercept. This can be done by using a robust location estimate of the residuals. An important advantage of R-estimators compared to M-estimators is that they are automatically scale equivariant, so they do not depend on a simultaneous scale estimator. Nevertheless, Jurecková (1977) showed that (under certain conditions) R-estimators are asymptotically equivalent to M-estimators. Heiler and Willers (1979) prove the asymptotic normality of R-estimators under weaker conditions than those imposed by Jurecková. Lecher (1980) developed (and implemented) an algorithm for R-estimators, in which the minimization (6.9) is carried out by a direct-search algorithm of Rosenbrock (1960). Cheng and Hettmansperger (1983) proposed an iteratively reweighted least squares algorithm for solving (6.9).

The class of *L-estimators* also plays a prominent role in robust univariate location. They are based on linear combinations of order statistics, and their popularity rests mainly on their simple computation. Bickel (1973) has proposed a class of one-step L-estimators for regression, which depend on a preliminary estimate of $\boldsymbol{\theta}$.

Koenker and Bassett (1978) formulated another proposal for L-estimators, making use of analogs of sample quantiles for linear regression. They defined the α-*regression quantile* $(0 < \alpha < 1)$ as the solution $\hat{\boldsymbol{\theta}}_\alpha$ of

$$\text{Minimize} \sum_{i=1}^{n} \rho_\alpha(r_i),$$
$$\hat{\boldsymbol{\theta}}_\alpha$$

where

$$\rho_\alpha(r_i) = \begin{cases} \alpha r_i & \text{if } r_i \geq 0 \\ (\alpha - 1)r_i & \text{if } r_i \leq 0. \end{cases}$$

(For $\alpha = 0.5$, one obtains the L_1 estimator.) Koenker and Bassett then proposed to compute linear combinations of these $\hat{\boldsymbol{\theta}}_\alpha$. Portnoy (1983) proved some asymptotic properties of these estimators. However, note that their breakdown point is still zero.

Also the trimmed least squares estimators of Ruppert and Carroll (1980) are L-estimators. They are counterparts of trimmed means, which are well-known L-estimators of location. (Note that they are not related to the LTS discussed in Section 4.) Ruppert and Carroll proposed two ways to select observations to be trimmed: one of these uses the concept of regression quantiles, whereas the other employs residuals from a preliminary estimator.

Heiler (1981) and Kühlmeyer (1983) describe the results of a simulation study about M-, L-, and R-estimators for linear regression. The behavior of these classes of estimators for different designs and error distributions was compared. It is important to note that the generated samples were rather small, $n \leq 40$ and $p \leq 3$, and that no leverage points were constructed. The conclusions from this study can be summarized as follows: LS is very poor, even for mild deviations from normality. M-estimators with redescending ψ-function turn out to work quite well. R-estimates with Wilcoxon scores, which have the advantage of being scale equivariant (and of being simple to use because no parameter has to be fixed in advance), are a good alternative. L-estimates achieved less satisfactory results.

The first regression estimator with maximal breakdown point is the repeated median due to Siegel (1982). Like the proposal of Oja and Niinimaa (1984) discussed above, it is based on all subsets of p points. Any p observations $(\mathbf{x}_{i_1}, y_{i_1}), \ldots, (\mathbf{x}_{i_p}, y_{i_p})$ determine a unique parameter vector, the jth coordinate of which is denoted by $\theta_j(i_1, \ldots, i_p)$. The repeated median is then defined coordinatewise as

$$\hat{\theta}_j = \underset{i_1}{\text{med}} (\ldots (\underset{i_{p-1}}{\text{med}} (\underset{i_p}{\text{med}} \, \theta_j(i_1, \ldots, i_p))) \ldots). \tag{6.10}$$

This estimator can be calculated explicitly. The fact that the medians are computed sequentially (instead of one median over all subsets) gives the estimator a 50% breakdown point, in which respect it is vastly superior to the Oja–Niinimaa proposal. (The asymptotic efficiency of the repeated median, as well as its influence function, appear to be unknown as yet.) Unfortunately, two disadvantages remain: the absence of affine equivariance and the large number of subsets. However, the second problem might be avoided by selecting some subsets at random instead of using all C_n^p of them.

Recently, Yohai (1985) introduced a new improvement toward higher efficiency for high-breakdown estimators like LMS and LTS. He called this new class *MM-estimators*. (Note that they are not related to the MM-estimators considered in Chapter 9 of Rey 1983.) Yohai's estimators are defined in three stages. In the first stage, a high-breakdown estimate θ^* is calculated, such as LMS or LTS. For this purpose, the robust estimator does not need to be efficient. Then, an M-estimate of scale s_n with 50% breakdown is computed on the residuals $r_i(\theta^*)$ from the robust fit. Finally, the MM-estimator $\hat{\theta}$ is defined as any solution of

$$\sum_{i=1}^{n} \psi(r_i(\theta)/s_n)\mathbf{x}_i = \mathbf{0} \,,$$

which satisfies

$$S(\theta) \leq S(\theta^*) \,,$$

where

$$S(\theta) = \sum_{i=1}^{n} \rho(r_i(\theta)/s_n) \,.$$

The function ρ must be like those used in the construction of S-estimators in Section 4; in particular, it must satisfy conditions (S1) and (S2). This implies that $\psi = \rho'$ has to be properly redescending: Some possibilities are three-part redescenders (6.8), Tukey's biweight (4.17), or the hyperbolic tangent ψ (4.16). The trick is that this ρ may be quite different from that of the scale estimate s_n of the second stage, because the first and the second stage must achieve the high breakdown point whereas the third stage is allowed to aim for a high efficiency. Indeed, Yohai showed that MM-estimators inherit the 50% breakdown point of the first stage and that they also possess the exact fit property. Moreover, he proved that MM-estimators are highly efficient when the errors are normally distributed. In a small numerical study, he showed that they compare favorably with GM-estimators.

In the same spirit of combining high breakdown with high efficiency, Yohai and Zamar (1986) consider "τ-estimators" defined by

$$\underset{\hat{\theta}}{\text{Minimize}} \; s_n^2 \, \frac{1}{n} \sum_{i=1}^{n} \rho\!\left(\frac{r_i}{s_n}\right),$$

where again ρ may be different from that of the scale estimate s_n, which is applied to the residuals $r_i(\theta)$. Asymptotically, a τ-estimator behaves like an M-estimator with a ρ-function that is a weighted average of the two ρ-functions used in this construction.

Another possibility, which we have not yet investigated, would be to minimize an objective function of the type

$$\left[\frac{1}{n} \sum_{i=1}^{n} r_i^2(\theta)\right] \wedge \left[k^2 S^2(\theta)\right], \tag{6.11}$$

where $\wedge$ denotes the minimum of two numbers, $k > 1$, and $S(\theta)$ is a high-breakdown estimator of scale on the residuals $r_1(\theta), \ldots, r_n(\theta)$, which is consistent under normality. For instance, $S^2(\theta)$ may be a multiple of $\text{med}_i \, r_i^2(\theta)$ or $(1/n) \sum_{i=1}^{h} \left(r^2(\theta)\right)_{i:n}$, or $S(\theta)$ may be a suitable M-estimator of scale. It seems that the minimization (over $\hat{\theta}$) of (6.11) would combine high asymptotic efficiency with a high breakdown point, because most often the first part (LS) would be used at "good" configurations whereas the second part protects from "bad" configurations. However, it remains to be verified whether the actual (finite-sample) behavior of this estimator would be good enough to compete with simple but effective methods like the combination of LMS with a one-step improvement.

REMARK. Suppose that we apply a weighted LS with weights given by

$$w_i = \begin{cases} 1 & \text{if } |r_i/\sigma^*| \le c \\ 0 & \text{otherwise}, \end{cases}$$

where r_i is the LMS residual of y_i, and σ^* is the corresponding LMS scale estimate. For each constant $c \ge 1$, this estimator has breakdown point 50%, whereas for $c \to \infty$ it becomes more and more efficient, and tends to LS. This paradox can be explained by understanding that the breakdown point is only a crude qualitative notion. Indeed, the above estimator with *large* c will not become unbounded for less than 50% of contamination, but it will not be very good either. [The same is true for the univariate M-estimator of location (6.7) with large b.] One should not forget that the breakdown point is only one out of several robustness criteria, so a high breakdown point alone is not a *sufficient* condition for a good method. We personally consider a good breakdown point as a *necessary* condition, because we do not want estimators that can become arbitrarily bad as a result of a small fraction of contamination. Indeed, Murphy's Law guarantees us that such contamination is bound to occur in practice.

EXERCISES AND PROBLEMS

Sections 1–3
1. Table 22 was taken from Gray (1985). It deals with 23 single-engine aircraft built over the years 1947–1979. The dependent variable is cost (in units of $100,000), and the explanatory variables are aspect

Table 22. Aircraft Data

Index	Aspect Ratio	Lift-to-Drag Ratio	Weight	Thrust	Cost
1	6.3	1.7	8,176	4,500	2.76
2	6.0	1.9	6,699	3,120	4.76
3	5.9	1.5	9,663	6,300	8.75
4	3.0	1.2	12,837	9,800	7.78
5	5.0	1.8	10,205	4,900	6.18
6	6.3	2.0	14,890	6,500	9.50
7	5.6	1.6	13,836	8,920	5.14
8	3.6	1.2	11,628	14,500	4.76
9	2.0	1.4	15,225	14,800	16.70
10	2.9	2.3	18,691	10,900	27.68
11	2.2	1.9	19,350	16,000	26.64
12	3.9	2.6	20,638	16,000	13.71
13	4.5	2.0	12,843	7,800	12.31
14	4.3	9.7	13,384	17,900	15.73
15	4.0	2.9	13,307	10,500	13.59
16	3.2	4.3	29,855	24,500	51.90
17	4.3	4.3	29,277	30,000	20.78
18	2.4	2.6	24,651	24,500	29.82
19	2.8	3.7	28,539	34,000	32.78
20	3.9	3.3	8,085	8,160	10.12
21	2.8	3.9	30,328	35,800	27.84
22	1.6	4.1	46,172	37,000	107.10
23	3.4	2.5	17,836	19,600	11.19

Source: Office of Naval Research.

ratio, lift-to-drag ratio, weight of the plane (in pounds), and maximal thrust. Run PROGRESS on these data. Do you find any outliers in the standardized observations? Does LS identify any regression outliers? How many outliers are identified by LMS and RLS, and of what type are they? Is there a good leverage point in the data?

2. Table 23 lists the delivery time data of Montgomery and Peck (1982, p. 116). We want to explain the time required to service a vending machine (y) by means of the number of products stocked (x_1) and the distance walked by the route driver (x_2). Run PROGRESS on these data. The standardized observations reveal two leverage points. Look at the LMS or RLS results to decide which of these is good and

Table 23. Delivery Time Data

Index (i)	Number of Products (x_1)	Distance (x_2)	Delivery Time (y)
1	7	560	16.68
2	3	220	11.50
3	3	340	12.03
4	4	80	14.88
5	6	150	13.75
6	7	330	18.11
7	2	110	8.00
8	7	210	17.83
9	30	1460	79.24
10	5	605	21.50
11	16	688	40.33
12	10	215	21.00
13	4	255	13.50
14	6	462	19.75
15	9	448	24.00
16	10	776	29.00
17	6	200	15.35
18	7	132	19.00
19	3	36	9.50
20	17	770	35.10
21	10	140	17.90
22	26	810	52.32
23	9	450	18.75
24	8	635	19.83
25	4	150	10.75

Source: Montgomery and Peck (1982).

which is bad. How does deleting the bad leverage point (as done by RLS) affect the significance of the regression coefficients?

3. Table 24 was taken from Prescott (1975), who investigated the effect of the concentration of inorganic phosphorus (x_1) and organic phosphorus (x_2) in the soil upon the phosphorus content (y) of the corn grown in this soil. Carry out a multiple regression analysis of y on x_1 and x_2 by means of PROGRESS. Are there extreme standardized observations? Which outliers are identified by LMS and RLS? In view of the fact that one of the explanatory variables has a very insignificant coefficient in both LS and RLS, it is recommended to run the analysis again without that variable. Indicate the previously discovered outliers in the scatterplot of this simple regression. Judging from the statistics and p-values in both models, do you think that switching to the smaller model is justified?

4. When fitting a multiplicative model

$$y_i = x_{i1}^{\theta_1} x_{i2}^{\theta_2} \dots x_{ip}^{\theta_p} ,$$

Table 24. Phosphorus Content Data

Index (i)	Inorganic Phosphorus (x_1)	Organic Phosphorus (x_2)	Plant Phosphorus (y)
1	0.4	53	64
2	0.4	23	60
3	3.1	19	71
4	0.6	34	61
5	4.7	24	54
6	1.7	65	77
7	9.4	44	81
8	10.1	31	93
9	11.6	29	93
10	12.6	58	51
11	10.9	37	76
12	23.1	46	96
13	23.1	50	77
14	21.6	44	93
15	23.1	56	95
16	1.9	36	54
17	26.8	58	168
18	29.9	51	99

Source: Prescott (1975).

it is natural to logarithmize the variables. However, in economics it happens that some observations are zero (typically in one of the explanatory variables). It is then customary to put the transformed observation equal to a very negative value. Would you rather use LS or a robust regression method on these transformed data? Why?

5. (Research problem) Is it possible to develop collinearity diagnostics that are not so much affected by outliers?

Sections 4–6

6. Show that the repeated median estimator is regression and scale equivariant, and give a counterexample to show that it is not affine equivariant.

7. (Research problem) It would be interesting to know the asymptotic behavior of the repeated median regression estimator and to obtain its influence function.

8. Show that the variant of the LMS given by formula (4.9) has breakdown point $([(n-p)/2]+1)/n$.

9. Explain why the breakdown point of the least quantile of squares estimator (4.10) is approximately equal to α. Why doesn't the breakdown point become higher than 50% when $\alpha > \frac{1}{2}$?

10. The L_∞ fit is determined by the narrowest band covering *all* the data. Consider again some of the simple regression examples of the preceding chapters to illustrate that the L_∞ line is even less robust than LS.

11. Prove that the LWS estimator (4.27) is regression, scale, and affine equivariant, and show that its breakdown point equals the desired value.

12. (Research problem) What is the maximal asymptotic efficiency of an S-estimator defined by means of function ρ satisfying (S1), (S2), and (S3) with $K = E_\Phi[\rho]$?

CHAPTER 4

The Special Case of
One-Dimensional Location

1. LOCATION AS A SPECIAL CASE OF REGRESSION

The estimation of a location and a scale parameter of a single variable can be considered as a particular case of the general regression model, obtained by putting $p = 1$ and $x_i = 1$ for all i. Then the sample reduces to $(y_i)_{i=1,\ldots,n}$ and the underlying model states

$$y_i = \theta + e_i,$$

where θ is an unknown one-dimensional parameter and e_i is normally distributed with mean zero and standard deviation σ. In this situation, an estimator T_n of θ is called a *univariate location estimator*, and an estimator S_n of σ is called a *scale estimator*.

From the equivariance properties in Section 4 of Chapter 3, which form an integral part of regression estimation, we can derive equivariance and invariance properties for location and scale estimators. First, regression equivariance reduces to so-called *location equivariance*. Indeed, when a constant v is added to the whole sample, then any reasonable location estimator should be increased by the same amount. More formally, T_n is a location (or translation) equivariant estimator if

$$T_n(y_1 + v, \ldots, y_n + v) = T_n(y_1, \ldots, y_n) + v \qquad (1.1)$$

for any constant v. Also, the scale equivariance of regression estimators can be formulated in the one-dimensional case. Indeed, a location estimator should follow a change of scale of the sample, which is obtained by multiplying all the observations by a constant c (not equal to zero). To be precise, a location estimator T_n satisfying

$$T_n(cy_1, \ldots, cy_n) = cT_n(y_1, \ldots, y_n) \qquad (1.2)$$

is said to be *scale equivariant*. A location estimator that is both location and scale equivariant is called *affine equivariant*, because it transforms well when all y_i are replaced by $cy_i + v$. (Note that the affine equivariance of *regression* estimators is different, and it has no counterpart here because we cannot transform any x-variables.)

Let us now look at the behavior of a scale estimator S_n under such transformations. First, adding a constant v to all observations should not change S_n. In other words,

$$S_n(y_1 + v, \ldots, y_n + v) = S_n(y_1, \ldots, y_n) \qquad (1.3)$$

for all v. Any S_n satisfying (1.3) is said to be *location invariant*. However, S_n does have to follow a change of scale:

$$S_n(cy_1, \ldots, cy_n) = |c| S_n(y_1, \ldots, y_n), \qquad (1.4)$$

that is, S_n is *scale equivariant*. Note that the absolute value is needed because scale estimators are always positive.

An extensive Monte Carlo study of robust estimators of location has been carried out by Andrews et al. (1972). A recent treatment of one-dimensional estimators can be found in Chapter 2 of Hampel et al. (1986). We will devote some attention to one-dimensional estimation for several reasons. First of all, location estimators can often be used for estimating the constant term of a robust fit when handling a regression model with intercept. Indeed, once $\hat{\theta}_1, \ldots, \hat{\theta}_{p-1}$ are found, it is possible to compute an appropriate $\hat{\theta}_p$ as an estimate of location on the sample consisting of all

$$d_i = y_i - x_{i,1}\hat{\theta}_1 - \cdots - x_{i,p-1}\hat{\theta}_{p-1} \qquad (i = 1, \ldots, n).$$

Furthermore, in any multivariate analysis, it is always useful to be able to study the different variables of the data set in a one-dimensional way before setting up a model. For this purpose, program PROGRESS can list the standardized data. In general, the standardization of a variable is obtained by first subtracting a location estimate of the one-dimensional sample associated with this variable and then by dividing by a scale estimate of the same sample. As such, the standardized data are dimensionless. Moreover, they possess a zero location and a unit scale (as measured by the same one-dimensional estimators that were used for the standardization). The investigation of these standardized measurements can aid the analyst in discovering some pattern in the distribution of each variable in the data set. For instance, the presence of skewness may be detected. The shape of the observed distribution of a certain variable may then suggest a suitable transformation for this variable, to be applied before carrying out the actual regression.

In PROGRESS, the standardization is performed in a robust way. When there is a constant term in the model, the program starts by subtracting the median of each variable, and then divides this result by the median of absolute deviations from the median (multiplied by the usual correction factor $1/\Phi^{-1}(0.75) \approx 1.4826$). More formally, each observation is transformed to

$$z_{ij} = \frac{x_{ij} - \operatorname*{med}_k x_{kj}}{1.4826 \operatorname*{med}_f \left| x_{fj} - \operatorname*{med}_k x_{kj} \right|}, \qquad j = 1, \dots, p-1, \qquad (1.5)$$

and we put $z_{ip} \equiv x_{ip} \equiv 1$ for the intercept term. Moreover, $z_{i,p+1}$ is the standardized version of the response y_i, computed in the same way as in (1.5) but with x_{ij} replaced by y_i. When there is no intercept term,

$$z_{ij} = \frac{x_{ij}}{1.4826 \operatorname*{med}_f \left| x_{fj} \right|}, \qquad j = 1, \dots, p \qquad (1.6)$$

and again $z_{i,p+1}$ corresponds to the standardization of the response but with x_{ij} replaced by y_i in (1.6).

Problems may occur if the denominator of (1.5) or (1.6) equals zero. If this happens, the standardization in (1.5) is substituted by

$$z_{ij} = \frac{x_{ij} - \operatorname*{med}_k x_{kj}}{1.2533 \dfrac{1}{n} \sum_{f=1}^{n} \left| x_{fj} - \operatorname*{med}_k x_{kj} \right|}, \qquad j = 1, \dots, p-1 \qquad (1.7)$$

and (1.6) becomes

$$z_{ij} = \frac{x_{ij}}{1.2533 \dfrac{1}{n} \sum_{f=1}^{n} \left| x_{fj} \right|}, \qquad j = 1, \dots, p, \qquad (1.8)$$

where $1.2533 = \sqrt{\pi/2}$ is another correction factor. (This is because the mean absolute deviation tends to $\sqrt{2/\pi}$ times the standard deviation when the x_{ij} are normally distributed.)

A denominator in (1.7) or (1.8) which equals zero would mean that the variable in question is constant over all observations. When this occurs, the program stops and recommends that the user start the analysis again without that variable, because it does not contain any information.

In Chapter 5 we will explain why standardization is also useful from a numerical point of view.

Finally, standardized observations also provide a simple scheme to diagnose univariate outliers in any single variable (of course, it is supposed here that the standardization has been performed in a robust way). By means of the standardized measurements, one introduces a so-called *rejection rule*, which consists of identifying all observations with a standardized value larger than some bound. (This is just a special case of looking at the standardized residuals from a robust regression in order to identify regression outliers.) When the target is one-dimensional estimation, one can then estimate location and scale in a classical way on the "cleaned" sample in which the extreme observations are removed. This idea of using rejection as a method of producing a robust estimator has been widely used. Hampel (1985) made a critical study of such rejection rules followed by computing the mean of the remaining data. One of the best rules is X84, corresponding to the removal of observations for which the standardization (1.5) yields an absolute value larger than some bound. This rule leads to a 50% breakdown point. On the other hand, Hampel showed that many classical rejection rules (Barnett and Lewis 1978, Hawkins 1980) yield a low breakdown point, which corresponds to their vulnerability to the masking effect. This is not surprising, because most of these methods are based on classical non-robust estimation.

We will now discuss the one-dimensional versions of some of the regression estimators defined in Chapters 2 and 3. The least squares (LS) estimator reduces to the minimization of

$$\sum_{i=1}^{n} (y_i - \hat{\theta})^2 , \tag{1.9}$$

which leads to the well-known arithmetic mean. As before, this estimator has a poor performance in the presence of contamination. Therefore, Huber (1964) has lowered the sensitivity of the LS objective function by replacing the square in (1.9) by a suitable function ρ. This leads to location M-estimators, defined by

$$\text{Minimize}_{\hat{\theta}} \sum_{i=1}^{n} \rho(y_i - \hat{\theta}), \tag{1.10}$$

which satisfy the necessary condition

$$\sum_{i=1}^{n} \psi(y_i - \hat{\theta}) = 0, \tag{1.11}$$

where ψ is the derivative of ρ. [Of course (1.10) and (1.11) are not always equivalent.] In order to make an M-estimator scale equivariant, Huber considered the simultaneous estimation of location θ and scale σ by solving the system of equations

$$\begin{cases} \sum_{i=1}^{n} \psi\left(\dfrac{y_i - \hat{\theta}}{\hat{\sigma}}\right) = 0 \\ \sum_{i=1}^{n} \chi\left(\dfrac{y_i - \hat{\theta}}{\hat{\sigma}}\right) = 0. \end{cases} \tag{1.12}$$

These equations have to be solved iteratively. A possible choice of ρ in (1.10) is the absolute value, yielding the L_1-estimator that reduces to the sample median $\hat{\theta} = \text{med}_i\, y_i$. The corresponding scale estimate is obtained by taking $\chi(y) = \text{sgn}\,(|y| - \Phi^{-1}\,(0.75))$ in (1.12), leading to the normalized median absolute deviation $\hat{\sigma} = 1.4826\,\text{med}_j\,|y_j - \text{med}_i\,y_i|$. Huber (1964) himself used the functions

$$\psi(y) = \begin{cases} -k, & y \le -k \\ y, & -k \le y \le k \\ k, & y \ge k \end{cases} \tag{1.13}$$

and

$$\chi(y) = \psi^2(y) - \int \psi^2(u)\,d\Phi(u).$$

A simplified variant is the one-step M-estimator (Bickel 1975). It results from applying Newton's iterative algorithm for solving (1.12) just once, with initial estimates T^0 for location and S^0 for scale. This means that the one-step estimate for θ equals

$$T^1 = T^0 + \frac{\dfrac{1}{n} \sum_{i=1}^{n} \psi\{(y_i - T^0)/S^0\}}{\dfrac{1}{n} \sum_{i=1}^{n} \psi'\{(y_i - T^0)/S^0\}}\, S^0. \tag{1.14}$$

For nonmonotone ψ the one-step variant is much safer than the fully iterated version, because it avoids the problem of multiple solutions of (1.12). However, one-step M-estimators should only be used with very robust starting values, preferably $T^0 = \text{med}_i \, y_i$ and $S^0 = 1.4826 \, \text{med}_j \, |y_j - \text{med}_i \, y_i|$. One problem can still occur: It may happen that the denominator of (1.14) becomes zero for certain nonmonotone functions ψ and certain samples. However, this pitfall can easily be avoided by replacing this denominator by $\int \psi'(u) \, d\Phi(u)$. Another way out is to switch to a *one-step W-estimator*

$$T_W = \frac{\sum\limits_{i=1}^{n} y_i w\{(y_i - T^0)/S^0\}}{\sum\limits_{i=1}^{n} w\{(y_i - T^0)/S^0\}} , \tag{1.15}$$

where the nonnegative weight function w is related to ψ through $w(u) = \psi(u)/u$. The computation of T_W is very simple, and it does not lead to problems because the denominator is nonzero, provided $w(u)$ is strictly positive for $0 \le |u| \le 0.68$ because then at least half of the y_i possess a strictly positive weight. Note that T_W is a one-step reweighted least squares (RLS) estimator, because it minimizes

$$\sum_{i=1}^{n} (y_i - T_W)^2 w\{(y_i - T^0)/S^0\} .$$

Also *L-estimators* have their location counterparts. If $y_{1:n}, \ldots, y_{n:n}$ represent the order statistics, then an L-estimate of location is given by

$$T = \sum_{i=1}^{n} a_i y_{i:n} , \tag{1.16}$$

where the a_i are appropriate weights. The sum of the a_i must be equal to 1 in order to satisfy location equivariance. For instance, the median can be expressed as an L-estimator. Another widely used L-estimator is the *α-trimmed mean*,

$$\frac{1}{n - 2[n\alpha]} \sum_{i=[n\alpha]+1}^{n-[n\alpha]} y_{i:n} .$$

This type of estimator can be considered to be the LS solution after discarding the $[n\alpha]$ smallest and the $[n\alpha]$ largest observations. For α equal to $\frac{1}{4}$, for instance, one finds the average of the middle 50% of the order statistics, which is sometimes referred to as the *midmean*. *L*-estimators have been investigated by, among others, Bickel and Hodges (1967), Dixon and Tukey (1968), Jaeckel (1971), Helmers (1980), and Serfling (1980).

An *R-estimate* of location is derived from a rank test statistic in the following way. Consider on the one hand the original sample $\{y_1, \ldots, y_n\}$ and on the other hand the sample $\{2T - y_1, \ldots, 2T - y_n\}$ which is obtained by reflecting each y_i with respect to some T. Let R_i be the rank of the observation y_i in the combined sample. Then an *R*-estimator of location is defined as the value T_n that makes the expression

$$\sum_{i=1}^{n} J\left(\frac{R_i}{2n + 1}\right) \tag{1.17}$$

as close to zero as possible, where J is a scores function satisfying $\int_0^1 J(t) = 0$. This means that the rank test based on the statistic (1.17) sees no shift between the true and the reflected sample. The first such construction made use of the Wilcoxon or Mann–Whitney test given by $J(t) = t - \frac{1}{2}$, leading to the Hodges–Lehmann (1963) estimator:

$$T = \underset{1 \le i < j \le n}{\text{med}} \frac{y_i + y_j}{2}. \tag{1.18}$$

This estimator equals the median of all pairwise means of the sample, which is analogous to the slope estimator of Sen in the case of simple regression [Chapter 2, equation (7.2)]. By means of the reasoning in formula (7.3) of Chapter 2, we find that its breakdown point is $\varepsilon^* = 1 - (\frac{1}{2})^{1/2} \approx 0.293$.

An estimator that is also worth noting in the framework of robustness is the *shorth*. This estimator is defined as the arithmetic mean of the shortest subsample with half of the observations. The shorth was introduced in the Princeton Robustness Study. Below, we shall see the resemblance between the shorth and the least median of squares (LMS) location estimator. We will also discuss the least trimmed squares (LTS) estimator.

2. THE LMS AND THE LTS IN ONE DIMENSION

In one-dimensional location, the LMS reduces to

$$\underset{\hat{\theta}}{\text{Minimize}} \underset{i}{\text{med}} \, (y_i - \hat{\theta})^2. \tag{2.1}$$

Before we investigate the properties of this estimator, we will first compare it with some well-known univariate estimators on a real data set. In Rosner (1977) we found a sample containing 10 monthly diastolic blood pressure measurements:

$$90, 93, 86, 92, 95, 83, 75, 40, 88, 80 , \qquad (2.2)$$

which are plotted in Figure 1. A glance at these measurements reveals that 40 stands out compared to the other values.

Figure 1 also contains some location estimates applied to this sample. The arithmetic mean, 82.2, is attracted by the outlier. On the other hand, the robust estimates range from 86.9 to 90.5. Indeed, the sample median equals 87.0, the LMS estimate is 90.5, and the reweighted mean (1.15) with weights based on the LMS becomes 86.9. If the outlier were to move even further away, none of these robust estimates would change any more.

It is striking that the LMS location estimate does not stand in the center of the "good" observations but rather in a position where the points lie very close together. The reason for this phenomenon is that this sample is not symmetric. The LMS looks for a concentration in the sample, which means that it might be considered as some kind of *mode estimator*. In order to illustrate this, we have sketched a possible underlying distribution from which this sample might have been drawn, judging from the observations. This density looks skewed, with the mode situated at the right-hand side in the neighborhood of the LMS location estimate. For such asymmetric samples the LMS lies closer to the mode than do the mean and the median. In fact, a classical measure for skewness is the ratio

$$\frac{\text{mean} - \text{mode}}{\text{scale estimate}} , \qquad (2.3)$$

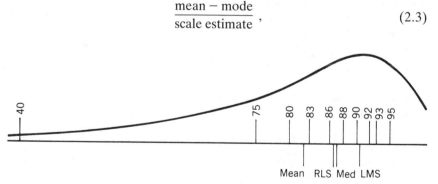

Figure 1. Rosner data: (above) Ten one-dimensional measurements; (below) some location estimates: Mean, arithmetic mean; Med, median; LMS, least median of squares estimator; RLS, reweighted mean with weights based on the LMS.

which becomes zero for symmetric situations, positive for skewed densities with their tail to the right, and negative if the tail is to the left.

It is important to note that this different behavior of the LMS does not only show up when the sampling distribution is actually skewed, but even when it is symmetric, because experience has taught us that:

*Small samples drawn from a symmetric distribution
tend to look asymmetric.*

This explains the larger variability of mode estimators (such as the LMS).

Note that we do not claim that the LMS would be particularly good for one-dimensional location, and indeed there are many robust estimators with smaller variability in this situation, such as the sample median. However, most of these competitors do not generalize appropriately to multiple regression (such as the usual L_1 estimator, which is very vulnerable to leverage points). The main motivation for the LMS (and its relatives) comes from the affine equivariant high-breakdown multiple regression setting, where it is not so easy to construct good tools.

Let us now compare the behavior of the objective function of the LMS and the LS on the one-dimensional data set of Rosner. In Figure 2, the dashed line represents the LS objective $\sum_{i=1}^{n} (y_i - t)^2$ as a function of t, whereas the solid line corresponds to the function $\text{med}_i (y_i - t)^2$, where $\{y_1, \ldots, y_{10}\}$ is the sample (2.2). The LMS objective function attains its minimum in some T, which equals 90.5 for this sample. The resulting $\text{med}_i (y_i - T)^2$ is denoted by m_T^2. The LS objective function attains its minimum more to the left, that is, closer to the outlying value 40. The values $T - m_T$ and $T + m_T$ are also indicated on the horizontal axis containing the measurements. Since m_T equals 4.5, this interval ranges from 86 to 95. Both of these boundary values exactly coincide with an observation. Moreover, it appears that this interval contains half of the sample. These properties are generally true, as stated in Theorem 1 below. We will, however, first show that there exists a solution to (2.1).

Lemma 1. For any univariate sample $\{y_1, y_2, \ldots, y_n\}$, there exists a solution to (2.1).

Proof. Since the convex hull of the sample is compact and the objective function $\text{med}_i (y_i - t)^2$ is continuous in t, it follows immediately that there exists a minimum.   □

Theorem 1. If $m_T^2 := \text{med}_i (y_i - T)^2$ equals $\min_\theta \text{med}_i (y_i - \theta)^2$, then both $T - m_T$ and $T + m_T$ are observations in the sample $\{y_1, y_2, \ldots, y_n\}$.

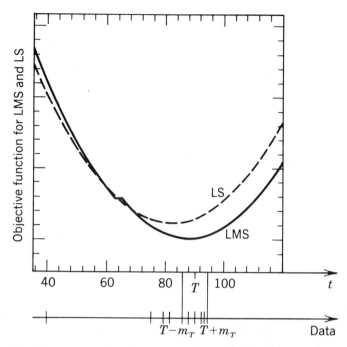

Figure 2. Plot of the objective function corresponding to the LS (dashed line) and the LMS (solid line) for the same data set as in Figure 1. These data are also indicated below the plot, which shows that $T - m_T$ and $T + m_T$ are observed values.

Proof. (a) *First suppose that n is odd*, with $n = 2k - 1$. Then the median is reached by the kth squared residual $r^2_{k:n}$. A residual is the difference between a sample point and the location estimate. Therefore, at least one of the points $T - m_T$ or $T + m_T$ is an observation in the sample. Without loss of generality, suppose that $T + m_T$ is and $T - m_T$ is not. There exists a partition of the $r^2_{i:n}$ in three parts:

$$(k - 1) \text{ squared residuals that are } \leq m^2_T,$$

$$1 \text{ squared residual that is } = m^2_T, \text{ and}$$

$$(k - 1) \text{ squared residuals that are } \geq m^2_T.$$

Suppose $y_{j:n}$ is the smallest observation that is larger than $T - m_T$ (choose one if there are several), and define $T' = \frac{1}{2}((T + m_T) + y_{j:n})$ and $m'^2 = \frac{1}{2}(|(T + m_T) - y_{j:n}|)^2$. It is clear that $m'^2 < m^2_T$. Then there exists a partition of the new residuals $(r'_i)^2 = (y_i - T')^2$ into

$(k-2)$ squared residuals $\leq m'^2$

(corresponding to the same points as before, except $y_{j:n}$),

2 squared residuals $= m'^2$

(corresponding to $y_{j:n}$ and $T + m_T$), and

$(k-1)$ squared residuals $\geq m'^2$

(corresponding to the same points as before).

Finally, $\text{med}_i (r_i')^2 = m'^2 < m_T^2$, a contradiction.

(b) *Suppose that n is even*, with $n = 2k$. If the ordered squares are denoted by $r_{1:n}^2 \leq \cdots \leq r_{n:n}^2$, then

$$m_T^2 = (r_{k:n}^2 + r_{k+1:n}^2)/2 .$$

There is a partition of the squared residuals into $(k-1)$ squares $\leq r_{k:n}^2$, $r_{k:n}^2$ itself, $r_{k+1:n}^2$ itself, and $(k-1)$ squares $\geq r_{k+1:n}^2$.

If $T + m_T$ is an observation and $T - m_T$ is not (or conversely), then m_T^2 is a squared residual and one can repeat the reasoning of case (a), yielding the desired contradiction. Now suppose that neither $T + m_T$ nor $T - m_T$ are observations, which implies $r_{k:n}^2 < r_{k+1:n}^2$ because otherwise $r_{k:n}^2 = m_T^2 = r_{k+1:n}^2$. Therefore, at least $r_{k+1:n}^2 > 0$. Now two situations can occur:

(b.1) *First Situation:* Assume that $r_{k:n}^2 = 0$. In that case, T coincides with exactly k observations, the nearest other observation (call it y_j) being at a distance $d = |r|_{k+1:n}$. Putting $T' = (T + y_j)/2$, however, we find

$$\text{med}_i (y_i - T')^2 = \tfrac{1}{2}\{(\tfrac{1}{2}d)^2 + (\tfrac{1}{2}d)^2\} = \tfrac{1}{4}d^2 < \tfrac{1}{2}d^2 = m_T^2 ,$$

a contradiction.

(b.2) *Second Situation:* Assume that $r_{k:n}^2 > 0$. Denote by y_j some observation corresponding to $r_{k:n}^2$ and by y_f some observation corresponding to $r_{k+1:n}^2$. If the observations leading to $r_{k:n}^2$ and $r_{k+1:n}^2$ are all larger than T or all smaller than T, one can again repeat the reasoning of case (a). Therefore, one may assume without loss of generality that $y_j < T < y_f$. Putting $T' = (y_j + y_f)/2$, we find

$$\text{med}_i (y_i - T')^2 = \tfrac{1}{2}\{(y_j - T')^2 + (y_f - T')^2\} < \tfrac{1}{2}\{(y_j - T)^2 + (y_f - T)^2\}$$

because the function $g(t) = (a - t)^2 + (b - t)^2$ attains its unique minimum at $t = (a + b)/2$. □

Theorem 1 makes it easy to compute the LMS in the location case. One only has to determine the *shortest half* of the sample. This is done by finding the smallest of the differences

$$y_{h:n} - y_{1:n}, \; y_{h+1:n} - y_{2:n}, \ldots, \; y_{n:n} - y_{n-h+1:n}, \quad (2.4)$$

where $h = [n/2] + 1$ and $y_{1:n} \leq \cdots \leq y_{n:n}$ are the ordered observations. The corresponding subsample containing h observations is then called the shortest half because of all the possible subsets with h elements it possesses the shortest range. Then using Theorem 1, T simply equals the midpoint of this shortest interval. (In case there are several shortest halves, which happens with probability zero when the distribution is continuous, one could take the average of their midpoints.) This reminds us of the shorth, where the *mean* of all observations in the shortest half is taken.

Let us now illustrate the LMS estimate on the sample in (2.2). We first order the observations as follows:

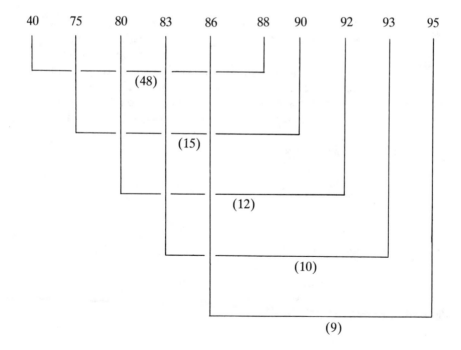

The horizontal lines indicate the halves of the sample. We now consider the differences defined in (2.4), yielding the values 48, 15, 12, 10, 9. The smallest of them corresponds to the distance between 86 and 95. Finally, the midpoint of this interval, namely 90.5, is the LMS estimate of location. Note that the presence of the aberrant value 40 does not affect this result.

REMARK.   Note that in the special case of $n = 3$ the LMS coincides with the mean of the closest pair. Lieblein (1952) investigated the properties of this estimator but concluded that it is not very satisfactory because in such small samples it often happens that two points are very close to each other by chance, making the third one look like an outlier, even when the observations come from a Gaussian distribution. Stefanski and Meredith (1986) recently gave an example in which the mean of the closest pair performs better than the sample median.

We are now in a position to explain why we have considered the median of the *squared* residuals in (2.1), instead of the median of the absolute residuals, corresponding to

$$\text{Minimize} \operatorname*{med}_{\hat{\theta}} | y_i - \hat{\theta} | \, . \tag{2.5}$$

It turns out that Theorem 1 no longer holds for the latter estimator when n is even. Indeed, consider the sample $\{1, 3, 4, 6, 7, 10\}$. The LMS solution is unique and equals $T = 5$ with $m_T = 2$. On the other hand, *all* values of T between 4.5 and 5.5 minimize (2.5) with $\operatorname{med}_i | y_i - T | = 2$, yielding counterexamples to Theorem 1. Therefore the square in (2.1) is not superfluous. In the general multiple regression model with intercept term (and thus also in the case of location), the following theorem holds.

Theorem 2.   Each solution of

$$\text{Minimize} \operatorname*{med}_{\hat{\theta}} (y_i - x_{i,1}\hat{\theta}_1 - \cdots - x_{i,p-1}\hat{\theta}_{p-1} - \hat{\theta}_p)^2 \tag{2.6}$$

is a solution of

$$\text{Minimize} \operatorname*{med}_{\hat{\theta}} | y_i - x_{i,1}\hat{\theta}_1 - \cdots - x_{i,p-1}\hat{\theta}_{p-1} - \hat{\theta}_p | \, . \tag{2.7}$$

Proof. (a) The implication is obvious when n is odd, $n = 2k - 1$. In that case, $\operatorname{med}_i r_i^2$ equals the kth squared residual $r_{k:n}^2$, where $r_{1:n}^2 \leq \cdots \leq r_{n:n}^2$ are the ordered squared residuals. Since $r_{k:n}^2 = (|r|_{k:n})^2$, it follows that $\operatorname{med}_i (r_i^2) = (\operatorname{med}_i |r_i|)^2$.

(b) When n is even, n can be written as $2k$. Let T be a solution to (2.6) and put $m_T^2 = (r_{k:n}^2 + r_{k+1:n}^2)/2 = \min_\theta \text{med}_i (r_i^2)$. From the proof of Theorem 1, it follows that $r_{k:n}^2 = m_T^2 = r_{k+1:n}^2$, so

$$\text{med}_i |r_i| = \tfrac{1}{2}(|r|_{k:n} + |r|_{k+1:n}) = m_T.$$

Now suppose that T would not minimize $\text{med}_i |r_i|$. Then there would exist a T' such that

$$\text{med}_i |r_i'| = m_{T'} < m_T.$$

Repeating case (b) of the proof of Theorem 1, replacing r_i^2 by $|r_i'|$, one can show that there exists a T'' such that $\text{med}_i |r_i''| \le m_{T'}$ and $|r''|_{k:n} = |r''|_{k+1:n}$. Therefore,

$$\text{med}_i (r_i'')^2 = \tfrac{1}{2}((r''^2)_{k:n} + (r''^2)_{k+1:n})$$
$$\le \tfrac{1}{2}(m_{T'}^2 + m_{T'}^2) = m_{T'}^2 < m_T^2,$$

which is a contradiction. □

Let us now look at the LTS estimator in the univariate case, given by

$$\underset{\hat{\theta}}{\text{Minimize}} \sum_{i=1}^{h} (r^2)_{i:n}, \tag{2.8}$$

where $h = [n/2] + 1$ and $(r^2)_{i:n}$ are the ordered squared residuals.

In order to determine the LTS location estimate one has to consider the following $n - h + 1$ subsamples:

$$\{y_{1:n}, \ldots, y_{h:n}\}, \{y_{2:n}, \ldots, y_{h+1:n}\}, \ldots, \{y_{n-h+1:n}, \ldots, y_{n:n}\}. \tag{2.9}$$

Each of these subsamples contains h observations, so we call it a (contiguous) half. For each half, one calculates the mean

$$\bar{y}^{(1)} = \frac{1}{h} \sum_{i=1}^{h} y_{i:n}$$
$$\vdots$$
$$\bar{y}^{(n-h+1)} = \frac{1}{h} \sum_{i=n-h+1}^{n} y_{i:n}$$

and the corresponding sum of squares

$$SQ^{(1)} = \sum_{i=1}^{h} \{y_{i:n} - \bar{y}^{(1)}\}^2$$

$$\vdots$$

$$SQ^{(n-h+1)} = \sum_{i=n-h+1}^{n} \{y_{i:n} - \bar{y}^{(n-h+1)}\}^2 .$$

The LTS solution then corresponds to the mean $\bar{y}^{(j)}$ with the smallest associated sum of squares $SQ^{(j)}$.

Proceeding in this way, one has to go twice through the observations of each half. The number of operations can be drastically reduced by using the mean of the preceding half. This means that the mean of the jth half ($j > 1$) is determined by the recursive formula

$$\bar{y}^{(j)} = \frac{h\bar{y}^{(j-1)} - y_{j-1:n} + y_{j+h-1:n}}{h} ,$$

making use of the mean of the ($j-1$)th half. (It is even better to update $h\bar{y}^{(j)}$ in order to avoid rounding errors caused by successive division and multiplication by the factor h.) The sum of squares can also be calculated in a recursive way, namely

$$SQ^{(j)} = SQ^{(j-1)} - (y_{j-1:n})^2 + (y_{j+h-1:n})^2 - h(\bar{y}^{(j)})^2 + h(\bar{y}^{(j-1)})^2 .$$

Applying this algorithm to the sample (2.2), one finds the value 90.67 for the LTS location estimator, which corresponds to the mean of the half $\{y_{5:10}, \ldots, y_{10:10}\}$.

The *least winsorized squares* (LWS) estimator in the case of location is defined by

$$\text{Minimize}_{\theta} \sum_{i=1}^{h} (r^2)_{i:n} + (n-h)(r^2)_{h:n} , \tag{2.10}$$

with $h = [n/2] + 1$. As for the LMS and the LTS, the LWS location estimate can be calculated explicitly.

Lemma 2. The LWS location estimate is given by the midpoint of the contiguous half which leads to the lowest objective function value in (2.10).

Proof. It is obvious that the selected half has to consist of successive observations, since if there were gaps, then one could make the objective function smaller by moving observations. We will now show that the LWS estimate (denoted by T) corresponds to the *midpoint* of this half. If $T < (y_{i:n} + y_{i+h-1:n})/2$, then the largest squared residual comes from the observation $y_{i+h-1:n}$. We are minimizing the sum of n squared residuals, so the objective function will decrease when T is moved in the direction of T', the *mean* of these n points (in which $y_{i+h-1:n}$ is counted $n - h + 1$ times):

$$T' = \frac{y_{i:n} + \cdots + y_{i+h-1:n} + (n - h)y_{i+h-1:n}}{n}$$

$$= \frac{y_{i:n} + \cdots + y_{i+h-2:n} + (n - h + 1)y_{i+h-1:n}}{n}.$$

If n is even, then $n - h + 1 = h - 1 = n/2$, so

$$T' \geq \frac{(h - 1)y_{i:n} + (h - 1)y_{i+h-1:n}}{n} = \frac{y_{i:n} + y_{i+h-1:n}}{2}.$$

If n is odd, it holds that $n = 2h - 1$ so that

$$T' \geq \frac{(h - 1)y_{i:n} + (2h - 1 - h + 1)y_{i+h-1:n}}{(2h - 1)}$$

$$= \frac{(h - 1)y_{i:n} + hy_{i+h-1:n}}{(h - 1) + h}$$

$$\geq \frac{y_{i:n} + y_{i+h-1:n}}{2}.$$

In both cases we find $T' \geq (y_{i:n} + y_{i+h-1:n})/2$, a contradiction. In an analogous way, one can contradict $T > (y_{i:n} + y_{i+h-1:n})/2$. Consequently, the LWS estimate equals the midpoint of a half. □

However, the selected half is not necessarily the *shortest* half as in the case of the LMS estimator. Take, for instance, the sample $\{0, 0, 2, 2, 3 + \varepsilon, 4 + 2\varepsilon\}$ with $\varepsilon > 0$. The midpoint of the shortest half $\{0, 0, 2, 2\}$ equals 1 and yields the value 6 for the LWS objective function, whereas the midpoint of the half $\{2, 2, 3 + \varepsilon, 4 + 2\varepsilon\}$ is $3 + \varepsilon$, which produces the value $5(1 + \varepsilon)^2$. It is possible to make $5(1 + \varepsilon)^2$ smaller than 6, for small values of ε. Similar examples with odd sample size can also be constructed, such as $\{0, 0, 2, 2, 3 + \varepsilon, 4 + 2\varepsilon, 4 + 2\varepsilon\}$ for $n = 7$. But still, Lemma 2 shows that the LWS is more similar to the LMS than to the LTS, and it will often be equal to the LMS. Indeed, for the Rosner data (2.2) the LWS estimate equals the LMS estimate.

We will now discuss the S-estimators of location. In Chapter 3, equations (4.28)–(4.30) defined this class of estimators for the multiple regression model. This definition is still valid in the case of location, keeping in mind that θ becomes one-dimensional and that r_i simply equals $y_i - \theta$. Therefore, we want to

$$\underset{\hat{\theta}}{\text{Minimize}}\ s(y_1 - \hat{\theta}, \ldots, y_n - \hat{\theta}),\qquad\qquad (2.11)$$

where the dispersion $s(r_1, \ldots, r_n)$ is defined as the solution of

$$\frac{1}{n}\sum_{i=1}^{n}\rho(r_i/s) = K.$$

The last formula defines an M-estimator of scale, which is usually computed in an iterative way starting with the initial value

$$S^0 = 1.4826 \underset{i}{\text{med}}\,|r_i|.$$

Because of constraints on computation time, it may be appropriate to use a simple and fast approximation of the objective function, such as the one-step estimate

$$S^1 = S^0 \sqrt{\frac{1}{K}\left\{\frac{1}{n}\sum_{i=1}^{n}\rho\!\left(\frac{r_i}{S^0}\right)\right\}},\qquad\qquad (2.12)$$

where ρ may, for instance, be given by formula (4.31) of Chapter 3. One then makes use of a numerical algorithm for minimizing S^1 as a function of $\hat{\theta}$.

3. USE OF THE PROGRAM PROGRESS

The same program that we used in the previous chapters for the computation of robust regression coefficients also provides one-dimensional estimators. The output of PROGRESS contains the arithmetic mean, the LMS, and the reweighted mean based on the LMS. For each of these location estimates, a corresponding scale estimate is given. As in the multivariate case, the output includes tables of residuals and index plots (at least if the print and plot options are chosen accordingly).

There are two ways to tell PROGRESS to execute the one-dimensional algorithm. The first is to answer 1 to the question

WHAT IS THE TOTAL NUMBER OF VARIABLES IN YOUR DATA SET?

PLEASE GIVE A NUMBER BETWEEN 1 AND 50:

Then, PROGRESS answers:

THE PROBLEM IS REDUCED TO ESTIMATING A LOCATION PARAMETER.

On the other hand, when you want to study the location and scale of a certain variable in a larger data set, you must choose that variable as your response variable, and then the question

HOW MANY EXPLANATORY VARIABLES DO YOU WANT TO USE IN THE ANALYSIS?

(AT MOST 9):

must be answered with 0.

Once PROGRESS is executing the one-dimensional case, some of the questions of the interactive part (as discussed in the case of more general regression) no longer apply, so they are not asked. Also, it does not matter whether you have answered YES or NO to the question about whether the regression model contains a constant term.

We will illustrate the program on a one-dimensional data set published by Barnett and Lewis (1978). The sample consists of the numbers

$$3, 4, 7, 8, 10, 949, 951 . \qquad (3.1)$$

Barnett and Lewis used this example for discussing the masking effect when applying outlier tests based on statistics of the type

$$\frac{y_{n:n} - y_{n-1:n}}{y_{n:n} - y_{1:n}} . \qquad (3.2)$$

For the data (3.1), this test does not detect $y_{n:n}$ as an outlier because of the small difference between $y_{n-1:n}$ and $y_{n:n}$. Repeated application of this single outlier test will not remedy the problem. Also, nonrobust estimators of location and scale will be strongly affected by these aberrant values. The output of PROGRESS, which is printed below, confirms the need for a robust estimator. Indeed, it appears that the mean (=276) is pulled in the direction of the outliers. Moreover, the standard deviation $\hat{\sigma}$ is inflated accordingly. The standardized residuals are displayed in an index plot. Note that both outliers fall inside the classical tolerance band, because $\hat{\sigma}$ has exploded! In contrast, the LMS and the LMS-reweighted mean both diagnose the values 949 and 951 as outliers in a very convincing way and provide good estimates for the majority of the data.

```
*************************************
* ROBUST ESTIMATION OF LOCATION. *
*************************************

NUMBER OF CASES =      7

DATA SET = ONE-DIMENSIONAL DATA SET OF BARNETT AND LEWIS (1978)

THE DESIRED OUTPUT IS LARGE.

AN INDEX PLOT WILL BE DRAWN.

THE OBSERVATIONS ARE:
     1       3.0000
     2       4.0000
     3       7.0000
     4       8.0000
     5      10.0000
     6     949.0000
     7     951.0000
```

```
THE MEDIAN          =       8.0000   THE MAD (* 1.4826)   =       5.9304
*****************************************************************************
THE MEAN            =     276.0000   STANDARD DEVIATION   =     460.4360

              OBSERVED            RESIDUAL    NO   RES/SC
                 Y

               3.00000         -273.00000     1    -.59
               4.00000         -272.00000     2    -.59
               7.00000         -269.00000     3    -.58
               8.00000         -268.00000     4    -.58
              10.00000         -266.00000     5    -.58
             949.00000          673.00000     6    1.46
             951.00000          675.00000     7    1.47
```

ONE-DIMENSIONAL DATA SET OF BARNETT AND LEWIS (1978)

--- L E A S T   S Q U A R E S ---

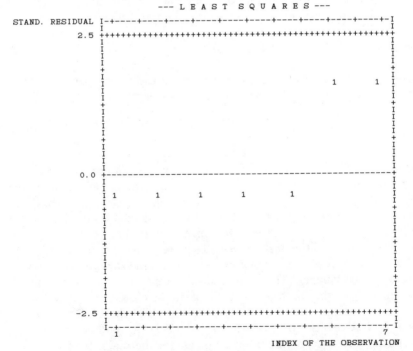

INDEX OF THE OBSERVATION

```
*****************************************************************************
```

```
LMS                    =      5.5000    CORRESPONDING SCALE   =      6.7952


            OBSERVED                RESIDUAL    NO   RES/SC
                  Y

            3.00000                -2.50000     1     -.37
            4.00000                -1.50000     2     -.22
            7.00000                 1.50000     3      .22
            8.00000                 2.50000     4      .37
           10.00000                 4.50000     5      .66
          949.00000               943.50000     6   138.85
          951.00000               945.50000     7   139.14

              ONE-DIMENSIONAL DATA SET OF BARNETT AND LEWIS (1978)
                --- L E A S T   M E D I A N   O F   S Q U A R E S ---
```

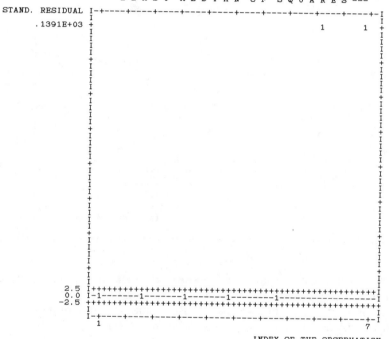

```
                                                       INDEX OF THE OBSERVATION


****************************************************************************************

REWEIGHTING USING THE WEIGHTS BASED ON THE LMS.
************************************************

WEIGHTED MEAN         =      6.4000    WEIGHTED STAND. DEV.   =      2.8810

THERE ARE     5 POINTS WITH NON-ZERO WEIGHT.

AVERAGE WEIGHT        =           .714286


            OBSERVED                RESIDUAL    NO   RES/SC WEIGHT
                  Y

            3.00000                -3.40000     1    -1.18   1.0
            4.00000                -2.40000     2     -.83   1.0
            7.00000                  .60000     3      .21   1.0
            8.00000                 1.60000     4      .56   1.0
           10.00000                 3.60000     5     1.25   1.0
          949.00000               942.60000     6   327.18    .0
          951.00000               944.60000     7   327.88    .0
```

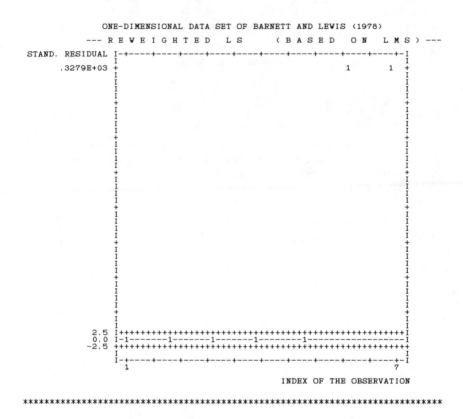

```
            ONE-DIMENSIONAL DATA SET OF BARNETT AND LEWIS (1978)
             --- R E W E I G H T E D   L S     ( B A S E D   O N   L M S ) ---
STAND. RESIDUAL I-+----+----+----+----+----+----+----+----+----+----+-I
               I                                                      I
     .3279E+03 +                                  1          1        +
               I                                                      I
               I                                                      I
               I                                                      I
               +                                                      +
               I                                                      I
               I                                                      I
               I                                                      I
               +                                                      +
               I                                                      I
               I                                                      I
               I                                                      I
               +                                                      +
               I                                                      I
               I                                                      I
               I                                                      I
               +                                                      +
               I                                                      I
               I                                                      I
               I                                                      I
               I                                                      I
               +                                                      +
               I                                                      I
               I                                                      I
               I                                                      I
               +                                                      +
               I                                                      I
               I                                                      I
               I                                                      I
               +                                                      +
           2.5 I+++++++++++++++++++++++++++++++++++++++++++++++++++++++I
           0.0 I-1-------1-------1-------1--------1-------------------I
          -2.5 +++++++++++++++++++++++++++++++++++++++++++++++++++++++++
               I                                                      I
               I-+----+----+----+----+----+----+----+----+----+-I
                 1                                              7
                                       INDEX OF THE OBSERVATION
```

****

REMARK. The scale estimate σ^* associated with the LMS location estimate is defined as in the regression case [see (3.10) in Chapter 2]. This σ^* is itself a high-breakdown robust estimator, some properties of which are discussed by Rousseeuw and Leroy (1987).

*4. ASYMPTOTIC PROPERTIES

We have seen that the one-dimensional LMS resembles the "shorth" discussed in the Princeton Robustness Study (Andrews et al. 1972; see also Shorack and Wellner 1986, pp. 767–771). Like the shorth, the LMS also converges as $n^{-1/3}$, which is abnormally slow. On the other hand, both the LTS and S-estimators are shown to converge at the usual $n^{-1/2}$ rate. However, the LMS is more easily computable than either the LTS or S-estimators. The fact that the LMS converges like $n^{-1/3}$ does not trouble us very much, because we consider it mainly as a data analytic tool, which may be followed by a one-step improvement to obtain higher efficiency (if necessary).

Theorem 3. If the observations $y_1, \ldots, y_n$ are i.i.d. according to $F(y - \theta)$, where F has a symmetric and strongly unimodal density f, then the distribution of the LMS estimator T_n converges weakly as

$$\mathcal{L}(n^{1/3}(T_n - \theta)) \to \mathcal{L}(C\tau) \,,$$

where $C = (f(q)f'(q)^2/2)^{-1/3}$, $q = F^{-1}(0.75)$; and τ is the random time s for which $W(s) + s^2$ attains its minimum, where $W(s)$ is a standard two-sided Brownian motion originating from zero.

Proof. This result is obtained by repeating parts 1, 2, and 3 of the heuristic reasoning of Andrews et al. (1972, p. 51). Using their notation, putting $\alpha = 0.25$ yields the constant

$$A = \{\sqrt{2}f(F^{-1}(0.75))/ -f'(F^{-1}(0.75))\}^{2/3} \,.$$

The remaining part of the calculation is slightly adapted. If $\hat{t}$ is the minimizing value of t, then the main asymptotic variability of T_n is given by

$$\tfrac{1}{2}[F_n^{-1}(\hat{t} + 0.75) + F_n^{-1}(\hat{t} + 0.25)] \approx \hat{t}(F^{-1})'(0.75)$$

$$= \hat{t}/f(F^{-1}(0.75)) \,,$$

where $n^{1/3}\hat{t}$ behaves asymptotically like $A\tau$. By putting $C := A/f(F^{-1}(0.75))$, we obtain the desired result.   □

REMARK. In Section 4 of Chapter 3 we considered variants of the LMS where it is not the median of the squared residuals that is minimized, but instead a larger quantile $(r^2)_{h:n}$ (or equivalently $|r|_{h:n}$). One puts h approximately equal to $n(1 - \alpha)$, where $0 < \alpha < 50\%$, so we must look for the shortest subsample containing a fraction of $(1 - \alpha)$ of the observations. In this case, Theorem 3 is still valid, but with $q = F^{-1}(1 - \alpha/2)$.

The minimization of $W(s) + s^2$ (or equivalently, the maximization of $W(s) - s^2$) occurs in many places in probability and statistics. For instance, the distribution of Chernoff's (1964) mode estimator (based on an interval of fixed length, shifted along the line to a position where it contains the highest number of observations) also converges (after standardization) to the distribution of the location of the maximum of $W(s) - s^2$. (See also Venter 1967.) Perhaps this is not unexpected, because we have already seen that the LMS behaves somewhat like a mode estimator. Groeneboom (1987) lists many other applications of these distributions, namely in epidemiology, density estimation, empirical Bayes tests, and other fields. In his article, Groeneboom obtains the (symmetric) density f_Z of the location of the maximum of $W(s) - cs^2$, which is given by

$$f_Z(s) = \tfrac{1}{2} g_c(s) g_c(-s),$$

where the function g_c has Fourier transform

$$\hat{g}_c(t) = (2/c)^{1/3} / \mathrm{Ai}(i(2c^2)^{-1/3}t)$$

in which Ai is the Airy function (see Abrahamowicz and Stegun 1964, p. 446). For $c = 1$ this is the distribution we are interested in. Groeneboom also shows that f_Z has very thin tails, and therefore the moments of all orders exist.

In the case of the LTS estimator, one can even show asymptotic normality:

Theorem 4. If the observations $y_1, \ldots, y_n$ are i.i.d. according to $F(y - \theta)$, where F has a symmetric density f, then the LTS estimator T_n is asymptotically normal:

$$\mathcal{L}(n^{1/2}(T_n - \theta)) \to N(0, V(\mathrm{LTS}, F)),$$

where the asymptotic variance equals

$$V(\mathrm{LTS}, F) = \frac{2 \displaystyle\int_0^q y^2 \, dF(y)}{(2F(q) - 1 - 2qf(q))^2},$$

with $q = F^{-1}(0.75)$.

Proof. For each t, denote by $R_i(t)$ the rank of $|y_i - t|$. Moreover, the function a on $[0, 1]$ is defined by

$$a(u) = \begin{cases} (2F(q) - 1)^{-1} & \text{if } u \le 2F(q) - 1 \\ 0 & \text{otherwise}. \end{cases}$$

Then the LTS estimator corresponds to the minimization of $\Sigma_{i=1}^{n} a(R_i(T_n)/n)(y_i - t)^2$. Because there exist only a finite number of points t where $R_i(t)$ changes, we may assume without loss of generality that all $R_i(t)$ are locally constant around the solution $t = T_n$. On this neighborhood $(T_n - \delta, T_n + \delta)$ we put

$$a_i = a(R_i(T_n)/n) \quad \text{and} \quad S_n = \frac{1}{n} \sum_{i=1}^{n} a(R_i(T_n)/n)y_i \, .$$

Now T_n minimizes $\Sigma_{i=1}^{n} a_i((y_i - S_n) + (S_n - t))^2 = \Sigma_{i=1}^{n} a_i(y_i - S_n)^2 + \Sigma_{i=1}^{n} a_i(S_n - t)^2$ for all t in $(T_n - \delta, T_n + \delta)$. Differentiating this expression with respect to t at T_n, we find that $T_n = S_n$. Therefore T_n is a solution of the equation

$$\frac{1}{n} \sum_{i=1}^{n} a(R_i(t)/n)y_i = t \, .$$

Estimators satisfying this equation with this function a are due to Gnanadesikan and Kettenring (1972). Their asymptotic normality was shown by Yohai and Maronna (1976), giving the desired result. □

An immediate consequence of this theorem is that the LTS has the same asymptotic behavior at F as the Huber-type "skipped" mean, which is the M-estimator of location corresponding to the ψ-function

$$\psi(y) = \begin{cases} y & \text{if } |y| \le q \\ 0 & \text{otherwise} \, . \end{cases} \tag{4.1}$$

REMARK. In the previous chapter we also discussed variants of the LTS where approximately $n(1 - \alpha)$ of the squared residuals are used, for $0 < \alpha < 50\%$. In this case, Theorem 4 and formula (4.1) can still be applied, but with $q = F^{-1}(1 - \alpha/2)$.

The asymptotic normality of S-estimators at the Gaussian model has already been proven in the general regression case in Chapter 3, Theorems 9 and 10. Of course, these results still hold in the special case of location.

At this point it may be noted that the asymptotic normality of both the LTS and S-estimators basically rests on (variants of) the central limit theorem (CLT). Unfortunately, there exists a fairly widespread misunderstanding about the interpretation of the CLT, leading some people to claim that "robust statistics are not necessary because data are normally distributed by the central limit theorem." However, the CLT says

Table 1. Some Statistics on the LMS, LWS, LTS, and Median Estimators, Based on 500 Generated Samples for Each n

n	Mean of				Standard Deviation of				Interquartile Range of			
	LMS	LWS	LTS	Med	LMS	LWS	LTS	Med	LMS	LWS	LTS	Med
10	-0.034	-0.026	-0.025	-0.027	0.486	0.491	0.495	0.340	0.669	0.643	0.711	0.463
20	0.005	0.003	0.006	0.012	0.416	0.428	0.435	0.255	0.546	0.574	0.602	0.358
30	0.039	0.030	0.030	0.022	0.368	0.359	0.371	0.202	0.485	0.468	0.517	0.282
40	0.043	0.041	0.043	0.020	0.348	0.353	0.374	0.186	0.459	0.487	0.534	0.245
50	0.008	0.028	0.036	0.016	0.323	0.329	0.347	0.165	0.413	0.459	0.499	0.219
60	0.005	0.008	0.034	0.009	0.313	0.312	0.327	0.150	0.412	0.418	0.437	0.202
80	0.011	0.011	0.017	0.010	0.284	0.289	0.309	0.136	0.385	0.387	0.427	0.179
100	0.024	0.015	0.017	0.010	0.275	0.267	0.285	0.122	0.377	0.362	0.405	0.169
200	-0.002	0.005	0.004	0.007	0.217	0.210	0.221	0.086	0.290	0.283	0.314	0.118
300	0.012	0.013	0.010	0.005	0.195	0.187	0.185	0.071	0.280	0.265	0.273	0.098
400	0.015	0.014	0.003	0.002	0.173	0.162	0.161	0.061	0.234	0.224	0.219	0.079
500	-0.003	0.001	0.000	0.001	0.158	0.151	0.146	0.053	0.213	0.216	0.201	0.073
600	-0.006	-0.003	-0.006	0.001	0.150	0.141	0.140	0.049	0.208	0.191	0.200	0.072
800	-0.004	-0.002	0.007	0.002	0.130	0.125	0.122	0.042	0.183	0.175	0.179	0.057
900	-0.008	-0.007	0.004	0.002	0.125	0.118	0.117	0.039	0.169	0.162	0.163	0.054
1000	-0.004	-0.004	0.002	0.001	0.124	0.116	0.114	0.038	0.175	0.157	0.166	0.052
1100	-0.009	-0.005	-0.003	0.001	0.121	0.115	0.110	0.036	0.166	0.161	0.158	0.048
1200	-0.003	-0.002	0.003	0.001	0.120	0.113	0.102	0.034	0.167	0.155	0.150	0.047
1300	-0.004	-0.003	0.002	0.001	0.115	0.112	0.100	0.032	0.159	0.153	0.137	0.043
1400	-0.002	-0.001	0.003	0.001	0.111	0.105	0.093	0.031	0.153	0.141	0.132	0.043
1500	-0.002	-0.005	0.003	0.001	0.113	0.102	0.092	0.031	0.145	0.138	0.129	0.040
1600	0.001	-0.004	0.004	0.002	0.110	0.101	0.091	0.030	0.152	0.141	0.131	0.041
1700	-0.003	-0.001	0.002	0.001	0.106	0.097	0.089	0.030	0.146	0.137	0.122	0.041
2000	-0.004	0.000	0.002	0.001	0.098	0.089	0.080	0.027	0.130	0.119	0.113	0.036

essentially that most *estimators* are asymptotically normal. For instance, a robust S-estimator applied to Cauchy-distributed observations will be asymptotically normal, but of course this does not imply that the observations themselves would look at all normal!

Let us now compare these asymptotic properties with simulation results. For different values of n, we generated 500 samples $\{y_1, \ldots, y_n\}$ from the standard normal distribution. On each sample the LMS, the LWS, the LTS, and the median were calculated. Some statistics on these 500 replications are listed in Table 1. The first column contains the average estimate over the 500 replications and shows that all four estimators approach the true value 0 with more accuracy when the sample size n increases. The standard deviation over the 500 replications is contained in the second column. For most values of n, the LMS possesses the largest standard deviation, followed by the LWS, the LTS, and the median. The same can be said for the interquartile range. Judging from the LMS and LTS columns, the difference between $n^{-1/3}$ and $n^{-1/2}$ convergence rates does not yet appear to be visible for n up to 2000.

*5. BREAKDOWN POINTS AND AVERAGED SENSITIVITY CURVES

The breakdown point ε^* for affine equivariant estimators of location is at most 50%, just as in the case of general regression models. Indeed, if ε^* were larger than 50%, then one could build a configuration of outliers which is just a translation image of the "good" data points.

The sample median enjoys a 50% breakdown point. On the other hand, the arithmetic mean breaks down in the presence of one outlying observation and thus has $\varepsilon^* = 0\%$. Theorems 2, 6, and 8 of Chapter 3 stated that the breakdown points of the LMS, the LTS, and S-estimators for regression (with $p > 1$) reached 50%. The same can be said for their location counterparts.

Theorem 5. At any univariate sample $Y = \{y_1, \ldots, y_n\}$, the breakdown point of the LMS estimator equals

$$\varepsilon_n^*(T, Y) = \frac{[(n+1)/2]}{n} ,$$

which converges to 50% as $n \to \infty$.

Proof. Without loss of generality, let $T(Y) = 0$. We put $M := \max_i |y_i|$. Let us first show that $\varepsilon_n^*(T, Y) \le [(n+1)/2]/n$. For this purpose we replace $[(n + 1)/2]$ observations of Y by the same number of points, which are all equal to the value y with $|y| > M$. This new sample Y' then contains only $h - 1$ of the "good" points, where $h = [n/2] + 1$. Therefore, each half of Y' comprises at least one such large value. Consequently, the midpoint of each half becomes arbitrarily large when $|y| \to \infty$, which entails that $|T(Y')| \to \infty$.

On the other hand, $\varepsilon_n^*(T, Y) \geq [(n + 1)/2]/n$. Indeed, construct a sample Y'' where less than $[(n + 1)/2]$ observations are replaced by arbitrary values. Because there are still at least h "good" points present in Y'', there exists a "good" half with length at most $2M$. Therefore, the length of the shortest half of Y'' is also at most $2M$. Now suppose that $|T(Y'')| > 2M$. But then all observations in the shortest half around $T(Y'')$ have absolute values strictly larger than M. This is a contradiction because $h > [(n + 1)/2] - 1$, which implies that each half contains at least one "good" point, with absolute value $\leq M$. To conclude, we have shown that $|T(Y'')| \leq 2M$ for any such contaminated data set Y''. ☐

Note that we were not able to apply Theorem 2 of Chapter 3 here because its proof rested on the assumption that $p > 1$. If we insert $p = 1$ in the old expression for ε_n^*, we do not find the exact value for the location estimator.

The location counterpart of "exact fit" is a configuration where a certain percentage of the observations in the sample are equal. For the LMS, one finds (as a consequence of the preceding theorem) that if at least $[n/2] + 1$ of the observations coincide, then the LMS estimate will also be equal to that value, whatever the other observations are.

Theorem 6. At any univariate sample $Y = \{y_1, \ldots, y_n\}$, the breakdown point of the LTS estimator equals

$$\varepsilon_n^*(T, Y) = \frac{[(n + 1)/2]}{n} ,$$

which converges to 50% as $n \to \infty$.

Proof. We retrace the proof of the preceding theorem. For $\varepsilon_n^*(T, Y) \leq [(n + 1)/2]/n$, it suffices to note that also the *mean* of each half of Y' becomes arbitrarily large when $|y| \to \infty$, and the LTS is just one of those. Also, the proof of $\varepsilon_n^*(T, Y) \geq [(n + 1)/2]/n$ looks very much the same. Because there are still at least h "good" points present in Y'', it is clear that $\Sigma_{i=1}^h (y''^2)_{i:n} \leq hM^2$. Therefore, the residuals $r_i'' = y_i - T(Y'')$ must also satisfy $\Sigma_{i=1}^h (r''^2)_{i:n} \leq hM^2$. Suppose now that $|T(Y'')| > (1 + \sqrt{h})M$.

But $h > [(n + 1)/2] - 1$ implies that each half of Y''' contains at least one "good" point, with absolute value $\leq M$, so that

$$\Sigma_{i=1}^h (r''^2)_{i:n} > \{(1 + \sqrt{h})M - M\}^2 = hM^2 ,$$

which is a contradiction. To conclude, we have shown that $|T(Y''')| \leq (1 + \sqrt{h})M$ for any contaminated data set Y'''. □

As a consequence, the LTS estimator of location on a sample where at least $[n/2] + 1$ of the points are equal will yield that value as its estimate, whatever the other observations are.

The same reasoning as in the proofs of Theorems 5 and 6 can be repeated for the LWS estimator of location. Also, for S-estimators we obtain the same breakdown point by applying Lemma 4 of Chapter 3.

We will now show that this breakdown point is indeed the highest possible value for any location equivariant estimator.

Theorem 7. The breakdown point of any location equivariant estimator satisfies

$$\varepsilon_n^*(T, Y) \leq \frac{[(n + 1)/2]}{n}$$

at any sample $Y = \{y_1, \ldots, y_n\}$.

Proof. Put $q = n - [(n + 1)/2]$, which also equals $[n/2]$. Suppose that the breakdown point is strictly larger than $[(n + 1)/2]/n$. This would mean that there exists a finite constant b such that $T(Y')$ lies in the interval $[T(Y) - b, T(Y) + b]$ for all samples Y' containing at least q points of Y. Take, for instance,

$$Y' = \{y_1, \ldots, y_q, y_{q+1} + v, \ldots, y_n + v\} ,$$

which satisfies $T(Y') \in [T(Y) - b, T(Y) + b]$. However, note that also the contaminated sample

$$Y'' = \{y_1 - v, \ldots, y_q - v, y_{q+1}, \ldots, y_n\}$$

contains at least q points of Y, so that also $T(Y'') \in [T(Y) - b, T(Y) + b]$. (Indeed, $2q = n$ if n is even, and $2q = n - 1$ if n is odd, so always $2q \leq n$.) However, $Y' = Y'' + v$ so by location equivariance we find that $T(Y') = T(Y'') + v$, and hence

$$T(Y') \in [T(Y) - b + v, T(Y) + b + v].$$

But this is a contradiction, because the intersection of $[T(Y) - b,$ $T(Y) + b]$ and $[T(Y) - b + v, T(Y) + b + v]$ is empty for large enough values of v. □

To conclude this discussion, we list some location estimators with their breakdown point in Table 2.

Some other robustness concepts can be found in Hampel et al. (1986). Whereas the breakdown point is a global measure of reliability, they also stress the importance of the *influence function* (IF), which measures the effect of infinitesimal perturbations on the estimator. The IF of an estimator T at a distribution F is given by

$$\text{IF}\,(y; T, F) = \lim_{\varepsilon \to 0} \frac{T((1 - \varepsilon)F + \varepsilon\Delta_y) - T(F)}{\varepsilon} \tag{5.1}$$

in those points y of the sample space where the limit exists. (Here Δ_y is the probability distribution that puts all its mass in the point y.) The IF reflects the bias caused by adding a few outliers at the point y, standardized by the amount ε of contamination. The IF provides a linear approximation of the estimator at such contaminated distributions:

$$T((1 - \varepsilon)F + \varepsilon\Delta_y) \approx T(F) + \varepsilon\,\text{IF}\,(y; T, F)\,. \tag{5.2}$$

When working with finite samples we do not use $T(F)$ but rather $T(F_n)$, where F_n is the empirical distribution generated by the sample $\{y_1, \ldots, y_n\}$. If we replace one observation by an outlier at y, we find

$$T(y_1, \ldots, y_{n-1}, y) \approx T(F_n) + \frac{1}{n}\,\text{IF}\,(y; T, F)\,, \tag{5.3}$$

Table 2. Breakdown Points of Some Location Estimators

Estimator	Breakdown Point (%)
Mean	0
α-Trimmed mean	α
Median	50
Hodges–Lehmann	29
Shorth	50
LMS, LTS, LWS	50
S-estimators	50

and it is even possible to replace a small fraction $\varepsilon = m/n$ of observations:

$$T(y_1, \ldots, y_{n-m}, y, \ldots, y) \approx T(F_n) + \frac{m}{n} \text{ IF }(y; T, F) .$$

The breakdown point ε_n^* provides us with a rough upper bound on the fraction of outliers for which such a linear approximation might be useful.

To understand better the difference between the concepts of breakdown point and influence function, let us resort to a physical analogy. Figure 3 was taken from a treatise of Galilei (1638) on applied mechanics. In this work he considered a beam which is fixed at one end, and to which a stone is attached. To clarify the relation with our statistical subject, we denote the weight of the stone by ε and the position where it exerts its force by y. Two questions naturally arise, which are still equally important—and nontrivial—in contemporary mechanics. First, how heavy does the stone have to be to make the beam break? (In his treatise, Galilei provided a reasoning showing that this breakdown point is propor-

Figure 3. Illustration taken from Galilei (1638), for showing the difference between the concepts of breakdown point and influence function.

tional to the width of the beam multiplied by the square of its height.) A second question is, how much does the beam bend for smaller weights? The answer to this question came only around 1825, when Navier gave a mathematical formulation of what is known today as Hooke's law, stating the proportionality of stress and strain. Nowadays one realizes that Hooke's law is only a first-order approximation: for small weights ε, the deformation of the beam is indeed linear in ε. The ratio of deformation divided by weight (or, more accurately, the slope of this relation at $\varepsilon = 0$) roughly corresponds to Young's elasticity modulus. Our influence function is just a plot of the elasticity as a function of y, the position of the external force. For small weights, one can predict the effect by means of a linear approximation such as (5.2). However, the actual computations in present-day mechanics have become much more laborious. As soon as the weight ε is larger than "infinitesimal," the deformations inside the beam become much more complex and Hooke's law is no longer valid. One resorts to systems of differential equations instead, which are solved by computer simulations (the so-called *finite element method*). These intricate models give a more accurate description of reality and are necessary to compute the breakdown point of complicated structures such as large buildings or bridges. In the same spirit, the study of influence functions of estimators is more and more being complemented by the consideration of breakdown points and simulations.

All estimators possess a breakdown point, but not all of them have an influence function. However, estimators that do are often asymptotically normal:

$$\sqrt{n}(T_n - T(F)) \sim N(0, V(T, F)) , \qquad (5.4)$$

where the asymptotic variance is related to the IF through

$$V(T, F) = \int \mathrm{IF}(y; T, F)^2 \, dF(y) \qquad (5.5)$$

(see Hampel et al. 1986, Section 2.1). Typical examples of this are M-, L-, and R-estimators.

From Theorem 3, we have learned that the LMS converges merely like $n^{-1/3}$. Therefore, its influence function is not well defined. Nevertheless, we find that statistical efficiency is not the most important criterion. As a data analytic tool, the LMS is very satisfactory. Moreover, it is not so difficult to construct variants with higher efficiency, as has already been discussed in Chapter 3 in the general regression case. In order to get some idea of the local robustness properties of the LMS, we resort to a finite-sample version of the IF, namely the *stylized sensitivity curve* SC_n (see Andrews et al. 1972). We start with an artificial sample containing n points

$$\{y_1, \ldots, y_n\} = \{\Phi^{-1}(i/(n+1)); i = 1, \ldots, n\}, \tag{5.6}$$

where Φ is the standard normal cumulative distribution function. Then we plot the curve

$$SC_n(y) = \frac{T_{n+1}(y_1, \ldots, y_n, y) - T_n(y_1, \ldots, y_n)}{1/n} \tag{5.7}$$

for y ranging over an interval containing the points y_i. By means of this curve we can also compute

$$V_n = \frac{1}{n} \sum_{i=1}^{n} [SC_n(y_i)]^2. \tag{5.8}$$

If the estimator T_n is really asymptotically normal, then the sensitivity curve should converge to the influence function as $n \to \infty$, and V_n to the asymptotic variance. We have investigated SC_n and V_n for samples as constructed in (5.6). Some LMS sensitivity curves appear in Figure 4. From this plot, we see that for larger sample sizes the upward and downward peaks become thinner and higher. Combining this information with heuristic arguments, one can conclude that the sensitivity curve of the LMS converges to zero, except in $-\Phi^{-1}(0.75)$ and $\Phi^{-1}(0.75)$ where the function tends to $-\infty$ and ∞.

The values of V_n for the LMS are contained in Table 3. For increasing n, the associated values of V_n are ever-growing numbers, indicating that $n \operatorname{Var}(T_n)$ does not tend to a finite limit. In contrast, the quantity V_n for the median (last column in Table 3) seems to stabilize around 1.57 for large n. This corresponds to the theoretical asymptotic variance because the influence function of the median at the normal is

$$IF(y; \text{median}, \Phi) = \operatorname{sgn}(y)/(2\phi(0)),$$

from which it follows that

$$V(\text{median}, \Phi) = (2\phi(0))^{-2} = \pi/2 \approx 1.571.$$

In Theorem 4 we have seen that the LTS converges like $n^{-1/2}$, with the same asymptotic behavior at the normal distribution as the Huber-type

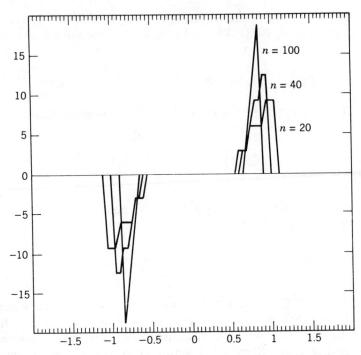

Figure 4. Stylized sensitivity curves of the LMS location estimator, for different sample sizes.

skipped mean [see (4.1)]. The influence function of such M-estimators of location at symmetric distributions F is given by

$$\text{IF}(y; T, F) = \frac{\psi(y)}{\int \psi' \, dF} = \frac{\psi(y)}{-\int \psi(z)f'(z) \, dz}, \qquad (5.9)$$

and their asymptotic variance is therefore

$$V(\psi, F) = \frac{\int \psi^2 \, dF}{\left(\int \psi' \, dF\right)^2}. \qquad (5.10)$$

For the LTS, this suggests that

Table 3. Value of V_n for the LMS, LWS, LTS, and Median (for Different n)

n	V_n for LMS	V_n for LWS	V_n for LTS	V_n for Median
10	4.213	5.005	10.206	1.578
20	13.182	6.247	14.702	1.573
40	14.373	15.759	14.588	1.571
100	17.738	20.795	14.267	1.571
300	38.048	23.178	13.545	1.560
600	54.074	38.340	13.596	1.566
1000	71.546	47.591	13.967	1.569

$$\mathrm{IF}(y;\mathrm{LTS},\Phi) = \frac{y1_{[-q,q]}(y)}{2\Phi(q) - 1 - 2q\phi(q)}, \tag{5.11}$$

which can also be rewritten as

$$\mathrm{IF}(y;\mathrm{LTS},\Phi) = \begin{cases} 14.021y & \text{if } |y| \le q \\ 0 & \text{otherwise,} \end{cases}$$

where $q = \Phi^{-1}(0.75) = 0.6745$. The central part of this influence function is therefore linear, as is the IF of the arithmetic mean. However, the IF of the LTS becomes exactly zero outside the middle half of the data, because outliers in that region do not have any influence on the LTS. Therefore, the LTS is said to be redescending. Using (5.10) or Theorem 4 we find that the asymptotic variance equals $V(\mathrm{LTS}, \Phi) = 14.021$, which corresponds to Table 3. This asymptotic variance is very large, but of course we will typically perform a one-step improvement (e.g., re-weighted least squares) which greatly reduces the asymptotic variance, making it smaller than that of the sample median.

An important robustness measure derived from the IF is the gross-error sensitivity γ^*, defined as

$$\gamma^* = \sup_y |\mathrm{IF}(y; T, F)|. \tag{5.12}$$

This γ^* describes the maximal effect on the estimate induced by a small contamination. Therefore, one aims for a finite gross-error sensitivity. For instance, the Huber estimator given by (1.13) has maximal asymptotic efficiency subject to an upper bound on γ^* (Hampel 1974). It is also the most efficient subject to an upper bound on the so-called change-of-variance sensitivity κ^* (Rousseeuw 1981). Moreover, the median mini-

mizes γ^* among all M-estimators (Hampel 1974), and it also minimizes
κ^* (Rousseeuw 1982). From (5.11), it follows that the LTS gross-error
sensitivity is equal to 9.457. This value is again very large, but also
diminishes when a one-step improvement is carried out.

Let us now look at the stylized sensitivity curves for the LTS in Figure
5. The different curves present the same characteristic jumps. For large
values of n, the peaks are only stretched out, but the jumps do not
disappear. We have found that they are a result of the generation of the
artificial sample (5.6). For small changes of y the selected half (for
calculating the LTS) remains the same, so the estimated location changes
as if it was the arithmetic mean on a sample with only $n/2$ observations.
If, however, y changes a little too much, then the selected half changes
and jumps to the next y_i. The total amount of change in T can roughly be
computed as follows: Put D equal to the distance between two of the y_i
around $q = \Phi^{-1}(0.75)$. Then one has that

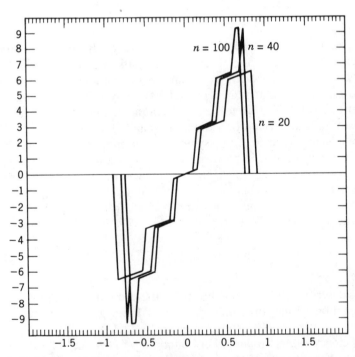

Figure 5. Stylized sensitivity curves of the LTS location estimator, for different sample
sizes.

$$\Phi'(q)D \text{ is approximately equal to } 1/n,$$

or that

$$D \text{ is about } 1/(n\phi(q)).$$

Since the sensitivity curve is n times the difference between $T(\{y_i\}, y)$ and $T(\{y_i\})$, it follows that the jump in the sensitivity curve is roughly equal to $nD \approx 1/\phi(q) \approx 3.1$, which is independent of n.

This "jump" effect can be avoided by considering the *averaged sensitivity curves* (ASC). These curves are defined as an average of some sensitivity curves as follows:

$$\text{ASC}_n(y) = \frac{1}{m} \sum_{j=1}^{m} \text{SC}_n^{(j)}(y) \qquad (5.13)$$

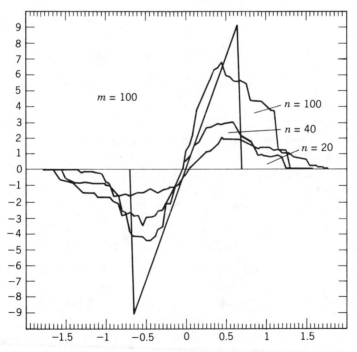

Figure 6. The influence function and averaged sensitivity curves (for different sample sizes) of the LTS location estimator.

where $SC_n^{(j)}$ is defined as in (5.7), but calculated on a real sample $\{y_1^{(j)}, \ldots, y_n^{(j)}\}$ drawn from a standard normal distribution. Such ASC curves of the LTS are shown in Figure 6 for some different sample sizes. Their basic idea is that the expected value $E[SC_n(y)]$ over all random samples $\{y_1, \ldots, y_n\}$ should be fairly close to the IF, and at least the systematic jumps do not occur because the distances between the observations now obey the true random variability. [Another possibility would be to compute the median of the m values in (5.13).] When both n and m increase, the ASC approaches the theoretical influence function, which has also been plotted.

EXERCISES AND PROBLEMS

Section 1

1. Assuming that the observations y_i are normally distributed with standard deviation σ, show that both $1.4826 \, \text{med}_f |y_f - \text{med}_k \, y_k|$ and $1.2533 \frac{1}{n} \sum_{f=1}^{n} |y_f - \text{med}_k \, y_k|$ tend to σ when n increases.
2. Show that the minimization of $\sum_{i=1}^{n} w_i (y_i - T_W)^2$ yields the solution

$$
T_W = \frac{\sum_{i=1}^{n} w_i y_i}{\sum_{i=1}^{n} w_i},
$$

which proves that the one-step W-estimator (1.15) is a reweighted least squares procedure.
3. Show that the sum of the a_i in (1.16) must be equal to 1 to make the corresponding L-estimator location equivariant. How about scale equivariance? Can the median be written as an L-estimator?

Section 2

4. Generate four random numbers according to the standard normal distribution $N(0, 1)$. Does this sample look symmetric? Calculate the arithmetic mean, sample median, and LMS, and compute the classical third moment for testing symmetry.
5. Make a plot of the objective functions of (2.1) and (2.5) as a function of θ for the sample $\{1, 3, 4, 6, 7, 10\}$.
6. Compute the LTS and the LWS for the Rosner data (2.2).

7. Show that the LMS of three observations equals the midpoint of the closest pair. What happens to the shorth and the LTS in this situation?

8. Show that for any affine equivariant location estimator T, it holds that $T(y_1) = y_1$ and $T(y_1, y_2) = \frac{1}{2}(y_1 + y_2)$. Things become more interesting when there are three observations (assuming that they are not all equal to the same value). By equivariance, we may assume that $y_1 = -1$ and $y_2 = 1$, so the whole estimator may be described by means of the univariate function $h(y) = T(-1, 1, y)$. Draw the function h for the arithmetic mean, the sample median, and the LMS.

Section 3

9. Analyze the monthly payments of exercise 8 of Chapter 1 (i.e., the response variable of Table 1) in a one-dimensional way by means of PROGRESS. (Motivation: In exercise 6 of Chapter 2 this variable was regressed against time, and the RLS slope was not significantly different from zero.) How many outliers do you find now?

10. Apply PROGRESS to the one-dimensional data set of Cushny and Peebles (1905), first analyzed by Student (1908), consisting of the following 10 observations:

$$0.0,\ 0.8,\ 1.0,\ 1.2,\ 1.3,\ 1.3,\ 1.4,\ 1.8,\ 2.4,\ 4.6\ .$$

Explain why the arithmetic mean is larger than the robust estimates of location. What happens to the standard deviation? Are there any points with standardized LS residuals outside the ± 2.5 region? Compare this with the residuals of LMS and RLS.

11. Grubbs (1969) considered the following one-dimensional data sets.
(a) Strengths of hard-drawn copper wire ($n = 10$):

$$568,\ 570,\ 570,\ 570,\ 572,\ 572,\ 572,\ 578,\ 584,\ 596$$

(b) Residuals of observed vertical semidiameters of Venus ($n = 15$):

$$-1.40,\ -0.44,\ -0.30,\ -0.24,\ -0.22,\ -0.13,\ -0.05\ ,$$
$$0.06,\ 0.10,\ 0.18,\ 0.20,\ 0.39,\ 0.48,\ 0.63,\ 1.01$$

(c) Percent elongations of plastic material ($n = 10$):

$$2.02,\ 2.22,\ 3.04,\ 3.23,\ 3.59,\ 3.73,\ 3.94,\ 4.05,\ 4.11,\ 4.13$$

(d) Ranges of projectiles $(n = 8)$:

4420, 4549, 4730, 4765, 4782, 4803, 4833, 4838.

Analyze these data sets by means of PROGRESS. How many outliers are identified?

Section 4

12. Write down the proof of Theorem 4 for the variant of the LTS in which $n(1 - \alpha)$ of the squared residuals are used, with $0 < \alpha < 50\%$.

13. (Research problem) In Table 1 it seems that the spread of the LWS lies between the LTS (which is known to converge like $n^{-1/2}$) and the LMS (which converges like $n^{-1/3}$). We do not yet know the convergence rate of the LWS estimator.

Section 5

14. Is there a relation between the function h of exercise 8 and sensitivity curves?

CHAPTER 5

Algorithms

1. STRUCTURE OF THE ALGORITHM USED IN PROGRESS

The computation of the least median of squares (LMS) regression coefficients is not obvious at all. It is probably impossible to write down a straightforward formula for the LMS estimator. In fact, it appears that this computational complexity is inherent to all (known) affine equivariant high-breakdown regression estimators, because they are closely related to projection pursuit methods (see Section 5 of Chapter 3). The algorithm implemented in PROGRESS (Leroy and Rousseeuw 1984) is similar in spirit to the bootstrap (Diaconis and Efron 1983). Some other possible algorithms will be described in Section 2.

The algorithm in PROGRESS proceeds by repeatedly drawing subsamples of p different observations. (We will use the same notation as in Section 6 of Chapter 3.) For such a subsample, indexed by $J = \{i_1, \ldots, i_p\}$, one can determine the regression surface through the p points and denote the corresponding vector of coefficients by $\boldsymbol{\theta}_J$. (This step amounts to the solution of a system of p linear equations in p unknowns.) We will call such a solution $\boldsymbol{\theta}_J$ a trial estimate. For each $\boldsymbol{\theta}_J$, one also determines the corresponding LMS objective function with respect to the whole data set. This means that the value

$$\operatorname*{med}_{i=1,\ldots,n} (y_i - \mathbf{x}_i \boldsymbol{\theta}_J)^2 \tag{1.1}$$

is calculated. Finally, one will retain the trial estimate for which this value is minimal. (Note that this algorithm is affine equivariant, as it should be.)

But how many subsamples $J_1, J_2, \ldots, J_m$ should we consider? In principle, one could repeat the above procedure for all possible subsamples of size p, of which there are C_n^p. Unfortunately, C_n^p increases very fast with n and p. In many applications, this would become infeasible. In such cases, one performs a certain number of random selections, such that the probability that at least one of the m subsamples is "good" is almost 1. A subsample is "good" if it consists of p good observations of the sample, which may contain up to a fraction ε of bad observations. The expression for this probability, assuming that n/p is large, is

$$1 - (1 - (1 - \varepsilon)^p)^m .\tag{1.2}$$

By requiring that this probability must be near 1 (say, at least 0.95 or 0.99), one can determine m for given values of p and ε. This has been done in Table 1 for $p \le 10$ and for a percentage of contamination varying between 5% and 50%. For fixed p, the number of subsamples increases with ε in order to maintain the same probability (1.2). This permits us to reduce the computation time of PROGRESS in situations where there are reasons to believe that there are at most 25% of bad data. This choice can be made by answering the question:

```
WHICH VERSION OF THE ALGORITHM WOULD YOU LIKE TO USE?
------------------------------------------
Q=QUICK VERSION
E=EXTENSIVE SEARCH

ENTER YOUR CHOICE PLEASE (Q OR E):
```

Table 1. Number m of Random Subsamples, Determined in Function of p and ε by Requiring That the Probability of at Least One Good Subsample Is 95% or More

Dimension p	Fraction ε of Contaminated Data						
	5%	10%	20%	25%	30%	40%	50%
1	1	2	2	3	3	4	5
2	2	2	3	4	5	7	11
3	2	3	5	6	8	13	23
4	2	3	6	8	11	22	47
5	3	4	8	12	17	38	95
6	3	4	10	16	24	63	191
7	3	5	13	21	35	106	382
8	3	6	17	29	51	177	766
9	4	7	21	36	73	296	1533
10	4	7	27	52	105	494	3067

The number of subsamples computed from (1.2) becomes tremendous for large p, at least in the most extreme case of about 50% of bad observations. At a certain stage, one can no longer increase the number m of replications, because the computation time must remain feasible. One then incurs a larger risk (the probability of which is known) of not encountering any "good" subsample during the m replications. Table 2 lists the values of m that are used in PROGRESS. For small p we have systematically taken m larger than the value needed in formula (1.2), in order to increase the expected number of good subsamples, so the objective function (1.1) may become even smaller.

Several elements of this algorithm can also be found in other statistical techniques. For instance, Oja and Niinimaa (1984) and Hawkins et al. (1985) have used trial estimates such as $\boldsymbol{\theta}_J$ for different purposes. The idea of drawing many (but not necessarily all) subsamples of the data at hand is related to the bootstrap (Efron 1979), which is, however, mainly concerned with the estimation of bias and variability of existing (point) estimators at the actual data. Formula (1.2) was first used by Stahel

Table 2. Number m of Subsamples That Are Actually Used in the Program PROGRESS, According to the Options "Extensive" or "Quick" [a]

	Option "Extensive"	Option "Quick"
$p = 1$	$m = C_n^p$ if $n \leq 500$ $m = 500$ if $n > 500$	$m = C_n^p$ if $n \leq 150$ $m = 150$ if $n > 150$
$p = 2$	$m = C_n^p$ if $n \leq 50$ $m = 1000$ if $n > 50$	$m = C_n^p$ if $n \leq 25$ $m = 300$ if $n > 25$
$p = 3$	$m = C_n^p$ if $n \leq 22$ $m = 1500$ if $n > 22$	$m = C_n^p$ if $n \leq 15$ $m = 400$ if $n > 15$
$p = 4$	$m = C_n^p$ if $n \leq 17$ $m = 2000$ if $n > 17$	$m = C_n^p$ if $n \leq 12$ $m = 500$ if $n > 12$
$p = 5$	$m = C_n^p$ if $n \leq 15$ $m = 2500$ if $n > 15$	$m = C_n^p$ if $n \leq 11$ $m = 600$ if $n > 11$
$p = 6$	$m = C_n^p$ if $n \leq 14$ $m = 3000$ if $n > 14$	$m = 700$
$p = 7$	$m = 3000$	$m = 850$
$p = 8$	$m = 3000$	$m = 1250$
$p \geq 9$	$m = 3000$	$m = 1500$

[a]The notation $m = C_n^p$ means that *all* subsamples with p elements are being considered.

(1981) in the context of robust multivariate location and covariance estimators.

Let us now illustrate the basic idea of the algorithm used in PROGRESS by means of the artificial two-dimensional example in Figure 1. For this data set, n equals 9 and p equals 2. Because n is very small, PROGRESS considers *all* pairs of points. We will restrict the explanation to only three such combinations, namely (f, g), (f, h), and (g, h). Let us start with the points f and g. The regression surface (which is a line here) passing through f and g is found by solving the system of equations

$$y_f = \theta_1^0 x_f + \theta_2^0$$

$$y_g = \theta_1^0 x_g + \theta_2^0 \, ,$$

where (x_f, y_f) and (x_g, y_g) are the coordinates of the points f and g. The resulting trial estimate is $(\theta_1^0, \theta_2^0)^t$. Next, the residuals $y_i - \theta_1^0 x_i - \theta_2^0$ from this line are determined for all points i in the sample. The median of the squared residuals is then calculated, and compared with the smallest value found for previous pairs of points. Because we want to minimize this quantity, the trial estimate corresponding to f and g will be retained only when it leads to a strictly lower value. Examining the scatterplot in Figure 1, we see that (f, g) is better than either (f, h) or (g, h). Indeed, the majority of the observations have small residuals with respect to the

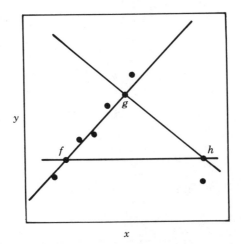

Figure 1. Illustration of the algorithm used in PROGRESS.

line passing through f and g. Repeating this procedure for each pair of points will finally yield the lowest objective function.

The entry $p = 1$ in Table 2 refers to simple regression through the origin (see Section 6 of Chapter 2) in which there is no intercept term. Note that in one-dimensional location we also have $p = 1$, but then we do not apply the resampling technique because an exact algorithm is available for that situation (see Section 2 of Chapter 4). Moreover, this exact algorithm is very useful in any regression model with intercept, because we can adjust our constant term with it. Indeed, suppose that the resampling algorithm has provided regression coefficients $\hat{\theta}_1, \ldots, \hat{\theta}_p$. Then we can replace $\hat{\theta}_p$ by the LMS location estimate of the n numbers

$$y_i - x_{i,1}\hat{\theta}_1 - \cdots - x_{i,p-1}\hat{\theta}_{p-1}, \qquad i = 1, \ldots, n$$

because we are certain that the resulting objective function (1.1) will become smaller, since the one-dimensional algorithm yields the optimal intercept term (conditional on the values of $\hat{\theta}_1, \ldots, \hat{\theta}_{p-1}$). Therefore, this last step has been incorporated in PROGRESS.

This argument also shows us how we could improve the result even further. Instead of adjusting the intercept term once (at the end), we could do the same thing at each replication, so that each trial estimate would contain an optimal intercept term. It follows that this slight modification of the algorithm could only lower the resulting objective function. However, we have decided not to include this version because it would consume more computation time. At present, we evaluate the median of the squared residuals (for each trial estimate) in $O(n)$ operations, whereas we would spend at least $O(n \log n)$ if we wanted to apply the one-dimensional algorithm each time. On the other hand, the present evolution of computers might soon allow this refinement.

REMARK. Note that the algorithm can be speeded up enormously by means of *parallel computing*, because it is very easy to split it up into parts that are carried out simultaneously. Indeed, each processing element can compute trial estimates θ_j and the associated median squared residual (1.1). To try this out, a parallelized version of PROGRESS was implemented on the ICAP 1 system at the IBM Research Center in Kingston, NY. (This happened in March 1987 in collaboration with Ph. Hopke and L. Kaufman.) This system contains an IBM 3081 host computer and 10 loosely coupled FPS-164 processors. PROGRESS was adapted by adding special statement lines for the precompiler used on the ICAP 1. Two versions of the program were constructed, one with and one without shared memory. In both cases, an important reduction of the computation time was achieved. This speed will, of course, still increase when a larger number of processors is used. Already systems with hundreds of processing elements are being developed (such as the DPP87 system in Delft), so eventually *all* trial values may be computed at the same time, making the LMS as fast as least squares.

Apart from the regression coefficients, also the scale parameter σ (the dispersion of the errors e_i) has to be estimated in a robust way. We begin by computing an initial scale estimate s^0. This s^0 is based on the minimal median and multiplied by a finite-sample correction factor (which depends on n and on p) for the case of normal errors:

$$s^0 = 1.4826(1 + 5/(n - p))\sqrt{\operatorname*{med}_i r_i^2(\hat{\theta})}. \tag{1.3}$$

The factor $1.4826 = 1/\Phi^{-1}(0.75)$ was introduced because $\operatorname{med}_i |z_i|/$ $\Phi^{-1}(0.75)$ is a consistent estimator of σ when the z_i are distributed like $N(0, \sigma^2)$. From an empirical study, it appeared that this factor 1.4826 alone was not enough, because the scale estimate became too small in regressions with normal errors, especially for small samples. Indeed, if n is only slightly larger than $2p$, then it easily happens that half of the data (which may mean only $p + 1$ or $p + 2$ points!) almost lie on a linear structure. As a consequence, also some good points may obtain relatively large standardized residuals. It was not obvious at all to find an appropriate factor to compensate for this effect. On the one hand, such a factor should not be too small because good points should not possess large standardized residuals, but on the other hand, a too-large factor would result in neglecting real outliers in contaminated samples. We have studied the behavior of the original scale estimate through simulation, both in normal error situations and in situations where there was contamination in the response or in the explanatory variables. It turned out that multiplication with the factor $1 + 5/(n - p)$ gave a satisfactory solution. (Of course, for large n this factor tends to 1.)

This preliminary scale estimate s^0 is then used to determine a weight w_i for the ith observation, namely

$$w_i = \begin{cases} 1 & \text{if } |r_i/s^0| \leq 2.5 \\ 0 & \text{otherwise}. \end{cases} \tag{1.4}$$

By means of these weights, the final scale estimate σ^* for LMS regression is calculated:

$$\sigma^* = \sqrt{\left(\sum_{i=1}^{n} w_i r_i^2\right) \Big/ \left(\sum_{i=1}^{n} w_i - p\right)}. \tag{1.5}$$

The advantage of this formula for σ^* is that outliers do not influence the scale estimate anymore. Moreover, at the classical model, σ^* would be a consistent estimator of σ if the weights w_i were independent of the data $(\mathbf{x}_i, y_i)$.

Like most regression programs, PROGRESS performs its actual computations on standardized data. The main motivation for this is to avoid numerical inaccuracies caused by different units of measurement. For instance, one of the explanatory variables might take values around 10^8, whereas another might be of the order of 10^{-11}. Therefore, the program first standardizes the data to make the variables dimensionless and of the same order of magnitude. The standardization used in PROGRESS is described in Section 1 of Chapter 4. The regression algorithm is then applied to these standardized data, but afterwards the results have to be transformed back in terms of the original variables. The regression coefficients of LS, LMS, and RLS are transformed in exactly the same way (by means of the subroutine RTRAN) because all three regression methods satisfy the equivariance properties listed in Section 4 of Chapter 3. For LS and RLS, the variance-covariance matrix between the coefficients is also transformed.

The running time of PROGRESS depends not only on the sample size n, the dimension p, and the required probability in (1.2), but, of course, also on the computer processing speed. To give an illustration, Table 3

Table 3. Computation Times on an Olivetti M24 Microcomputer Without 8087 Mathematical Coprocessor, for Some Data Sets Used in Chapters 2 and 3

Data Set	Computation Time (minutes)	
	Option "Extensive"	Option "Quick"
Pilot-Plant data	1.37	1.37
Hyperinflation in China	0.90	0.90
Fires data	0.83	0.83
Mickey data	1.32	1.32
Hertzsprung–Russell	3.88	1.93
Telephone data	1.45	1.45
Lactic acid concentration	1.00	1.00
Kootenay River	0.83	0.83
Stackloss data	7.50	2.58
Coleman data	19.13	5.23
Salinity data	9.72	3.05
Air quality data	8.93	2.18
Hawkins–Bradu–Kass data	16.23	4.77
Education expenditure	12.52	3.87

lists some typical running times on an IBM-PC compatible microcomputer. These times can be reduced substantially by means of an 8087
Mathematical Coprocessor.

*2. SPECIAL ALGORITHMS FOR SIMPLE REGRESSION

In simple regression, one only has to find the slope $\hat{\theta}_1$ and the intercept $\hat{\theta}_2$
of a line determined by n points in the plane. One can, of course, apply
the above resampling algorithm, as it is implemented in PROGRESS.
However, in this particular situation some other approaches are also
possible.

Our oldest algorithm for the LMS and the LTS was of the "scanning"
type (Rousseeuw et al. 1984b). As in Section 1, one starts by standardiz-
ing the observations. The idea of the algorithm can be understood most
easily when writing the definition of the LMS estimator in the following
way:

$$\min_{\theta_1} \left\{ \min_{\theta_2} \ \text{med} \ ((y_i - \theta_1 x_i) - \theta_2)^2 \right\} . \tag{2.1}$$

Indeed, one can treat the parts in (2.1) separately. The second portion of
the minimization is quite easy, because for any given θ_1 it becomes
essentially a one-dimensional problem, which can be solved explicitly as
we saw in Chapter 4. (This part even cancels when there is no constant
term in the model.) Then one has to find $\hat{\theta}_1$ for which

$$m^2(\theta_1) = \min_{\theta_2} \ \text{med} \ ((y_i - \theta_1 x_i) - \theta_2)^2$$

is minimal. This is just the minimization of a one-dimensional function
$m^2(\theta_1)$, which is continuous but not everywhere differentiable. In order to
find this minimum, one goes through all angles α from -1.55 rad to
1.55 rad with a stepsize of 0.02 rad and uses the slope $\theta_1 = \tan \alpha$ to
compute the corresponding value of $m^2(\theta_1)$. Then one scans with a
precision of 0.001 rad in the two most promising areas (as determined by
the local minima of the first step). In this way, the objective function does
not have to be calculated too often. In Figure 2, a part of the function
$m^2(\theta_1)$ is plotted for the telephone data in Table 3 of Chapter 2. Such a
plot may also be useful in keeping track of local minima. Indeed, a
prominent secondary minimum indicates a possible ambiguity in that
there may be two lines, each fitting the data reasonably well.

The generalization of this scanning-type algorithm to multivariate
models is not feasible. In such situations, the scanning would have to be

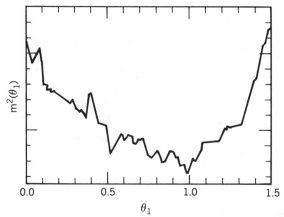

Figure 2. A plot of the function $m^2(\theta_1)$ for the telephone data, for θ_1 between 0.0 rad and 1.5 rad.

performed on higher-dimensional grids, which leads to a tremendous amount of computation.

Yohai (1986, personal communication) is presently experimenting with a multivariate algorithm based on repeated application of the estimate $\hat{\theta} = \text{med}_i \, y_i/\text{med}_i \, x_i$ for simple regression through the origin. At each step, the variable is selected which minimizes $\text{med}_i \, r_i^2$, and then its effect is swept from the response variable and the remaining carriers. The objective function $\text{med}_i \, r_i^2$ is also used to decide how many variables are used within a cycle and as a stopping rule for the iteration of cycles. For such an algorithm it is difficult to prove anything in the way of equivariance or overall breakdown point, but some preliminary experience indicates that the computation time is not too high, and the results have been satisfactory.

Recently, Steele and Steiger (1986) studied the numerical aspects of the LMS estimator in the framework of simple regression with intercept. Their first result states that not only are there two points that possess an LMS residual of $\pm m_T$ (as we have proved in Theorem 1 of Chapter 4) but that there are even *three* such observations. This property is one out of three (necessary and sufficient) conditions under which a pair (θ_1, θ_2) is a local optimizer of the LMS objective function. The two other conditions say that the three residuals of magnitude $\pm m_T$ alternate in sign, and secondly that the number of points with $|r_i| > m_T$ is at least one more than the number of points with $|r_i| < m_T$. As a consequence, this characterization of local minimizers reduces the continuous minimization to a discrete one, in the sense that only lines satisfying the above conditions have to be considered and that the one with the smallest objective function value has to be retained. Indeed, it implies that there exists (at least) one pair of data points such that the line joining them has the same slope as the LMS line. This provides a posteriori justification of the way we work in PROGRESS because it means that our algorithm, at least in the simple regression case and for small n, leads to an exact solution. (For large n, we only use random pairs of points to reduce the computation time.) Furthermore, Steele and Steiger also showed that the LMS objective function can have at most $O(n^2)$ local minima, and they provide some ideas on decreasing the computation time. Finally, they stated that the necessary and sufficient conditions for local minimizers can be generalized to p dimensions ($p > 2$). This corresponds with our own algorithm for multiple regression, where we consider hyperplanes passing through p points, which are again vertically adjusted by estimating the intercept term in a one-dimensional way.

Souvaine and Steele (1986) provided even more refined algorithms for LMS simple regression, making use of modern tools of numerical analysis and computer science such as heaps and hammocks. Their best computational complexity is again $O(n^2)$, and they also stated that everything can be generalized to p dimensions, yielding $O(n^p)$. It seems reasonable to conjecture that one can do no better than $O(n^p)$ for computing LMS or any other affine equivariant high-breakdown regression estimator.

*3.   OTHER HIGH-BREAKDOWN ESTIMATORS

In addition to the LMS estimator, we have also discussed the LTS estimator in Section 2 of Chapter 1 and in Section 4 of Chapter 3. In order to compute the LTS regression coefficients, one only has to change the objective function in either the resampling algorithm of Section 1 or the scanning algorithm described in Section 2. That is, we now have to minimize

$$\sum_{i=1}^{h} (r^2)_{i:n}$$

instead of $\operatorname{med}_i r_i^2$. Note, however, that the calculation of the LTS objective requires sorting of the squared residuals, which takes more computation time than the LMS goal function. In a model with a constant term, one also needs the LTS location estimator in order to adjust the intercept. For this purpose, Section 2 of Chapter 4 contains an exact algorithm, which again consumes more computation time than its LMS counterpart. Therefore, the overall LTS regression algorithm is somewhat more expensive than the LMS one. This effect is shown in Figure 3, where the computation times of LMS and LTS are plotted for different sample sizes. (Note that, according to Table 2, the number of trial estimates m increases until $n = 50$ and then remains equal to 1000.) Also, analogous changes in the LMS algorithm are sufficient for computing the least winsorized squares (LWS) estimator.

The calculation of S-estimators (Section 4 of Chapter 3) is somewhat more complicated. The objective function now must be replaced by the solution s of the equation

$$\frac{1}{n} \sum_{i=1}^{n} \rho\left(\frac{r_i}{s}\right) = K \,, \tag{3.1}$$

which may be computed in an iterative way. [A possible choice for ρ is given by (4.31) in Chapter 3.] The residuals r_i in (3.1) correspond to a trial estimate $\boldsymbol{\theta}_J$ in the resampling algorithm or to a certain slope and

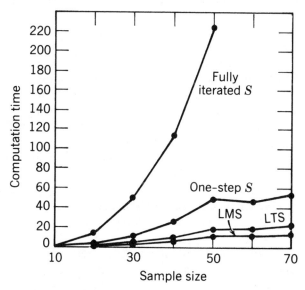

Figure 3. Computation times (in CP seconds on a VAX 11/750) of the LMS, the LTS, one-step S-, and fully iterated S-estimators, for $p = 2$ and sample sizes $n = 10, \ldots, 70$. The resampling algorithm was used, and the running times have been averaged over several simulated data sets. (Data from van Zomeren 1986.)

intercept in the scanning algorithm. It is possible to restrict our attention to the first step of the iteration [see (2.12) in Chapter 4] in order to reduce the computation time. The vector $\hat{\boldsymbol{\theta}}$ minimizing s is called an S-estimate of the regression coefficients. The final scale estimate is then given by

$$\hat{\sigma} = s(r_1(\hat{\boldsymbol{\theta}}), \ldots, r_n(\hat{\boldsymbol{\theta}})) .$$

From a computational point of view, Figure 3 shows that even the one-step version of the S-estimator is very time-consuming, not to mention the fully iterated version. Moreover, some preliminary experience (both for simple and multiple regression) indicates that S-estimators do not really perform better than the LMS, at least from a practical point of view, which is why we have focused our attention to the LMS as the most easily computable member of our high-breakdown family.

*4. SOME SIMULATION RESULTS

In Chapters 2 and 3 we compared the LS and LMS regressions by means of some real data sets. A shortcoming of this approach is that it may lead to disputes, because it is impossible to "prove" which analysis is best. A popular alternative is to carry out a simulation, because then one knows the true parameter values of the generated data.

In the well-known Princeton robustness study (Andrews et al. 1972), a large number of location estimators have been tried out in a variety of situations. This large-scale simulation has provided many new insights, resulting in a significant impact on robustness theory. We hope that, some day, funds may be raised to carry out a similar project devoted to regression estimators. In the meantime there have been some smaller comparative studies, such as Kühlmeyer's (1983) investigation of 24 regression estimators of the types M, L, and R, and the article by Johnstone and Velleman (1985b). In the present section we report on some of our own simulation results, which focus on the LMS and its variants.

As a preliminary step, we investigated the scale estimate $\hat{\sigma}$ associated with LMS regression. This was necessary because the naive estimate

$$1.4826\sqrt{\operatorname*{med}_i r_i^2(\hat{\boldsymbol{\theta}})} \tag{4.1}$$

underestimates σ in small-sample situations, as was already explained in

Section 1 above. To determine a suitable correction factor, we generated many samples according to the standard linear model. For each choice of n and p we considered 200 samples and compared the resulting values of (4.1) with the true σ. [It turned out that (4.1) behaved somewhat differently depending on whether n and p were even or odd. A first attempt to understand this phenomenon was restricted to $p = 1$, so only the effect of n had to be studied (Rousseeuw and Leroy 1987). The effect of parity—or of mod $(n, 4)$—persisted even in artificial samples of the type $\{\Phi^{-1}(i/(n + 1)), i = 1, \ldots, n\}$ and could be traced back by expanding Φ^{-1} by means of Taylor series in the endpoints of the shortest halves.] In order to determine a reasonable overall adjustment factor, we have plotted the average values of (4.1) versus $1/(n - p)$, yielding the factor $1 + 5/(n - p)$ in (1.3) above.

After this preliminary step, we set up the actual simulation experiment in order to investigate the LS and the LMS, as well as the one-step reweighted least squares (RLS) based on the LMS, and the one-step M-estimator (OSM) based on the LMS. For this purpose, we have resorted to three types of configurations. The first one is the normal situation,

$$y_i = x_{i,1} + \cdots + x_{i,p-1} + x_{i,p} + e_i \, ,$$

in which $e_i \sim N(0, 1)$ and the explanatory variables are generated as $x_{i,j} \sim N(0, 100)$ for $j = 1, \ldots, p$ (if there is no intercept term) or for $j = 1, \ldots, p - 1$ (if there is an intercept term, and then $x_{i,p} = 1$).

In the second situation we construct outliers in the y-direction. For this purpose, we generate samples where 80% of the cases are as in the first situation and 20% are contaminated by using an error term $e_i \sim N(10, 1)$.

Finally, in the third situation we introduce outliers in the x-direction. Eighty percent of the cases are again as in the first situation. In the remaining 20% the y_i are generated as before, but afterwards the $x_{i,1}$ are replaced by values that are now normally distributed with mean 100 and variance 100.

Figure 4 shows an example of each of these three configurations in the framework of simple regression through the origin, with $n = 50$.

The purpose of our simulation is to measure to what extent the estimates differ from the true values $\theta_1 = \cdots = \theta_p = \sigma = 1$. Some summary values over the $m = 200$ runs are computed, such as the *mean estimated value*

$$\bar{\theta}_j = \frac{1}{m} \sum_{k=1}^{m} \hat{\theta}_j^{(k)} \, , \tag{4.2}$$

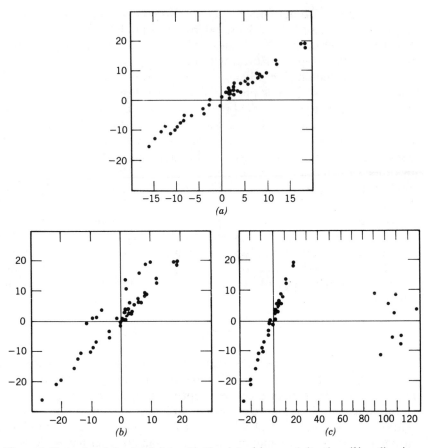

Figure 4. Examples of simulated data with 50 points: (*a*) normal situation, (*b*) outliers in *y*, and (*c*) outliers in *x*.

which yields the bias $\bar{\theta}_j - \theta_j$. Also, the *mean squared error*

$$\text{MSE}\,(\theta_j) = \frac{1}{m} \sum_{k=1}^{m} (\hat{\theta}_j^{(k)} - \theta_j)^2 \qquad (4.3)$$

is an important quantity, which can be decomposed into a sum of the squared bias and the variance:

$$\text{MSE}\,(\theta_j) = (\bar{\theta}_j - \theta_j)^2 + \frac{1}{m} \sum_{k=1}^{m} (\hat{\theta}_j^{(k)} - \bar{\theta}_j)^2 . \qquad (4.4)$$

The same summary statistics are also computed for the estimates of σ.

In our simulation we performed 200 replications for each value of p between 1 and 10; for n equal to 10, 20, 30, 40 and 50; and for all three sampling situations; this was repeated for models with and without intercept term. (This being only a modest study, we did not yet investigate the suitability of variance reduction techniques as described by Johnstone and Velleman 1985a.) Table 4 lists the results for regression with intercept term, for $n = 40$ and $p = 4$. (The tables for other values of n and p are all extremely similar, so we only display one of them to save space.)

In Table 4, we see that in the normal error situation the mean estimated values produced by the robust techniques are almost as good as the LS ones, whereas their MSE and variance do increase, but not dramatically. (The constant term produces the largest MSE and variance for all four techniques.) For all estimates, one also sees that the variance makes up most of the MSE, which means that the bias is negligible in this situation. To conclude, the LS provides the best results when the errors are normally distributed, but then the robust techniques also behave quite well.

However, the LS becomes very bad when there are outliers in y. In this situation the mean estimated intercept and scale differ considerably from the true value of 1, and they have a high MSE. This large MSE is mostly due to the bias, since the variances are rather small. This also means that the LS estimates behave in the same way in all 200 runs. In other words, the LS breaks down systematically at these contaminated samples. On the other hand, the robust regression estimates are scarcely altered when compared to the normal error situation. Only the scale estimate of the OSM is too large, which is because the outliers are not really rejected since the OSM scale is not defined by means of a redescending function. (Note that the LMS scale does contain a rejection step, given by (1.4) above.)

In the situation of outliers in x, we only have put contamination on $x_{i,1}$. The LS regressions are pulled in the direction of these leverage points, thereby affecting the mean estimated value of θ_1 and causing a large MSE. The LS estimates of intercept and scale also break down in a spectacular way. On the other hand, the robust regressions still yield good results for all parameters. (Note that the OSM again performs somewhat less than the RLS, especially for the scale parameter. This is one of the reasons why we have preferred to use the RLS in the previous chapters, some of the other reasons being intuitive appeal and ease of interpretation.)

It may be noted that the LMS columns in Table 4 are *exactly* equal in the situations of outliers in y and outliers in x, and the same holds for the RLS. This is not surprising since the samples in both situations share the

Table 4. Simulation Results of Regression with Intercept, for $n = 40$ and $p = 4$ (Including the Constant Term)[a]

Parameter	Normal				Outliers in y				Outliers in x			
	LS	LMS	RLS	OSM	LS	LMS	RLS	OSM	LS	LMS	RLS	OSM
θ_1	1.000365	1.001555	1.000796	1.000829	0.999226	0.999641	0.999238	0.999367	0.042766	0.999641	0.999238	0.999365
	0.000279	0.001350	0.000479	0.000619	0.004935	0.001262	0.000430	0.000556	0.918092	0.001262	0.000430	0.001224
	0.000279	0.001347	0.000479	0.000618	0.004934	0.001262	0.000429	0.000556	0.001794	0.001262	0.000429	0.001223
θ_2	1.000884	1.002744	1.001689	1.002117	1.004542	1.003353	1.001238	1.002902	1.006389	1.003353	1.001238	1.002395
	0.000268	0.001320	0.000481	0.000570	0.004563	0.001646	0.000533	0.000726	0.022725	0.001646	0.000533	0.000730
	0.000267	0.001313	0.000478	0.000565	0.004542	0.001635	0.000532	0.000717	0.022684	0.001635	0.000532	0.000724
θ_3	0.999125	0.996997	0.998000	0.997707	0.997524	0.997381	0.997542	0.997274	0.987313	0.997381	0.997542	0.996846
	0.000301	0.001598	0.000625	0.000785	0.005007	0.001500	0.000492	0.000694	0.031526	0.001500	0.000492	0.000731
	0.000300	0.001589	0.000621	0.000780	0.005001	0.001493	0.000486	0.000687	0.031365	0.001493	0.000486	0.000721
Constant	0.999892	1.007599	0.989971	0.990862	2.994885	0.981571	1.004995	0.996266	0.292098	0.981571	1.004995	0.997807
	0.023451	0.127050	0.034034	0.040130	4.041006	0.109773	0.033947	0.045180	3.089755	0.109773	0.033947	0.040871
	0.023451	0.126992	0.033934	0.040046	0.061441	0.109434	0.033922	0.045166	2.588630	0.109434	0.033922	0.040867
σ	0.993785	0.937518	0.861388	1.029884	4.156255	1.094557	0.937264	1.442410	9.680913	1.094557	0.937264	1.461946
	0.013573	0.043511	0.047492	0.038412	10.005019	0.047954	0.026198	0.258837	76.745529	0.047954	0.026198	0.280485
	0.013535	0.039607	0.028279	0.037519	0.043075	0.039012	0.022262	0.063110	1.387274	0.039012	0.022262	0.067090

[a]The entries in each cell are the mean estimated value, the mean squared error, and the variance over the 200 runs.

Table 5. Simulation Results of Regression Without Intercept, for $n = 50$ and $p = 8$[a]

Parameter	Normal				Outliers in y				Outliers in x			
	LS	LMS	RLS	OSM	LS	LMS	RLS	OSM	LS	LMS	RLS	OSM
θ_1	1.001943	1.002612	1.001572	1.002003	0.999899	1.000960	1.001128	1.001258	0.034819	1.000960	1.001182	1.000959
	0.000218	0.001252	0.000405	0.000531	0.004761	0.001377	0.000326	0.000519	0.932610	0.001377	0.000325	0.001322
	0.000214	0.001245	0.000403	0.000527	0.004761	0.001376	0.000325	0.000517	0.001035	0.001376	0.000324	0.001321
θ_2	0.999810	0.999127	0.999727	1.000177	1.002180	1.000647	1.000132	1.000959	1.005823	1.000647	1.000297	1.000575
	0.000287	0.001180	0.000445	0.000537	0.004794	0.001459	0.000457	0.000618	0.020716	0.001459	0.000448	0.000620
	0.000287	0.001179	0.000445	0.000537	0.004789	0.001458	0.000457	0.000617	0.020682	0.001458	0.000448	0.000619
θ_3	1.001059	1.000904	1.000546	1.000880	0.996228	1.004423	1.002371	1.001842	0.991331	1.004423	1.002275	1.003173
	0.000233	0.001217	0.000470	0.000608	0.004548	0.001153	0.000327	0.000449	0.026097	0.001153	0.000327	0.000454
	0.000232	0.001216	0.000469	0.000607	0.004534	0.001134	0.000321	0.000445	0.026022	0.001134	0.000322	0.000444
θ_4	1.000511	1.000979	1.000768	1.000216	1.001978	1.003547	1.001338	1.002251	1.002652	1.003547	1.001330	1.002426
	0.000254	0.001116	0.000419	0.000498	0.004536	0.001325	0.000334	0.000507	0.025608	0.001325	0.000334	0.000526
	0.000290	0.001115	0.000419	0.000498	0.004532	0.001312	0.000333	0.000502	0.025601	0.001312	0.000333	0.000520
θ_5	1.000265	1.001546	0.998959	0.998897	0.999070	1.001347	0.999485	1.000078	0.999605	1.001347	0.999607	0.999709
	0.000290	0.001185	0.000467	0.000640	0.006565	0.001339	0.000426	0.000645	0.024870	0.001339	0.000419	0.000641
	0.000290	0.001183	0.000466	0.000639	0.006564	0.001337	0.000426	0.000645	0.024870	0.001337	0.000419	0.000641
θ_6	0.999495	0.997944	0.998165	0.997982	1.001850	1.002854	0.999834	1.000624	0.990852	1.002854	0.999726	1.000547
	0.000257	0.001186	0.000456	0.000574	0.004388	0.001341	0.000406	0.000536	0.020759	0.001341	0.000401	0.000559
	0.000257	0.001182	0.000452	0.000570	0.004384	0.001333	0.000406	0.000536	0.020675	0.001333	0.000400	0.000558
θ_7	1.000390	1.000006	1.000231	1.000143	1.002043	0.998583	1.000299	0.999266	1.001794	0.998583	1.000298	0.999228
	0.000267	0.001237	0.000469	0.000603	0.006173	0.001446	0.000349	0.000557	0.022490	0.001446	0.000349	0.000585
	0.000267	0.001237	0.000469	0.000603	0.006168	0.001444	0.000349	0.000556	0.022487	0.001444	0.000349	0.000584
θ_8	1.001183	1.002922	1.002269	1.002989	1.007825	0.999898	1.001737	1.000080	0.989495	0.999898	1.001769	1.000586
	0.000285	0.001232	0.000474	0.000590	0.005454	0.001277	0.000423	0.000550	0.023626	0.001277	0.000424	0.000566
	0.000284	0.001223	0.000469	0.000581	0.005393	0.001277	0.000420	0.000550	0.023515	0.001277	0.000421	0.000566
σ	0.999105	1.030184	0.865524	0.920439	4.592490	1.270508	0.957984	1.302358	9.701845	1.268939	0.956662	1.324905
	0.013545	0.034845	0.039224	0.032424	12.959248	0.117720	0.022465	0.132850	76.718758	0.115622	0.021174	0.149173
	0.013544	0.033934	0.021140	0.026095	0.053265	0.044545	0.020700	0.041430	0.996646	0.043294	0.019296	0.043610

[a]The entries in each cell are the mean estimated value, the mean squared error, and the variance over the 200 runs.

213

80% of "good" observations, whereas the 20% of outliers are properly rejected by the LMS and the RLS. This is no longer true for the OSM, because there the outliers still have an (albeit very small) influence.

Table 5 contains the summary statistics for regression *without* intercept, with $n = 50$ and $p = 8$. (The results for other choices of n and p are very similar and hence are not listed here.) For the normal situation we arrive at the same conclusions as before, the robust techniques behaving almost as well as LS.

In the presence of outliers in the response variable, we find that the LS coefficients are still quite good. This is because the model now forces the fit to go through the origin, so the contamination (as in Figure 4b) cannot move up the fit as in a model with intercept. (Moreover, the contamination happens to be balanced about the origin, so the solution does not even tilt in this experiment.) However, the LS scale estimate clearly breaks down because all points are taken into account, including the bad ones. On the other hand, the robust estimators provide decent estimates of both the θ_j and σ. The simulation results also show that the RLS is really an improvement on the LMS as the mean squared errors are decreased even more, particularly that of the scale estimate. This is not true for the OSM, which again overestimates σ because it does not really reject the outliers.

The last columns of Table 5 deal with the outliers in $x_{i,1}$, which cause the LS estimate of θ_1 to break down. As was to be expected, also the corresponding scale estimate explodes. Fortunately, the robust techniques yield satisfactory estimates for all quantities. Moreover, the RLS again constitutes a slight improvement on the LMS results.

The overall conclusion of these simulations is that the LMS indeed achieves the goals for which it was constructed, because it gives reasonable results in the normal situation and is also able to withstand substantial amounts of outliers in x and in y.

EXERCISES AND PROBLEMS

Section 1

1. Illustrate that the LMS algorithm in PROGRESS is regression, scale, and affine equivariant by running one or more example(s) in which you transform the data as in (4.2), (4.3), and (4.4) of Section 4 in Chapter 3.
2. Derive the probability (1.2).

3. Write down formulas expressing the regression coefficients $\hat{\theta}_j$ in terms of the regression results on the standardized data. Why are these formulas the same for LS, LMS, and RLS regression?

Section 2

4. Why is it useful to look at more than one local minimum of the objective function (2.2)? Consider both computational and data-analytic aspects.

5. (Research problem) Is it true that one can do no better than $O(n^p)$ algorithms for computing affine equivariant high-breakdown regression estimators?

Section 3

6. In Section 3 the computation times of the LMS, the LTS, and S-estimators are discussed. What do you think would be the total cost of a combination of LMS with a subsequent one-step improvement (either a one-step RLS or a one-step M-estimator)? Indicate the computation time of such an estimator in Figure 3.

Section 4

7. (Research problem) Why does the behavior of (4.1) depend on n and p being even or odd?

CHAPTER 6

Outlier Diagnostics

1. INTRODUCTION

Outlier diagnostics are statistics that focus attention on observations having a large influence on the least squares (LS) estimator, which is known to be nonrobust. Several diagnostic measures have been designed to detect individual cases or groups of cases that may differ from the bulk of the data. The field of diagnostics consists of a combination of numerical and graphical tools.

Many diagnostics are based on the residuals resulting from LS. However, this starting point may lead to useless results because of the following reason. By definition, LS tries to avoid large residuals. Consequently, one outlying case may cause a poor fit for the majority of the data because the LS estimator tries to accommodate this case at the expense of the remaining observations. Therefore, an outlier may have a small LS residual, especially when it is a leverage point (see Figure 2 in Section 1 of Chapter 1). As a consequence, diagnostics based on LS residuals often fail to reveal such points.

Another class of diagnostics is based on the principle of deleting one case at a time. For example, denote by $\hat{\theta}(i)$ the estimate of θ computed from the sample without the ith case. Then the difference between $\hat{\theta}$ and $\hat{\theta}(i)$ gives the extent to which the presence of the ith case affects the regression coefficients. These are so-called *single-case diagnostics*, which are computed for each case i. It is possible to generalize this to *multiple-case diagnostics* for highlighting the simultaneous influence of several cases. Indeed, the deletion of one or more cases seems to be a logical way to examine their impact. However, it is not obvious at all which subset of cases should be deleted. It may happen that some points are jointly influential but the individual points are not! Moreover, the computations involved are often infeasible because of the large number of subsets that would have to be considered.

Some quantities that occur frequently in classical diagnostics are the diagonal elements of the LS projection matrix **H**. This matrix is well known under the name of *hat matrix*, because it puts a hat on the column vector $\mathbf{y} = (y_1, \ldots, y_n)^t$. This means that $\hat{\mathbf{y}} = \mathbf{H}\mathbf{y}$, where $\hat{\mathbf{y}}$ is the LS prediction for **y**. In fact, the diagonal elements of the hat matrix are often used as diagnostic tools in their own right. Section 2 is devoted to the hat matrix, because this matrix is a fundamental component of the classical diagnostics treated in Sections 3–5. We will discuss the advantages and disadvantages of these tools, which appear in statistical packages such as SPSS, BMDP, SAS, and others. In Section 6, we propose alternative diagnostics that avoid some shortcomings of the classical ones.

2. THE HAT MATRIX AND LS RESIDUALS

Consider the following situation. Each individual $i = 1, \ldots, n$ is observed on $p + 1$ variables (that is, p explanatory variables and one response variable). The n-by-1 vector of responses is denoted by $\mathbf{y} = (y_1, \ldots, y_n)^t$. The linear model states that

$$\mathbf{y} = \mathbf{X}\boldsymbol{\theta} + \mathbf{e}, \tag{2.1}$$

where **X** is the n-by-p matrix

$$\mathbf{X} = \begin{bmatrix} x_{11} & x_{12} & \cdots & x_{1p} \\ x_{21} & x_{22} & \cdots & x_{2p} \\ \vdots & \vdots & & \vdots \\ x_{n1} & x_{n2} & \cdots & x_{np} \end{bmatrix}, \tag{2.2}$$

$\boldsymbol{\theta}$ is the vector of unknown parameters, and **e** is the error vector, that is, $\mathbf{e} = (e_1, \ldots, e_n)^t$.

The hat matrix **H** mentioned above is defined by

$$\mathbf{H} = \mathbf{X}(\mathbf{X}^t\mathbf{X})^{-1}\mathbf{X}^t. \tag{2.3}$$

(We assume that $\mathbf{X}^t\mathbf{X}$ is invertible.) The n-by-n matrix **H** is called the hat matrix because it transforms the observed vector **y** into its LS estimate $\hat{\mathbf{y}} = \mathbf{H}\mathbf{y}$. It can easily be verified that **H** is idempotent ($\mathbf{H}\mathbf{H} = \mathbf{H}$) and symmetric ($\mathbf{H}^t = \mathbf{H}$). From (2.3), it follows that

$$\text{trace } \mathbf{H} = p \qquad \left(\text{i.e., } \sum_{i=1}^{n} h_{ii} = p\right) \tag{2.4}$$

because trace $\mathbf{H} = \text{trace } [\mathbf{X}^t \mathbf{X}(\mathbf{X}^t \mathbf{X})^{-1}] = \text{trace } \mathbf{I}_p = p$, where $\mathbf{I}_p$ is the p-by-p identity matrix. Moreover, the rank of $\mathbf{H}$ also equals p. The fact that $\mathbf{H}$ is idempotent and symmetric implies that

$$h_{ii} = (\mathbf{HH})_{ii} = \sum_{j=1}^{n} h_{ij} h_{ji} = \sum_{j=1}^{n} h_{ij} h_{ij},$$

hence

$$h_{ii} = \sum_{j=1}^{n} h_{ij}^2 \qquad \text{for all } i. \tag{2.5}$$

Furthermore, $\mathbf{H}$ is invariant under nonsingular linear transformations of the explanatory variables. Indeed, let $\mathbf{A}$ be a p-by-p matrix of full rank. If the explanatory variables are transformed as

$$\tilde{\mathbf{X}} = \mathbf{XA}, \tag{2.6}$$

it easily follows that

$$\tilde{\mathbf{H}} = \tilde{\mathbf{X}}(\tilde{\mathbf{X}}^t \tilde{\mathbf{X}})^{-1} \tilde{\mathbf{X}}^t = \mathbf{X}(\mathbf{X}^t \mathbf{X})^{-1} \mathbf{X}^t = \mathbf{H}. \tag{2.7}$$

In particular, it is possible to multiply all variables x_j by a constant a_j, corresponding to multiplication by the matrix

$$\mathbf{A} = \begin{bmatrix} a_1 & 0 & \cdots & 0 & 0 \\ 0 & a_2 & \cdots & 0 & 0 \\ \vdots & \vdots & & \vdots & \vdots \\ 0 & 0 & \cdots & 0 & a_p \end{bmatrix}, \tag{2.8}$$

without affecting $\mathbf{H}$. This is applied in the program PROGRESS in the case of regression through the origin (regression without a constant term), because there each explanatory variable x_j is standardized by dividing it by a robust measure of its spread.

In the case of regression *with* a constant term, the same matrix algebra can be used by means of an artificial explanatory variable x_p which is identical to 1. This means that

$$\mathbf{X} = \begin{bmatrix} x_{1,1} & \cdots & x_{1,p-1} & 1 \\ x_{2,1} & \cdots & x_{2,p-1} & 1 \\ \vdots & & \vdots & \vdots \\ x_{n,1} & \cdots & x_{n,p-1} & 1 \end{bmatrix} \tag{2.9}$$

and $\mathbf{H}$ is again defined by (2.3) with this $\mathbf{X}$. Then one can even transform the $p-1$ "real" explanatory variables by

$$x_j \rightarrow a_j x_j + b_j$$

without affecting $\mathbf{H}$, because this amounts to applying (2.6) and (2.7) with the matrix

$$\mathbf{A} = \begin{bmatrix} a_1 & 0 & \cdots & 0 & 0 \\ 0 & a_2 & \cdots & 0 & 0 \\ \vdots & \vdots & & \vdots & \vdots \\ 0 & 0 & \cdots & a_{p-1} & 0 \\ b_1 & b_2 & \cdots & b_{p-1} & 1 \end{bmatrix}. \tag{2.10}$$

Therefore, the standardization performed in the program PROGRESS (subtracting the median of the x_j and dividing by their spread) does not change $\mathbf{H}$. As a consequence, it was rather easy to compute $\mathbf{H}$ in PROGRESS. In order to obtain $\mathbf{H}$ in the output, one has to answer YES to the question

DO YOU WANT TO COMPUTE OUTLIER DIAGNOSTICS?
PLEASE ANSWER YES OR NO:

in the interactive input.

The hat matrix also plays a central role in the covariance matrices of $\hat{\mathbf{y}}$ and $\mathbf{r} = \mathbf{y} - \hat{\mathbf{y}}$ since

$$\text{cov}(\hat{\mathbf{y}}) = \sigma^2 \mathbf{H} \tag{2.11}$$

$$\text{cov}(\mathbf{r}) = \sigma^2 (\mathbf{I} - \mathbf{H}), \tag{2.12}$$

where $\mathbf{r} = (r_1, \ldots, r_n)^t$ is the n-by-1 vector of LS residuals.

An intuitive geometric argument reveals a link between the diagonal elements of $\mathbf{H}$ and the location of the points in the space of the explanatory variables. The set of p-dimensional points $\mathbf{x}$ (not necessarily observations) that satisfy

$$h_{\mathbf{x}} = \mathbf{x}(\mathbf{X}^t \mathbf{X})^{-1} \mathbf{x}^t \leq \max_i h_{ii} \tag{2.13}$$

determine an ellipsoid (see, e.g., Montgomery and Peck 1982, p. 143). This ellipsoid contains the smallest convex set enclosing the n observations. Consequently, one can say that the point $\mathbf{x}$ lies close to the bulk of the space formed by the explanatory variables if $h_{\mathbf{x}}$ is small.

The above properties of $\mathbf{H}$ are helpful in understanding and interpreting its elements h_{ij}. From $\hat{\mathbf{y}} = \mathbf{Hy}$ it follows that the element h_{ij} of $\mathbf{H}$ has a direct interpretation as the effect exerted by the jth observation on $\hat{y}_i$. Especially the diagonal elements h_{ii} are of interest, because they equal $\partial \hat{y}_i / \partial y_i$. Therefore, they measure the effect of the ith observation on its own prediction. A diagonal element h_{ii} of zero indicates a point with no influence on the fit. From (2.4) it is immediately clear that the average value of the h_{ii} is p/n. Moreover, (2.5) implies that $h_{ii} = h_{ii}^2 + \Sigma_{j \neq i} h_{ij}^2$, from which it follows that $0 \leq h_{ii} \leq 1$ for all i. These limits are useful for interpreting the size of the effect on the prediction but do not yet tell us when h_{ii} is "large." Most authors (Hoaglin and Welsch 1978, Henderson and Velleman 1981, Cook and Weisberg 1982, Hocking 1983, Paul 1983, Stevens 1984) determine potentially influential points by looking at the h_{ii} and paying particular attention to points for which $h_{ii} > 2p/n$ (some people also use the cut-off value $3p/n$).

From (2.5), it follows that $h_{ii} = 0$ implies $h_{ij} = 0$ for all j, and $h_{ii} = 1$ implies $h_{ij} = 0$ for all i except j. Both extremes can be interpreted in a data analytic way. A simple example of $h_{ii} = 0$ is provided by a point with all explanatory variables equal to zero in a regression model without intercept. The prediction $\hat{y}_i$ is then fixed at zero and is not affected by y_i or by any other y_j. On the other hand, when $h_{ii} = 1$ we have $\hat{y}_i$ equal to y_i and consequently $r_i = 0$. In other words, when $h_{ii} = 1$, the ith observation is fitted exactly by the regression hyperplane. When h_{ii} is large, that is, near 1, then (2.12) implies that the variance of the ith residual is almost zero. The main cause for this may be that the case in question has an unusually large influence on the LS regression coefficients. However, although the LS residual is near zero, one cannot identify that case as a bad or a good point.

Let us now investigate the diagnostic power of the h_{ii} on some examples. In Table 1, the diagonal elements of the hat matrix are printed for the Hertzsprung–Russell diagram data (see Table 3 of Chapter 2). From the LMS regression, we know that the outliers (the giant stars) correspond to the indices 11, 20, 30, and 34. In Table 1, the h_{ii} for these observations are larger than the cut-off value $2p/n$ (0.085 in this case). Indeed, these stars are the most outlying in x-space.

From the scatterplot of the telephone data (Figure 3 of Chapter 2), it was clear that the responses associated with the years 1964–1969 were contaminated. However, the diagonal elements of the hat matrix (shown

Table 1. Diagonal Elements of the Hat Matrix, Squared Mahalanobis Distance, and Standardized, Studentized, and Jackknifed LS Residuals for the Hertzsprung–Russell Diagram Data[a]

Index i	h_{ii} (0.085)	MD_i^2 (3.84)	r_i/s (2.50)	t_i (2.50)	$t(i)$ (2.50)
1	0.022	0.043	0.44	0.44	0.44
2	0.037	0.738	1.47	1.50	1.52
3	0.022	0.027	−0.18	−0.19	−0.18
4	0.037	0.738	1.47	1.50	1.52
5	0.021	0.001	0.30	0.31	0.31
6	0.027	0.271	0.90	0.91	0.91
7	0.078	2.592	−0.99	−1.03	−1.03
8	0.038	0.783	0.64	0.65	0.64
9	0.022	0.027	0.95	0.96	0.96
10	0.022	0.043	0.24	0.25	0.24
11	<u>0.195</u>	<u>7.994</u>	0.67	0.75	0.74
12	0.025	0.171	0.87	0.88	0.88
13	0.029	0.342	0.85	0.86	0.86
14	0.044	1.055	−1.92	−1.96	−2.03
15	0.021	0.005	−1.35	−1.36	−1.38
16	0.024	0.144	−0.69	−0.70	−0.69
17	0.023	0.084	−1.96	−1.98	−2.05
18	0.024	0.144	−1.40	−1.41	−1.43
19	0.023	0.084	−1.53	−1.55	−1.58
20	<u>0.195</u>	<u>7.994</u>	0.95	1.06	1.06
21	0.021	0.005	−1.13	−1.15	−1.15
22	0.021	0.005	−1.42	−1.43	−1.45
23	0.024	0.144	−0.97	−0.98	−0.98
24	0.030	0.383	−0.16	−0.16	−0.16
25	0.023	0.059	0.06	0.07	0.07
26	0.024	0.144	−0.54	−0.55	−0.55
27	0.021	0.005	−0.64	−0.65	−0.64
28	0.023	0.059	−0.15	−0.15	−0.15
29	0.023	0.094	−1.16	−1.17	−1.18
30	<u>0.196</u>	<u>8.031</u>	1.24	1.38	1.40
31	0.023	0.059	−1.00	−1.01	−1.01
32	0.037	0.738	0.34	0.34	0.34
33	0.026	0.232	0.47	0.48	0.47
34	<u>0.195</u>	<u>7.994</u>	1.66	1.85	1.91
35	0.023	0.084	−1.25	−1.26	−1.27
36	0.046	1.134	1.30	1.33	1.35
37	0.034	0.572	0.32	0.32	0.32
38	0.026	0.232	0.47	0.48	0.47
39	0.034	0.572	0.46	0.47	0.46
40	0.025	0.171	1.08	1.10	1.10
41	0.023	0.059	−0.64	−0.65	−0.65
42	0.026	0.232	0.19	0.19	0.19
43	0.031	0.427	0.72	0.73	0.73
44	0.026	0.232	0.68	0.69	0.69
45	0.036	0.681	1.11	1.13	1.13
46	0.026	0.232	0.04	0.05	0.05
47	0.024	0.144	−0.83	−0.84	−0.84

[a] The cut-off value for h_{ii} is $2p/n = 0.85$, and that for MD_i^2 is $\chi^2_{1.0.95} = 3.84$.

221

in Table 2) do not point to outlying observations. This example illustrates that outliers in the y-direction are totally neglected by the hat matrix, because $\mathbf{H}$ is only based on the $\mathbf{x}_i$. Further on, we will discuss classical diagnostics that were designed especially to detect such outlying observations.

An unfortunate property of the h_{ii} diagnostics is their vulnerability to the masking effect. To show this, let us consider the data set generated by Hawkins, Bradu, and Kass (1984), which was reproduced in Table 9 of Chapter 3. Since the cut-off value for the h_{ii} is 0.107, we would conclude from Table 3 that the cases with indices 12, 13, and 14 are suspect. However, from the construction of the data it is known that only the first

Table 2. Diagonal Elements of the Hat Matrix, Squared Mahalanobis Distance, and Standardized, Studentized, and Jackknifed LS Residuals for the Telephone Data[a]

Index i	Year x_i	h_{ii} (0.167)	MD_i^2 (3.84)	r_i/s (2.50)	t_i (2.50)	$t(i)$ (2.50)
1	50	0.157	2.653	0.22	0.24	0.24
2	51	0.138	2.216	0.14	0.15	0.14
3	52	0.120	1.802	0.05	0.05	0.05
4	53	0.105	1.457	−0.02	−0.02	−0.02
5	54	0.091	1.135	−0.10	−0.10	−0.10
6	55	0.078	0.836	−0.18	−0.18	−0.18
7	56	0.068	0.606	−0.25	−0.26	−0.26
8	57	0.059	0.399	−0.33	−0.34	−0.33
9	58	0.052	0.238	−0.39	−0.40	−0.39
10	59	0.047	0.123	−0.45	−0.46	−0.45
11	60	0.044	0.054	−0.51	−0.53	−0.52
12	61	0.042	0.008	−0.58	−0.59	−0.58
13	62	0.042	0.008	−0.65	−0.66	−0.65
14	63	0.044	0.054	−0.65	−0.66	−0.65
15	64	0.047	0.123	1.00	1.03	1.03
16	65	0.052	0.238	1.00	1.03	1.03
17	66	0.059	0.399	1.23	1.27	1.29
18	67	0.068	0.606	1.45	1.50	1.54
19	68	0.078	0.836	1.77	1.84	1.95
20	69	0.091	1.135	2.21	2.32	2.60
21	70	0.105	1.457	−0.89	−0.94	−0.93
22	71	0.120	1.802	−1.31	−1.40	−1.43
23	72	0.138	2.216	−1.35	−1.45	−1.49
24	73	0.157	2.653	−1.40	−1.53	−1.58

[a] The cut-off vaue for h_{ii} is $2p/n = 0.167$, and that for MD_i^2 is $\chi_{1,0.95}^2 = 3.84$.

Table 3. Diagonal Elements of the Hat Matrix, Squared Mahalanobis Distance, and Standardized, Studentized, and Jackknifed LS Residuals for the Hawkins–Bradu–Kass Data[a]

Index i	h_{ii} (0.107)	MD_i^2 (7.82)	r_i/s (2.50)	t_i (2.50)	$t(i)$ (2.50)
1	0.063	3.674	1.50	1.55	1.57
2	0.060	3.444	1.78	1.83	1.86
3	0.086	5.353	1.33	1.40	1.41
4	0.081	4.971	1.14	1.19	1.19
5	0.073	4.411	1.36	1.41	1.42
6	0.076	4.606	1.53	1.59	1.61
7	0.068	4.042	2.01	2.08	2.13
8	0.063	3.684	1.71	1.76	1.79
9	0.080	4.934	1.20	1.26	1.26
10	0.087	5.445	1.35	1.41	1.42
11	0.094	5.986	−3.48	−3.66	−4.03
12	0.144	9.662	−4.16	−4.50	−5.29
13	0.109	7.088	−2.72	−2.88	−3.04
14	0.564	40.725	−1.69	−2.56	−2.67

[a] Only the first 14 cases are listed. The cut-off value for h_{ii} is 0.107, and that for MD_i^2 is $\chi^2_{3,0.95} = 7.82$.

10 observations are leverage outliers, whereas the next four cases (11, 12, 13, and 14) are good leverage points. Therefore, the real outliers are masked by the effect of good leverage points. As far as the hat matrix is concerned, some leverage points may be masked by others.

In the case of regression with a constant, one also measures leverage by means of the *Mahalanobis distance*. Let us first split up the $\mathbf{x}_i$ into the essential part $\mathbf{v}_i$ and the last coordinate 1:

$$\mathbf{x}_i = (x_{i,1} \cdots x_{i,p-1} \ 1) = (\mathbf{v}_i \ 1),$$

where

$$\mathbf{v}_i = (x_{i,1} \cdots x_{i,p-1})$$

is a $(p-1)$-dimensional row vector. One computes the arithmetic mean $\bar{\mathbf{v}}$ and the covariance matrix $\mathbf{C}$ of these $\mathbf{v}_i$:

$$\bar{\mathbf{v}} = \frac{1}{n} \sum_{i=1}^{n} \mathbf{v}_i$$

$$C = \frac{1}{n-1} \sum_{i=1}^{n} (\mathbf{v}_i - \bar{\mathbf{v}})^t (\mathbf{v}_i - \bar{\mathbf{v}}) .$$

One then measures how far $\mathbf{v}_i$ is from $\bar{\mathbf{v}}$ in the metric defined by C, yielding

$$MD_i^2 = (\mathbf{v}_i - \bar{\mathbf{v}}) C^{-1} (\mathbf{v}_i - \bar{\mathbf{v}})^t , \tag{2.14}$$

which is called the squared Mahalanobis distance of the ith case. The purpose of these MD_i^2 is to point to observations for which the explanatory part lies far from that of the bulk of the data. For this reason, several packages (notably BMDP9R, SAS, and SPSS) compute the squared Mahalanobis distance. The values of MD_i^2 may be compared with quantiles of the chi-squared distribution with $p - 1$ degrees of freedom (in Tables 1–3, the MD_i^2 are compared with the 95% quantiles).

It turns out that there is a one-to-one relationship between the squared Mahalanobis distance and the diagonal elements of the hat matrix, given by

$$MD_i^2 = (n-1)\left[h_{ii} - \frac{1}{n} \right], \tag{2.15}$$

which provides another clarification of the meaning of the h_{ii}. To prove (2.15), we first note that we may subtract the average of each of the first $p - 1$ explanatory variables, because this changes neither the hat matrix nor the MD_i^2. Therefore, we may assume without loss of generality that $(1/n) \sum_{i=1}^{n} x_{ij} = 0$ for each variable $j = 1, \ldots, p - 1$, hence $\bar{\mathbf{v}} = \mathbf{0}$. Therefore,

$$\mathbf{X}'\mathbf{X} = \begin{bmatrix} x_{11} & \cdots & x_{i1} & \cdots & x_{n1} \\ \vdots & & \vdots & & \vdots \\ 1 & \cdots & 1 & \cdots & 1 \end{bmatrix} \begin{bmatrix} x_{11} & \cdots & 1 \\ \vdots & & \vdots \\ x_{i1} & \cdots & 1 \\ \vdots & & \vdots \\ x_{n1} & \cdots & 1 \end{bmatrix}$$

$$= \begin{bmatrix} \sum_{i=1}^{n} x_{i1}x_{i1} & \cdots & \sum_{i=1}^{n} x_{i1}x_{i,p-1} & \sum_{i=1}^{n} x_{i1} \\ \vdots & & \vdots & \vdots \\ \sum_{i=1}^{n} x_{i,p-1}x_{i1} & \cdots & \sum_{i=1}^{n} x_{i,p-1}x_{i,p-1} & \sum_{i=1}^{n} x_{i,p-1} \\ \sum_{i=1}^{n} x_{i1} & \cdots & \sum_{i=1}^{n} x_{i,p-1} & n \end{bmatrix}$$

$$= \begin{bmatrix} \boxed{(n-1)\mathbf{C}} & \begin{matrix} 0 \\ \vdots \\ 0 \end{matrix} \\ 0 \quad \cdots \quad 0 & n \end{bmatrix},$$

from which it follows that

$$h_{ii} = \mathbf{x}_i (\mathbf{X}'\mathbf{X})^{-1} \mathbf{x}_i^t$$

$$= (\mathbf{v}_i \ 1) \begin{bmatrix} (1/(n-1))\mathbf{C}^{-1} & 0 \\ 0 & 1/n \end{bmatrix} \begin{bmatrix} \mathbf{v}_i \\ 1 \end{bmatrix}$$

$$= \frac{1}{n-1} \mathrm{MD}_i^2 + \frac{1}{n} ,$$

which proves (2.15).

The hat matrix is useful when the data set contains only a single observation with outlying $\mathbf{x}_i$. However, when there are several such observations, it may easily happen that this does not show up in the corresponding h_{ii}. The reason for this is that (2.14) is based on a classical covariance matrix, which is not robust. Therefore, there may be multiple outliers in $\mathbf{x}$ which do not show up through the hat matrix. To conclude, we think that it is not appropriate to define "leverage" in terms of h_{ii}. Rather, one should say that leverage points are any outliers in $\mathbf{x}$-space (which may be more difficult to identify), because such points have a potentially large influence on the fit. In order to discover such leverage points, it is better to insert a robust covariance estimator in (2.14). More details on this can be found in Section 1 of Chapter 7 (two examples on the identification of leverage points are given in Subsection 1d of Chapter 7).

At any rate, examining the h_{ii} alone does not suffice when attempting to discover outliers in regression analysis, because they do not take y_i into account. For a long time, one has thought that the LS residuals possess all the complementary information. We will restrict our attention to the standardized, the studentized, and the jackknifed residuals. The packages SPSS and SAS provide these statistics in their regression programs. However, these packages do not use a uniform terminology.

In the LS part of the PROGRESS output, we computed the *standardized residuals* defined as

$$\frac{r_i}{s} , \tag{2.16}$$

where s^2 is given by

$$s^2 = \frac{1}{n-p} \sum_{j=1}^{n} r_j^2 \, . \tag{2.17}$$

When the measurement errors on y_i are independent and normally distributed with mean zero and standard deviation σ, it is known that s^2 is an unbiased estimator of σ^2.

Many authors (Cook 1977, Hoaglin and Welsch 1978, Draper and John 1981, Atkinson 1982, Cook and Weisberg 1982, Hocking 1983, Hocking and Pendleton 1983, Paul 1983, Stevens 1984) have taken (2.12) into account for the scaling of the residuals. They recommend the *studentized residuals*

$$t_i = \frac{r_i}{s\sqrt{1 - h_{ii}}} \, . \tag{2.18}$$

In the literature these are sometimes called "standardized residuals". Then, the term "studentized residual" is mostly applied to

$$t(i) = \frac{r_i}{s(i)\sqrt{1 - h_{ii}}} \, , \tag{2.19}$$

where $s(i)$ is the estimate of σ when the entire regression is run again without the ith case. We will call the $t(i)$ *jackknifed residuals*. Velleman and Welsch (1981) and Cook and Weisberg (1982) call t_i an "internally studentized residual" and $t(i)$ an "externally studentized residual", whereas Belsley et al. (1980) used the term "RSTUDENT" for $t(i)$. Atkinson (1983) referred to $t(i)$ as "cross-validatory residual".

We will now illustrate these diagnostics on the three examples for which we discussed the diagonal elements of the hat matrix earlier. Table 1 gives the squared Mahalanobis distance and the standardized, studentized, and jackknifed LS residuals for the Hertzsprung–Russell data. The squared Mahalanobis distance confirms the information that was already apparent from the h_{ii}. On the other hand, none of the residuals shows the presence of outliers, as they are all very small (much smaller than 2.5). This phenomenon is typical for data sets containing multiple outliers in the x-direction: neither the standardized residuals nor the transformed residuals detect the outliers because the latter have pulled the LS fit in their direction.

Let us now look at these residuals for the telephone data (Table 2), which is a sample containing outliers in the y-direction. Although at least six observations (the years 1964–1969) contained a wrongly recorded response variable, none of the residuals in Table 2 are really large. This is because s and $s(i)$ explode in the presence of outliers.

The effectiveness of diagnostics becomes much more important in multiple regression models. Therefore, let us investigate the squared Mahalanobis distance and the standardized, studentized, and jackknifed residuals for the Hawkins–Bradu–Kass data set. Table 3 is restricted to the first 14 observations because these are the interesting ones (all the other cases yield small values anyway). Because of (2.15), the observations with a high MD_i^2 also have a high h_{ii}. The three types of residuals are small everywhere, except for the observations 11, 12, and 13. Unfortunately, these observations should have small residuals because they were not generated as outliers. None of the table entries indicates the true outliers in this data set.

3. SINGLE-CASE DIAGNOSTICS

In order to assert the influence of the ith observation, it is useful to run the regression both with and without that observation. For this purpose, Cook's squared distance (Cook 1977) measures the change in the regression coefficients that would occur if a case was omitted. It is defined by

$$CD^2(i) = \frac{(\hat{\boldsymbol{\theta}} - \hat{\boldsymbol{\theta}}(i))'\mathbf{M}(\hat{\boldsymbol{\theta}} - \hat{\boldsymbol{\theta}}(i))}{c} , \qquad (3.1)$$

where $\hat{\boldsymbol{\theta}}$ is the LS estimate of $\boldsymbol{\theta}$, and $\hat{\boldsymbol{\theta}}(i)$ is the LS estimate of $\boldsymbol{\theta}$ on the data set without case i. Usually, one chooses $\mathbf{M} = \mathbf{X}'\mathbf{X}$ and $c = ps^2$. Alternative choices for $\mathbf{M}$ and c lead to other diagnostics. In words, $CD^2(i)$ is the squared standardized distance over which $\boldsymbol{\theta}$ moves when estimated without the ith case. A large value of $CD^2(i)$ indicates that the ith observation has a considerable influence on the determination of $\hat{\boldsymbol{\theta}}$. An approximate $(1 - \alpha) \times 100\%$ confidence ellipsoid for $\boldsymbol{\theta}$ centered at $\hat{\boldsymbol{\theta}}$ is given by the set of all $\boldsymbol{\theta}^*$ such that

$$\frac{(\boldsymbol{\theta}^* - \hat{\boldsymbol{\theta}})'(\mathbf{X}'\mathbf{X})(\boldsymbol{\theta}^* - \hat{\boldsymbol{\theta}})}{ps^2} \qquad (3.2)$$

is less than the $(1 - \alpha)$-quantile of an F-distribution with p and $n - p$ degrees of freedom. So, the removal of the ith data point, which yields $\hat{\boldsymbol{\theta}}(i)$, would lead to the shift of the LS estimate to the edge of an $(1 - \alpha) \times 100\%$ confidence region based on the complete data. However, $CD^2(i)$ is not really distributed as an F-statistic. The F-distribution has merely been introduced by analogy to the theory of a confidence region for the LS estimate of $\boldsymbol{\theta}$ because it allows one to assess the magnitude of

$CD^2(i)$. Most authors (Cook and Weisberg 1982, Montgomery and Peck 1982) indicate that a $CD^2(i)$ of about 1.0 would generally be considered large.

Making use of $\hat{\mathbf{y}} = \mathbf{X}\hat{\boldsymbol{\theta}}$, Cook's squared distance can also be written as

$$CD^2(i) = \frac{(\hat{\mathbf{y}} - \hat{\mathbf{y}}(i))'(\hat{\mathbf{y}} - \hat{\mathbf{y}}(i))}{ps^2} . \tag{3.3}$$

This equivalent form allows us to say that $CD^2(i)$ measures the effect of deleting the ith observation on the fitted value vector. Another revealing form is

$$CD^2(i) = \frac{1}{p} t_i^2 \frac{h_{ii}}{1 - h_{ii}} . \tag{3.4}$$

From this equation it follows immediately that $CD^2(i)$ depends on three relevant quantities, all of which are related to the full data set. These quantities are the number of coefficients p, the ith studentized residual, and the ratio of the variance of y_i [i.e., $\sigma^2 h_{ii}$, see (2.11)] to the variance of the ith residual [i.e., $\sigma^2(1 - h_{ii})$, see (2.12)]. Thus $CD^2(i)$ is made up of a component that reflects how well the model fits y_i and a component that measures how far $\mathbf{x}_i$ is from the rest of the $\mathbf{x}_j$. Both may contribute to a large value of $CD^2(i)$. Formula (3.4) is very similar to one of the diagnostics of Belsley et al. (1980), namely

$$DFFITS(i) = \frac{r_i}{s(i)} \frac{(h_{ii})^{1/2}}{1 - h_{ii}} \tag{3.5}$$

(the only essential difference is the use of $s(i)$ instead of s). In fact, $DFFITS(i)$ originates from the standardization of the ith component of $\hat{\mathbf{y}} - \hat{\mathbf{y}}(i)$, so this diagnostic measures the influence on the prediction when an observation is deleted. Observations for which DFFITS exceeds $2(p/n)^{1/2}$ should be scrutinized. In an analogous way, Belsley et al. (1980) defined a diagnostic based on the change in the jth regression coefficient, namely

$$DFBETAS_j(i) = \frac{c_{ji}}{[\Sigma_{k=1}^n c_{jk}^2]^{1/2}} \frac{r_i}{s(i)(1 - h_{ii})} , \tag{3.6}$$

where $\mathbf{C} = (\mathbf{X}'\mathbf{X})^{-1}\mathbf{X}'$ is sometimes called the catcher matrix. The cut-off value for DFBETAS is $2/\sqrt{n}$. (Polasek 1984 treated an extension to general error structures for the residuals.) Belsley et al. also examine some other quantities in order to investigate whether the conclusions of hypothesis testing could be affected by the deletion of a case. They consider for example the t-statistic (for testing whether the jth regression coefficient is significantly different from zero) and compute

$$\text{DFSTUD}_j(i) = \frac{\hat{\theta}_j}{s((\mathbf{X}'\mathbf{X})^{-1})_{jj}} - \frac{\hat{\theta}_j(i)}{s(i)([\mathbf{X}'(i)\mathbf{X}(i)]^{-1})_{jj}}. \quad (3.7)$$

This diagnostic measures the difference between the t-statistic corresponding to the whole data set and that corresponding to the data set without case i.

It is worth noting that all these diagnostics are characterized by some function of the LS residuals, the diagonal elements of the hat matrix, and/or elements of the catcher matrix. These building blocks contain two types of information: the LS residuals r_i are measures of poor fit, and the hat matrix and the catcher matrix reflect the remoteness of $\mathbf{x}_i$ in the space of the explanatory variables. Unfortunately, their interpretation is no longer reliable when the data contains more than one outlying case. These diagnostics are susceptible to the masking effect, because two or several outliers can act together in complicated ways to reinforce or to cancel each other's influence. The final result is that these single-case diagnostics often do not succeed in identifying the outliers.

Most standard statistical packages provide single-case diagnostics. From SPSS one can obtain Cook's squared distance, and SAS also yields DFFITS and DFBETAS. Velleman and Welsch (1981) presented a survey of the field and clarified some conflicting terminology. They gave formulas for the efficient computation of regression diagnostics and mentioned statistical programs from which some diagnostics can be obtained.

Let us now apply these single-case diagnostics to the data sets considered above. Cook's squared distance, DFFITS, and DFBETAS appear in Tables 4–6 for the Hertzsprung–Russell data, the telephone data, and Hawkins' data set. The cut-off value for each diagnostic is printed in parentheses at the top of the corresponding column. Values exceeding this cut-off are underlined.

For these data sets, we are well aware of the position of the outliers. This makes it easy to evaluate the information provided by these single-case diagnostics. For the Hertzsprung–Russell data, we see that the outliers are not noticed by $\text{CD}^2(i)$. The DFFITS and DFBETAS diagnostics for the cases 14, 20, 30, and 34 (also 11 for DFBETAS) are larger in absolute value than the cut-off value. Apart from 14, these cases were also considered as outlying by least median of squares (LMS) in Chapter 2.

The telephone data seem to be very tough for the single-case diagnostics. Cook's squared distance remains below the cut-off value 1.0

Table 4. Outlier Diagnostics (Including the Resistant Diagnostic RD_i, which will be explained in Section 6) for the Hertzsprung–Russell Diagram Data[a]

Index i	$CD^2(i)$ (1.000)	DFFITS (0.413)	DFBETAS (0.292) Intercept	DFBETAS (0.292) Slope	RD_i (2.500)
1	0.002	0.066	−0.009	0.014	0.841
2	0.044	0.300	−0.181	0.197	1.111
3	0.000	−0.027	−0.006	0.005	1.194
4	0.044	0.300	−0.181	0.197	1.111
5	0.001	0.045	0.005	−0.002	1.189
6	0.012	0.151	−0.061	0.071	0.772
7	0.044	−0.298	−0.264	0.254	<u>3.505</u>
8	0.008	0.128	−0.079	0.086	0.882
9	0.010	0.144	0.033	−0.024	1.885
10	0.001	0.037	−0.005	0.008	0.700
11	0.068	0.366	<u>0.353</u>	<u>−0.345</u>	<u>6.891</u>
12	0.010	0.141	−0.046	0.054	0.833
13	0.011	0.147	−0.066	0.075	0.769
14	0.089	<u>−0.437</u>	<u>−0.334</u>	<u>0.315</u>	2.290
15	0.020	−0.203	−0.027	0.014	1.422
16	0.006	−0.109	0.032	−0.039	0.977
17	0.046	−0.315	−0.109	0.089	1.819
18	0.025	−0.226	0.067	−0.081	1.421
19	0.028	−0.242	−0.083	0.068	1.553
20	0.137	<u>0.524</u>	<u>0.505</u>	<u>−0.495</u>	<u>7.163</u>
21	0.014	−0.170	−0.023	0.012	1.289
22	0.022	−0.215	−0.029	0.014	1.467
23	0.012	−0.155	0.046	−0.056	1.154
24	0.000	−0.028	0.013	−0.015	0.845
25	0.000	0.010	−0.002	0.002	0.599
26	0.004	−0.087	0.026	−0.031	0.888
27	0.005	−0.095	−0.013	0.006	0.979
28	0.000	−0.023	0.004	−0.005	0.650
29	0.016	−0.182	−0.065	0.054	1.320
30	0.233	<u>0.689</u>	<u>0.664</u>	<u>−0.651</u>	<u>7.364</u>
31	0.012	−0.154	0.026	−0.037	1.182
32	0.002	0.067	−0.041	0.044	1.000
33	0.003	0.078	−0.029	0.034	0.585
34	0.416	<u>0.939</u>	<u>0.905</u>	<u>−0.886</u>	<u>7.641</u>
35	0.019	−0.196	−0.067	0.055	1.376
36	0.043	0.295	−0.202	0.216	1.108
37	0.002	0.059	−0.033	0.036	0.806
38	0.003	0.078	−0.029	0.034	0.585
39	0.004	0.086	−0.048	0.052	0.725
40	0.015	0.176	−0.057	0.068	0.978
41	0.005	−0.098	0.017	−0.023	0.960
42	0.000	0.031	−0.012	0.014	0.516
43	0.008	0.129	−0.064	0.071	0.734
44	0.006	0.113	−0.043	0.050	0.682
45	0.024	0.219	−0.129	0.141	0.947
46	0.000	0.007	−0.003	0.003	0.522
47	0.009	−0.132	0.039	−0.047	1.066

[a] The cut-off value for DFFITS is $2(p/n)^{1/2} = 0.413$, and for DFBETAS it is $2/\sqrt{n} = 0.292$.

Table 5. Outlier Diagnostics (Including the Resistant Diagnostic RD$_i$) for the Telephone Data

Index	Year	CD$^2(i)$	DFFITS	DFBETAS (0.408)		RD$_i$
i	x_i	(1.000)	(0.577)	Intercept	Slope	(2.500)
1	50	0.005	0.101	0.092	−0.087	1.054
2	51	0.002	0.057	0.051	−0.048	0.946
3	52	0.000	0.018	0.016	−0.014	0.835
4	53	0.000	−0.008	−0.007	0.006	0.731
5	54	0.001	−0.032	−0.026	0.024	0.624
6	55	0.001	−0.052	−0.040	0.036	0.518
7	56	0.002	−0.069	−0.049	0.043	0.472
8	57	0.004	−0.083	−0.053	0.046	0.439
9	58	0.004	−0.092	−0.050	0.041	0.490
10	59	0.005	−0.101	−0.045	0.034	0.584
11	60	0.006	−0.111	−0.035	0.023	0.677
12	61	0.008	−0.122	−0.022	0.009	0.772
13	62	0.010	−0.137	−0.006	−0.010	0.867
14	63	0.010	−0.139	0.014	−0.030	1.566
15	64	0.026	0.229	−0.053	0.078	<u>34.566</u>
16	65	0.029	0.242	−0.085	0.109	<u>35.879</u>
17	66	0.051	0.324	−0.145	0.177	<u>41.632</u>
18	67	0.082	0.417	−0.221	0.259	<u>47.042</u>
19	68	0.144	0.570	−0.341	0.390	<u>54.502</u>
20	69	0.267	<u>0.821</u>	<u>−0.538</u>	<u>0.604</u>	<u>64.352</u>
21	70	0.051	−0.319	0.223	−0.247	<u>6.251</u>
22	71	0.134	−0.530	0.391	<u>−0.428</u>	1.754
23	72	0.169	<u>−0.597</u>	<u>0.458</u>	<u>−0.498</u>	1.835
24	73	0.217	<u>−0.681</u>	<u>0.541</u>	<u>−0.584</u>	1.924

everywhere. The contaminated years also go undiscovered by DFFITS and the DFBETAS. Worse still, these diagnostics point to "good" years as being outlying!

The Hawkins–Bradu–Kass data show that also multivariate outliers may be hard to find by means of single-case diagnostics (see Table 6). The cut-off value 1.0 for CD$^2(i)$ is only attained by observation 14, which is a good leverage point. Also DFFITS and DFBETAS do not succeed in separating the bad points from the good ones, whereas the LMS residuals did (see Section 3 of Chapter 3).

We will end this section with a discussion of some of the above diagnostics in connection with the data of Mickey et al. (1967) concerning the relation between age at first word and Gesell adaptive score. (We

Table 6. Outlier Diagnostics (Including the Resistant Diagnostic RD$_i$) for the First 14 Cases of the Hawkins–Bradu–Kass Data

Index i	CD$^2(i)$ (1.000)	DFFITS (0.462)	DFBETAS (0.231) Constant	θ_1	θ_2	θ_3	RD$_i$ (2.500)
1	0.040	0.407	−0.076	0.116	−0.062	0.039	12.999
2	0.053	0.470	0.006	−0.043	0.006	0.092	13.500
3	0.046	0.430	−0.038	0.080	−0.180	0.160	13.911
4	0.031	0.352	0.045	−0.085	−0.070	0.161	13.961
5	0.039	0.399	−0.007	−0.001	−0.090	0.135	13.982
6	0.052	0.459	−0.141	0.201	−0.050	−0.019	13.451
7	0.079	0.575	−0.156	0.189	0.027	−0.055	13.910
8	0.052	0.464	−0.034	0.060	−0.108	0.121	13.383
9	0.034	0.372	0.055	−0.086	−0.103	0.191	13.718
10	0.048	0.439	0.098	−0.126	−0.167	0.273	13.535
11	0.348	−1.300	−0.052	0.238	0.174	−0.491	11.730
12	0.851	−2.168	−0.024	−0.025	1.192	−1.262	12.004
13	0.254	−1.065	0.367	−0.257	−0.424	0.359	12.297
14	2.114	−3.030	0.559	0.337	−2.795	1.920	13.674

already treated this example briefly in Chapter 2.) Many articles dealing with diagnostics use this data set as an illustration, but the conclusions are not unanimous. Mickey et al. (1967) themselves indicated observation 19 as an outlier. Andrews and Pregibon (1978) stated that case 18 is most influential and that 19 is not an outlier. On the other hand, Draper and John (1981) maintained that case 19 is an outlier, and they called case 18 an influential point. Dempster and Gasko-Green (1981) found that the pair of points 2 and 18 only extend the domain of the model and that case 19 was an outlier that does not matter. Paul (1983) summarized his analysis by saying that 18 is a high leverage point and that 19 is not an outlier. The h_{ii}, LS residuals, CD$^2(i)$, DFFITS, DFBETAS, and the weights based on LMS are reproduced in Table 7.

The weights determined by LMS regression are 1 everywhere except for case 19. The diagnostics that are designed to detect leverage focus their attention on observation 18, because it lies farthest away in the factor space (look at h_{ii} and CD$^2(i)$). The standardized LS residual exceeds 2.5 (in absolute value) for observation 19 only. This is easy to understand. Since cases 2 and 18 are leverage points, they produce small LS residuals. Their joint position, isolated from the rest of the data, is the reason why the deletion of only one of them will not greatly affect the residuals. The single-case diagnostic DFFITS points at observation 18

Table 7. Diagnostics for the Data Set of Mickey et al. [h_{ii}; Standardized, Studentized, and Jackknifed LS Residuals; $CD^2(i)$; DFFITS; DFBETAS; weights based on LMS; and the resistant diagnostic RD_i.]

Index i	h_{ii} (0.190)	Based on LS						DFBETAS (0.436)		Robust	
		r_i/s (2.50)	t_i (2.50)	$t(i)$ (2.50)	$CD^2(i)$ (1.000)	DFFITS (0.617)		Intercept	Slope	w_i	RD_i (2.500)
1	0.048	0.18	0.19	0.18	0.001	0.041		0.017	0.003	1.0	0.786
2	0.155	-0.87	-0.94	-0.94	0.081	-0.403		0.189	-0.335	1.0	2.867
3	0.063	-1.42	-1.46	-1.51	0.072	-0.391		-0.331	0.192	1.0	1.419
4	0.071	-0.79	-0.82	-0.81	0.026	-0.224		-0.200	0.128	1.0	1.000
5	0.048	0.82	0.84	0.83	0.018	0.187		0.075	0.015	1.0	0.892
6	0.073	-0.03	-0.03	-0.03	0.000	-0.009		0.001	-0.005	1.0	1.688
7	0.058	0.31	0.32	0.31	0.003	0.077		0.005	0.033	1.0	1.288
8	0.057	0.23	0.24	0.23	0.002	0.056		0.044	-0.023	1.0	0.339
9	0.080	0.29	0.30	0.29	0.004	0.085		0.079	-0.054	1.0	0.618
10	0.073	0.60	0.63	0.62	0.015	0.173		-0.023	0.101	1.0	1.624
11	0.091	1.00	1.05	1.05	0.055	0.322		0.316	-0.229	1.0	0.887
12	0.071	-0.34	-0.35	-0.34	0.005	-0.094		-0.084	0.054	1.0	0.677
13	0.063	-1.42	-1.46	-1.51	0.072	-0.391		-0.331	0.192	1.0	1.419
14	0.057	-1.22	-1.26	-1.28	0.048	-0.314		-0.247	0.125	1.0	1.258
15	0.057	0.41	0.42	0.41	0.005	0.101		0.080	-0.041	1.0	0.387
16	0.063	0.13	0.13	0.13	0.001	0.033		0.028	-0.016	1.0	0.363
17	0.052	0.78	0.81	0.80	0.018	0.187		0.133	-0.055	1.0	0.534
18	0.652	-0.50	-0.85	-0.85	0.678	-1.156		0.831	-1.113	1.0	5.752
19	0.053	2.75	2.82	3.61	0.223	0.854		0.144	0.273	0.0	1.803
20	0.057	-1.04	-1.07	-1.08	0.035	-0.264		-0.208	0.105	1.0	1.129
21	0.063	0.13	0.13	0.13	0.001	0.033		0.028	-0.016	1.0	0.363

and, to a lesser extent, at 19. DFBETAS is larger than the cut-off value for case 18 only. For this sample, it is hard to draw an overall conclusion from the different diagnostics. This is because of the lack of observations between the pair of leverage points (2 and 18) and the remaining data points.

The present variety of single-case diagnostics appears rather overwhelming. Moreover, the single-case diagnostics, although simple from a computational point of view, often fail to reveal the impact of small groups of cases because the influence of one point could be masked by another.

4. MULTIPLE-CASE DIAGNOSTICS

Most of the single-case diagnostics of Section 3 can be extended in order to assess the changes caused by the deletion of more than one observation at the same time. These generalizations are called *multiple-case diagnostics*. Because of constraints of computation time, they are less popular than their single-case counterparts. The diagnostics described here are the most widely used.

Cook and Weisberg (1982) generalized formula (3.1) to measure the joint effect of deleting more than one case, yielding

$$CD^2(I) = \frac{(\hat{\boldsymbol{\theta}} - \hat{\boldsymbol{\theta}}(I))'M(\hat{\boldsymbol{\theta}} - \hat{\boldsymbol{\theta}}(I))}{c} , \qquad (4.1)$$

where I represents the indices corresponding to a subset of cases, and $\hat{\boldsymbol{\theta}}(I)$ is the LS estimate of $\boldsymbol{\theta}$ calculated without the cases indexed by I. Again, several possibilities for M and c can be used, such as $M = X'X$ and $c = ps^2$. The quantity $CD^2(I)$ can be interpreted in an analogous way as $CD^2(i)$. However, the selection of the cases to include in I is not obvious at all. It may happen that subsets of points are jointly influential but individual points are not. Therefore, single-case diagnostics do not indicate which subsets have to be considered. Moreover, the computation for all pairs, triples, and so on, leads to C_n^m runs, where $m = 1, \ldots, n/2$. This is usually infeasible. This fundamental problem is shared by all diagnostics that require consideration of all subsets of cases.

Andrews and Pregibon (1978) proposed the determinantal ratio

$$AP(I) = \frac{\det [Z^t(I)Z(I)]}{\det [Z'Z]} , \qquad (4.2)$$

where $\mathbf{Z}$ is the $\mathbf{X}$ matrix with the response variable $\mathbf{y}$ appended, that is,

$$\mathbf{Z} = \begin{bmatrix} x_{11} & \cdots & x_{1p} & y_1 \\ \vdots & & \vdots & \vdots \\ x_{n1} & \cdots & x_{np} & y_n \end{bmatrix}. \tag{4.3}$$

$\mathbf{Z}(I)$ is the part of $\mathbf{Z}$ obtained by deleting the rows corresponding to the indices in I. From (4.2) it is clear that $AP(I)$ is a dimensionless measure. Geometrically, one can interpret the quantity $AP(I)$ as a measure of the remoteness of the subset of cases with indices I. This interpretation is based on the fact that $1 - AP(I)^{1/2}$ corresponds to the relative change in the volume of an ellipsoid produced by $\mathbf{Z}'\mathbf{Z}$ when the cases corresponding to the index set I are omitted (Andrews and Pregibon 1978). Hence, it follows that small values of $AP(I)$ are associated with extreme cases, which may be influential. The conclusions drawn from Cook's squared distance and the Andrews–Pregibon statistic are not necessarily similar. Little (1985) decomposes $AP(I)$ into two terms, which shows a relation between the AP statistic and Cook's statistic. Both $CD^2(I)$ and $AP(I)$ are discussed in Draper and John (1981).

The extension of diagnostics to groups of cases appears to be conceptually straightforward. However, they involve prohibitive computations if all possible subsets of cases are considered. Some strategies exist for approaching the computational problems, but none of them guarantees to identify all the bad cases present in the data set.

5. RECENT DEVELOPMENTS

Recently, Gray and Ling (1984) proposed to perform a k-clustering on the modified hat matrix

$$\mathbf{H}^* = \mathbf{Z}(\mathbf{Z}'\mathbf{Z})^{-1}\mathbf{Z}' \tag{5.1}$$

(where $\mathbf{Z}$ is again the $\mathbf{X}$ matrix with the $\mathbf{y}$ vector appended), in order to identify subsets of cases which are similar with regard to the elements of $\mathbf{H}^*$. In particular, they consider the elements of $\mathbf{H}^*$ as similarity measures between the objects. Then they use the k-clustering algorithm in the hope of discovering a block-diagonal structure in $\mathbf{H}^*$. They perform the clustering for several values of the unknown parameter k. These clusterings yield (in a very ad hoc way) different candidate subsets that then have to be evaluated by one or more multiple-case diagnostics. The clustering results appear to be rather ambiguous, but in principle this approach avoids the treatment of all possible subsets of cases. However, the basic problem with the hat matrix remains: It breaks down easily in the presence of multiple outliers, after which it does not even contain the necessary information on the outliers anymore. Therefore, one cannot be sure that all bad subsets will be discovered.

Another step in the development of diagnostics is the use of the "elemental sets" that have been proposed by Hawkins et al. (1984). The underlying idea resembles the algorithm for the LMS. They also use random subsets J consisting of p observations. They call such a J an elemental set. The estimate determined by these p observations is called the "elemental regression" $\boldsymbol{\theta}_J$ and is equal to the "pseudo-observation" used by Oja and Niiminaa (see Section 6 of Chapter 3). Then they define the "elemental predicted residuals" $r_i(\boldsymbol{\theta}_J)$ resulting from the elemental regression $\boldsymbol{\theta}_J$ by means of

$$r_i(\boldsymbol{\theta}_J) = y_i - \mathbf{x}_i \boldsymbol{\theta}_J . \tag{5.2}$$

In order to summarize the information in all subsets, they look at the median of the elemental predicted residuals over all subsets J,

$$\tilde{r}_i = \operatorname*{med}_J r_i(\boldsymbol{\theta}_J) , \tag{5.3}$$

for each case i. An observation for which this statistic is large may be identified as outlying. Unfortunately, this technique does not have a very good breakdown point. Indeed, ε^* decreases fast with the dimension of the problem:

$$\varepsilon^* = 1 - (\tfrac{1}{2})^{1/p} . \tag{5.4}$$

This expression for the breakdown point is derived from the condition that the "good" subsets J have to be in the majority in order for the median in (5.3) to be meaningful. More formally, the probability of selecting a subset of p good points out of a sample that may contain a fraction ε of contamination must be greater than 50%. This yields

$$(1 - \varepsilon)^p \geq 50\% ,$$

which is equivalent to

$$\varepsilon \leq 1 - (\tfrac{1}{2})^{1/p} ,$$

giving the upper bound on the amount of contamination ε that may be tolerated. The same reasoning was already applied to the Theil–Sen estimator [formula (7.3) in Section 7 of Chapter 2] and its generalization due to Oja and Niinimaa [formula (6.6) in Section 6 of Chapter 3]. Table 21 in Section 6 of Chapter 3 lists the resulting breakdown point for some values of p.

Hawkins et al. (1984) tested the elemental set method on their artificial data set (which we discussed in Sections 2 and 3 of this chapter). The proportion of outliers in this data set is close to the breakdown point of the elemental set technique. Hawkins et al. applied four iterative algorithms, one of which uncovered the real outliers. On the other hand, we saw in Chapter 3 that the LMS easily finds them.

6. HIGH-BREAKDOWN DIAGNOSTICS

We need diagnostics that are able to cope with multiple outliers, without suffering from the masking effect. That is, we want diagnostics that have high breakdown. One way to formalize this objective would be to compute the breakdown point of the estimator defined by (a) deleting all observations for which the diagnostic exceeds its cutoff, and then (b) applying LS to the remainder. We shall look at two ways to construct such "sturdy" diagnostics.

a. Residuals from Robust Fits

As already indicated in Section 2, the residuals from an LS fit are not very useful as outlier diagnostics. On the other hand, the residuals computed from a very robust estimator embody powerful information for detecting all the outliers present in the data. In particular, one can look at the standardized LMS residuals r_i/σ^* (with σ^* as defined in Chapters 2 and 5), which are dimensionless. A plot of these standardized residuals versus the estimated value of y_i or versus the index i gives a visual portrayal of this quantitative information. Whatever the dimension of the original data set, these plots enable the data analyst to detect the bad points in a simple display. These tools (provided by PROGRESS) have been used extensively for the analysis of various examples in Chapters 2 and 3.

REMARK. While this book was already in press, a paper by Atkinson (1986) appeared in which the LMS residuals are used for the same purpose. Atkinson applies the LMS to "unmask" the outliers (i.e., to identify them), after which he carries out a second stage by means of LS diagnostics. This confirmatory stage is motivated by the fact that the LMS sometimes assigns a large residual to a good leverage point, as small variations in the parameter estimates may already have this effect. Atkinson also recommended the use of the LMS for identifying outliers in his invited comments on the discussion papers by Cook (1986) and Chatterjee and Hadi (1986). As a further extension, Atkinson (1987) uses LMS as a diagnostic method for inference about transformation of the response variable.

The standardized residuals resulting from a very robust fit such as LMS can be used to determine a weight w_i for the ith observation, by means of a nonincreasing function of the $|r_i/\sigma^*|$. In PROGRESS we use

$$w_i = \begin{cases} 1 & \text{if } |r_i/\sigma^*| \le 2.5 \\ 0 & \text{otherwise} \end{cases} \tag{6.1}$$

for the computation of the RLS estimates. This is a weight function of the "hard rejection" type. It produces a radical dichotomy between the accepted and the rejected points. It is also possible to use "smooth rejection" weight functions, such as (4.23) and (4.24) in Section 4 of Chapter 3. At any rate, the resulting w_i can also be considered as outlier diagnostics.

b. The Resistant Diagnostic

The resistant diagnostic RD_i proposed by Rousseeuw (1985) is aimed at identifying all points that are either outliers in the y-direction or leverage points (whether "good" or "bad"). The basic idea of this diagnostic is closely related to the definition of LMS regression. In our algorithm for the LMS, one *first minimizes* the median of the absolute residuals over the subsets J. This θ_J yields residuals $r_i(\theta_J)$. Then the standardized residuals are constructed by *dividing* the residuals $r_i(\theta_J)$ by

$$\operatorname*{med}_{j=1,\ldots,n} |r_j(\theta_J)| .$$

The resistant diagnostic essentially performs these operations in the inverse order. The "outlyingness" for case i is now defined by

$$u_i = \max_J \frac{|r_i(\theta_J)|}{\operatorname*{med}_{j=1,\ldots,n} |r_j(\theta_J)|} . \tag{6.2}$$

So, in (6.2) one *first divides* the residuals $r_i(\theta_J)$ by the median absolute residual. This ratio is then *maximized* over all subsets J. The maximum is taken over a large collection of random subsets J (in order to avoid complete enumeration). The notation is the same as that used in the "elemental sets" technique of Hawkins et al. (1984) in Section 5, but note that the median is now over j instead of J, which makes all the difference for the breakdown behavior. Also note that (6.2) is somewhat related to the outlyingness defined in a multivariate location context by Stahel (1981) and Donoho (1982).

As in the case of standardized LMS residuals, the idea is that observations with a large u_i are considered aberrant, or at least suspicious. But how can these u_i be computed? The algorithm designed for the LMS (see Chapter 5) permits us to calculate the values $|r_i(\boldsymbol{\theta}_J)| / \text{med}_j |r_j(\boldsymbol{\theta}_J)|$ at each replication. No excessive supplementary computations are needed. The only additional storage we need is an array of length n with the current u_i values. The preceding value for case i will be updated only if that of the current subset J is larger. Therefore, the u_i can be obtained as a relatively cheap by-product of the LMS algorithm.

The u_i defined in (6.2) is merely a computable variant of the theoretical formula

$$U_i = \sup_{\boldsymbol{\theta}} \frac{|r_i(\boldsymbol{\theta})|}{\underset{j=1,\ldots,n}{\text{med}} |r_j(\boldsymbol{\theta})|} , \qquad (6.3)$$

where the supremum is over all parameter vectors $\boldsymbol{\theta}$. In this general form, the U_i is clearly related to projection pursuit (Section 5 of Chapter 3), but hard to compute exactly. An obvious lower bound is

$$U_i \geq |r_i(\boldsymbol{\theta}_{\text{LMS}})| / \underset{j}{\text{med}} |r_j(\boldsymbol{\theta}_{\text{LMS}})| , \qquad (6.4)$$

where $\boldsymbol{\theta}_{\text{LMS}}$ minimizes $\text{med}_j |r_j(\boldsymbol{\theta})|$. On the other hand, if we restrict our attention to the computable version (6.2), we can still write

$$u_i \geq |r_i(\boldsymbol{\theta}_{\text{ALGO}})| / \underset{j}{\text{med}} |r_j(\boldsymbol{\theta}_{\text{ALGO}})| , \qquad (6.5)$$

where $\boldsymbol{\theta}_{\text{ALGO}}$ is the approximate LMS solution obtained by minimizing $\text{med}_j |r_j(\boldsymbol{\theta}_J)|$ over the same subsets J as those used for computing u_i. In either case, it follows that observations with large standardized LMS residuals will also possess large u_i or U_i.

Our experience with these u_i showed that they may become very large. (Theoretically, U_i may sometimes become infinite.) From an inferential point of view, a data point i might be identified as outlying if its u_i is greater than a critical value C_n, which could be derived from the distribution of the u_i under normality conditions. Unfortunately, this critical value is very hard to find and would depend on the distribution of the x_i also. Even in the simple case of one-dimensional location, the search for a number C_n such that

$$P[u_i \geq C_n] = \alpha \qquad (6.6)$$

(with α for example equal to 5%) looks difficult. On the other hand, a statistic for which one has no information about "how large is large" is not useful. A possible way to get more or less around this problem is to use a standardized version of the u_i. Therefore, we propose to define the *resistant diagnostic*

$$\mathrm{RD}_i = \frac{u_i}{\underset{j=1,\ldots,n}{\mathrm{med}}\ u_j} \qquad (6.7)$$

for each observation i. Of course, (6.7) can only be applied when the median of all the u_j is different from zero. However, this implies no real restriction because $\mathrm{med}_j\ u_j = 0$ happens only in the case of exact fit, when more than half of the observations lie on the same hyperplane (see Section 5 of Chapter 2). This event will be discovered by the robust regression algorithm. In that case, there is no need anymore for having diagnostics, because all the information is contained in the residuals resulting from the LMS solution.

From (6.7), one can conclude that 50% of the RD_i will be at most 1. Some preliminary experience leads us to say that the diagnostic measure RD_i is "large" if its value exceeds 2.5. Some simulation would be useful in order to obtain more detailed insight into the effects of the dimensionality p and the distribution of the x_i.

By construction, the diagnostic RD_i is not susceptible to masking. However, an observation that has a large RD_i is not necessarily outlying in the sense of producing a large residual. Indeed, RD_i will also be large in the case of leverage points that are only potentially influential. Therefore, the resistant diagnostic RD_i should be used and interpreted in conjunction with the standardized residuals resulting from a very robust fit. One can say that all points that have a large standardized residual also possess a large value for RD_i, but the reverse is not true. It may happen that one (or more) observations yield a resistant diagnostic which exceeds the cut-off value and simultaneously produce small standardized robust fit residuals. Such observations are precisely the harmless leverage points, which do not influence the fit. (To identify leverage points, we can use the technique described in Subsection 1d of Chapter 7.)

Furthermore, the resistant diagnostic can also be used for determining a weight for each observation by means of a weight function as in (6.1) but with $|r_i/\sigma^*|$ replaced by RD_i. These weights can then be used for performing a reweighted least squares (RLS). One is sure that at least half of the weights will be 1. In contrast with RLS with weights based on the LMS, this solution will always be unique. A situation where the problem is acute is sketched in Figure 1. The points are generated in the neighborhood of two different lines. Five points satisfy $y_i = x_i + 1 + e_i$, $e_i \sim N(0, 1)$ and the other five points satisfy $y_i = -x_i + 11 + e_i$, $e_i \sim N(0, 1)$. The RLS with weights based on the RD_i will yield an intermediate but unique solution, which almost bisects the two lines. On the other hand, the LMS will hesitate between both lines and will finally choose one of them. However, the other line also fits about half of the observations. This rather extreme situation in fact poses a new problem. The data set in Figure 1 is not well-fitted by *one* linear fit but rather by *two* lines. In this sense it has more to do with cluster analysis (in a framework of linear clusters) than with regression analysis. There exists some literature concerning this so-called two-line regression problem, namely Lenstra et al. (1982) and Kaufman et al. (1983). Because of such ambiguous situations, D. Hoaglin (personal communication) proposed to look also at the second best local minimum of $\text{med}_i|r_i(\boldsymbol{\theta})|$ when computing the LMS.

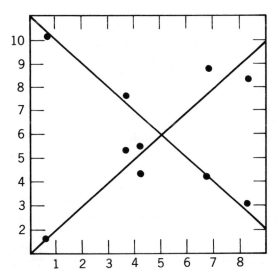

Figure 1. Data set illustrating the two-line regression problem.

The resistant diagnostic is just a by-product of our algorithm for robust regression, so it does not need additional computation time. Therefore, the RD_i is also incorporated in PROGRESS and can be obtained by answering yes to the question

```
DO YOU WANT TO COMPUTE OUTLIER DIAGNOSTICS?
PLEASE ANSWER YES OR NO:
```

in the interactive input.

The information provided by the resistant diagnostic will now be illustrated on some examples. The last column of Table 4 contains the RD_i values for the Hertzsprung–Russell data. We see that the large RD_i values are assigned to the giant stars, which also received a zero weight on the basis of the standardized LMS residuals.

The same observation can be made for the telephone data. In Table 5 we see that the magnitude of RD_i is strikingly large for the years 1964–1970.

Table 6 contains the resistant diagnostic values for the first 14 cases of the Hawkins–Bradu–Kass data. (The other cases all possess a small RD_i value.) It is not surprising that the 14 printed values are all rather large. The RD_i diagnostic measures the outlyingness in x-space as well as the deviation from the fit. Therefore, the good leverage points (11–14) also receive a large RD_i value, although they are not regression outliers. From the LMS standardized residuals we know that only the first 10 cases have to be considered as regression outliers.

The RD_i for the widely studied Mickey et al. data set are listed in Table 7. Our RD_i exceeds the cut-off value 2.50 for the cases 2 and 18. These observations possess small LMS residuals. From both diagnostics together we learn that 2 and 18 are *good* leverage points. Case 19 (which received a zero weight from LMS) also possesses a relatively large RD_i value (1.803) compared to the others.

Let us now briefly compare the diagnostics discussed in this chapter. It seems that an entirely real example with "messy" data might not be completely convincing because we might end up with different results without a conclusive way to decide which analysis is best, possibly causing some debates. Therefore, we start with a real data set that is rather well behaved and contaminate it by replacing a few observations. We selected a data set containing 20 points with six parameters to be estimated. The raw data came from Draper and Smith (1966, p. 227) and were used to determine the influence of anatomical factors on wood specific gravity, with five explanatory variables and an intercept. (Draper and Smith conclude that x_2 could be deleted from the model, but this matter is not

considered for the present purpose.) Table 8 lists a contaminated version
of these data, in which a few observations have been replaced by outliers.
A collection of results for these data is given in Table 9. Each column
shows a different diagnostic. Entries that exceed a cut-off value (possibly
in absolute value) are underlined. Applying LS regression to these data
yields

$$\hat{y} = 0.4407x_1 - 1.475x_2 - 0.2612x_3 + 0.0208x_4 + 0.1708x_5 + 0.4218 .$$

It is not easy to spot the outliers just by looking at the observations, and
the standardized LS residuals in Table 9 are of little help. Indeed, the
standardized LS residuals look very inconspicuous, except perhaps for
observation 11. (Unfortunately, this is a false trail!) Because of this,
many people would probably be satisfied with the LS fit (especially when
not expecting trouble). The studentized and jackknifed residuals in Table
9 (which are based on LS) also point to case 11. The diagonal elements of
the hat matrix are all smaller than the cut-off value $2p/n$ (which equals

Table 8. Modified Data on Wood Specific Gravity

Index	x_1	x_2	x_3	x_4	x_5	y
1	0.5730	0.1059	0.4650	0.5380	0.8410	0.5340
2	0.6510	0.1356	0.5270	0.5450	0.8870	0.5350
3	0.6060	0.1273	0.4940	0.5210	0.9200	0.5700
4	0.4370	0.1591	0.4460	0.4230	0.9920	0.4500
5	0.5470	0.1135	0.5310	0.5190	0.9150	0.5480
6	0.4440	0.1628	0.4290	0.4110	0.9840	0.4310
7	0.4890	0.1231	0.5620	0.4550	0.8240	0.4810
8	0.4130	0.1673	0.4180	0.4300	0.9780	0.4230
9	0.5360	0.1182	0.5920	0.4640	0.8540	0.4750
10	0.6850	0.1564	0.6310	0.5640	0.9140	0.4860
11	0.6640	0.1588	0.5060	0.4810	0.8670	0.5540
12	0.7030	0.1335	0.5190	0.4840	0.8120	0.5190
13	0.6530	0.1395	0.6250	0.5190	0.8920	0.4920
14	0.5860	0.1114	0.5050	0.5650	0.8890	0.5170
15	0.5340	0.1143	0.5210	0.5700	0.8890	0.5020
16	0.5230	0.1320	0.5050	0.6120	0.9190	0.5080
17	0.5800	0.1249	0.5460	0.6080	0.9540	0.5200
18	0.4480	0.1028	0.5220	0.5340	0.9180	0.5060
19	0.4170	0.1687	0.4050	0.4150	0.9810	0.4010
20	0.5280	0.1057	0.4240	0.5660	0.9090	0.5680

Source: Draper and Smith (1966) and Rousseeuw (1984).

Table 9. Diagnostics for the Data in Table 8 [h_{ii}; Squared Mahalanobis Distance; Standardized, Studentized, and Jackknifed LS Residuals; $CD^2(i)$, DFFITS, DFBETAS, Standardized LMS Residuals, and RD_i].[a]

| | | | | | | | | Based on LS | | | | | | | Robust | |
| | | | | | | | | DFBETAS (0.447) | | | | | | | | |
Index i	h_{ii} (0.600)	MD_i^2 (11.07)	r_i/s (2.50)	t_i (2.50)	$t(i)$ (2.50)	$CD^2(i)$ (1.000)	DFFITS (1.095)	Const.	θ_1	θ_2	θ_3	θ_4	θ_5	r_i/s (2.50)	RD_i (2.500)
1	0.278	4.327	-0.73	-0.85	-0.84	0.047	-0.524	-0.347	-0.004	0.055	0.328	-0.052	0.215	-0.16	0.798
2	0.132	1.552	0.05	0.05	0.05	0.000	0.019	-0.003	0.009	0.002	-0.005	0.002	0.000	0.00	0.701
3	0.220	3.224	1.24	1.41	1.46	0.093	0.776	-0.356	-0.651	-0.523	-0.206	-0.429	0.549	0.55	0.577
4	0.258	3.959	0.35	0.41	0.40	0.010	0.236	-0.074	0.035	-0.049	0.015	-0.105	0.118	-14.79	3.938
5	0.223	3.277	1.00	1.14	1.15	0.062	0.615	-0.244	0.286	-0.517	0.164	-0.388	0.437	1.75	0.605
6	0.259	3.974	-0.45	-0.53	-0.51	0.016	-0.302	0.050	-0.053	0.037	0.035	0.130	-0.113	-17.68	4.520
7	0.530	9.124	0.91	1.32	1.36	0.329	1.448	1.027	-0.956	0.424	0.521	0.133	-0.964	0.73	1.421
8	0.289	4.536	-0.03	-0.04	-0.04	0.000	-0.025	-0.005	0.011	-0.012	0.005	-0.005	0.006	-17.31	4.466
9	0.348	5.665	-0.40	-0.49	-0.48	0.021	-0.348	-0.075	0.052	0.105	-0.224	0.161	0.007	-0.73	1.243
10	0.449	7.588	-0.42	-0.56	-0.55	0.043	-0.492	0.257	-0.008	-0.198	-0.256	-0.137	-0.029	-0.40	1.267
11	0.317	5.075	1.99	2.40	3.02	0.447	2.059	0.521	0.425	0.970	0.748	0.198	-0.800	0.00	1.258
12	0.410	6.833	-1.20	-1.56	-1.65	0.281	-1.376	-0.566	-0.597	0.013	0.556	0.359	0.368	-1.88	1.030
13	0.287	4.506	-0.49	-0.58	-0.56	0.022	-0.356	0.180	-0.098	0.045	-0.251	0.106	-0.121	0.00	1.015
14	0.129	1.500	-1.26	-1.35	-1.40	0.045	-0.537	0.021	-0.169	0.228	0.178	-0.006	-0.103	-1.30	0.668
15	0.152	1.945	-0.59	-0.64	-0.62	0.012	-0.264	-0.073	0.148	-0.061	-0.011	-0.162	0.108	-0.34	0.465
16	0.526	9.049	0.52	0.76	0.75	0.107	0.789	0.122	-0.529	0.559	-0.052	0.745	-0.432	0.00	0.865
17	0.289	4.548	-0.25	-0.30	-0.29	0.006	-0.187	0.133	-0.019	0.019	-0.044	-0.055	-0.086	0.00	0.802
18	0.294	4.637	0.28	0.34	0.33	0.008	0.211	-0.002	-0.062	-0.096	0.081	-0.024	0.045	-0.21	0.985
19	0.292	4.599	-1.08	-1.29	-1.32	0.114	-0.849	-0.128	0.195	-0.287	0.231	-0.024	0.079	-20.84	5.201
20	0.318	5.084	0.55	0.66	0.65	0.034	0.441	0.064	0.092	-0.154	-0.305	0.037	0.046	0.00	0.816

[a]The cut-off value for h_{ii} is $2p/n = 0.600$, and for MD_i^2 it is $\chi^2_{5,0.95} = 11.07$.

0.6). Also, $CD^2(i)$ is less than 1 everywhere. The single-case diagnostics DFFITS, and in part also DFBETAS, point to observations 7, 11, and 12. The cases 3, 5, and 16 also show some values of DFBETAS which are larger than $2/\sqrt{n}$ (which equals 0.447). However, none of these diagnostics have succeeded in identifying the actual contaminated cases. Let us now examine the column with the standardized residuals associated with the LMS fit

$$\hat{y} = 0.2687x_1 - 0.2381x_2 - 0.5357x_3 - 0.2937x_4 + 0.4510x_5 + 0.4347 .$$

These standardized residuals make it easy to spot the four outliers. Indeed, the cases with indices 4, 6, 8, and 19 are the contaminated ones. The resistant diagnostic confirms the conclusions drawn from the standardized LMS residuals. In this example, the outliers are bad leverage points that are not outlying in any of the individual variables, as in Figure 4 of Section 1 in Chapter 1. Note that neither the hat matrix nor any other classical diagnostic is able to detect this fact, because they are susceptible to the masking effect (which is just another way of saying that they have a low breakdown point).

Our last example is the stackloss data set (Table 1 of Chapter 3). From Table 10, we see that this is relatively easy, because the different diagnostics lead to more or less the same conclusions. The LS residuals are largest (but not very large) for observations 1, 3, 4, and 21. The DFFITS and DFBETAS diagnostics exceed their cut-off values for case 21. The DFFITS value approaches the cut-off for cases 1, 3, and 4. The 17th diagonal element of the hat matrix slightly surpasses $2p/n = 0.381$, so it may be considered a good leverage point. The outlying cases 1, 3, 4, and 21 are detected by both the LMS residuals and the resistant diagnostic RD_i.

EXERCISES AND PROBLEMS

Section 2

1. Show that the hat matrix $\mathbf{H}$ defined in (2.3) is idempotent ($\mathbf{HH} = \mathbf{H}$) and symmetric ($\mathbf{H}' = \mathbf{H}$).

Section 3

2. Prove the alternative formulas (3.3) and (3.4) of Cook's squared distance.

Table 10. Diagnostics Associated with the Stackloss Data [h_{ii}; Squared Mahalanobis Distance; Standardized, Studentized, and Jackknifed LS Residuals; $CD^2(i)$; DFFITS; DFBETAS; Standardized LMS Residuals; RD_i]

Index i	h_{ii} (0.381)	MD_i^2 (7.82)	Based on LS									Robust	
			r_i/s (2.50)	t_i (2.50)	$t(i)$ (2.50)	$CD^2(i)$ (1.000)	DFFITS (0.873)	Const.	θ_1	θ_2	θ_3	r_i/s (2.50)	RD_i (2.500)
1	0.302	5.079	1.00	1.19	1.21	0.154	0.795	−0.085	0.400	0.103	−0.210	7.70	3.289
2	0.318	5.404	−0.59	−0.72	−0.71	0.060	−0.481	0.013	−0.250	−0.061	0.165	3.74	1.596
3	0.175	2.540	1.40	1.55	1.62	0.126	0.744	−0.188	0.391	−0.005	−0.047	7.14	3.047
4	0.129	1.618	1.76	1.88	2.05	0.131	0.788	−0.122	−0.415	0.619	0.027	7.64	3.265
5	0.052	0.092	−0.53	−0.54	−0.53	0.004	−0.125	0.012	0.012	−0.030	−0.007	0.28	0.617
6	0.078	0.597	−0.93	−0.97	−0.96	0.020	−0.279	0.039	0.106	−0.169	−0.012	0.00	0.821
7	0.219	3.432	−0.74	−0.83	−0.83	0.049	−0.438	0.296	0.268	−0.263	−0.282	0.51	1.000
8	0.219	3.432	−0.43	−0.49	−0.47	0.017	−0.251	0.169	0.154	−0.151	−0.162	1.30	0.892
9	0.140	1.851	−0.97	−1.05	−1.05	0.045	−0.423	0.077	0.307	−0.328	−0.078	−0.11	0.860
10	0.200	3.049	0.39	0.44	0.43	0.012	0.213	0.153	0.124	−0.132	−0.128	0.51	1.275
11	0.155	2.148	0.81	0.88	0.88	0.036	0.376	−0.068	0.108	−0.265	0.158	0.51	0.842
12	0.217	3.391	0.86	0.97	0.97	0.065	0.509	−0.011	0.233	−0.427	0.128	0.00	1.044
13	0.158	2.198	−0.44	−0.48	−0.47	0.011	−0.203	−0.119	−0.116	0.141	0.088	−1.87	1.018
14	0.206	3.164	−0.02	−0.02	−0.02	0.000	−0.009	0.005	0.001	0.003	−0.007	−1.36	0.902
15	0.191	2.857	0.73	0.81	0.80	0.039	0.388	−0.116	−0.195	−0.029	0.247	0.28	0.990
16	0.131	1.669	0.28	0.30	0.29	0.003	0.113	0.000	−0.053	−0.010	0.042	−0.51	0.700
17	0.412	7.290	−0.47	−0.61	−0.60	0.065	−0.502	−0.462	0.020	−0.063	0.424	0.00	2.137
18	0.161	2.260	−0.14	−0.15	−0.15	0.001	−0.065	−0.047	0.023	−0.013	0.033	0.00	1.146
19	0.175	2.538	−0.18	−0.20	−0.20	0.002	−0.091	−0.049	0.052	−0.042	0.034	0.51	0.941
20	0.080	0.651	0.44	0.45	0.44	0.004	0.131	0.085	−0.011	0.005	−0.067	1.87	0.806
21	0.285	4.738	−2.23	−2.64	−3.33	0.692	−2.100	0.402	−1.624	1.642	−0.363	−6.06	2.854

3. How many outliers does it take to make the diagnostics of Sections 2 and 3 unreliable? Consider the following procedure: First delete all points for which a classical diagnostic exceeds its cut-off, then apply LS on the remaining data. What is the breakdown point of this two-stage estimator?

4. There are basically three kinds of outlier diagnostics: those that aim at identifying outliers in y, those that try to find leverage points (whether they are good or bad), and the combined diagnostics that look for both leverage points and outliers in y. In which category are the h_{ii}, r_i/s, MD_i^2, $CD^2(i)$, and DFFITS?

Section 6

5. Try to find the leverage points of the data in Table 8 by making all scatterplots between any two explanatory variables. Do you think it is always possible to discover the leverage points in this way?

6. Consider again the aircraft data listed in Table 22 of Chapter 3. Run PROGRESS with the diagnostics option, and discuss the h_{ii} and RD_i together with the residuals of LS, LMS, and RLS to distinguish between good and bad leverage points. Compare these results with the analysis of Gray (1985), who uses diagnostic graphics, and with any other diagnostics that may be available in your local statistical software.

7. Analyze the delivery time data (exercise 2 of Chapter 3) by running PROGRESS with the diagnostics option. For which cases is $h_{ii} > 2p/n$? Explain the masking effect by comparing with the h_{ii} in the RLS output, where the bad leverage point has been deleted. Also discuss the robustly standardized residuals and the resistant diagnostic RD_i.

8. Use the h_{ii} to discover the leverage point in the phosphorus content data (Table 24 of Chapter 3). Should you look at the LS or the LMS residuals to find out whether it is a good or a bad leverage point? Also discuss the resistant diagnostic RD_i in this example, as well as any other diagnostics provided by your local software.

CHAPTER 7

Related Statistical Techniques

The preceding chapters all dealt with high-breakdown methods in regression analysis (including the special case of one-dimensional location). However, the same ideas can also be applied to other statistical procedures. In the first section of this chapter we shall focus our attention on multivariate location and covariance, a topic which is itself a key to various statistical techniques. Section 2 is about robust time series analysis, and Section 3 briefly discusses the merits of robustification in other situations.

1. ROBUST ESTIMATION OF MULTIVARIATE LOCATION AND COVARIANCE MATRICES, INCLUDING THE DETECTION OF LEVERAGE POINTS

Outliers are much harder to identify in multivariate data clouds than in the univariate case. Therefore, the construction of robust techniques becomes more difficult. We shall focus on the estimation of the "center" of a point cloud, in which all variables are treated in the same way (unlike regression analysis, where one tries to "explain" one variable by means of the remaining ones). We are also interested in the dispersion of the data about this "center."

Suppose we have a data set

$$X = \{\mathbf{x}_1, \ldots, \mathbf{x}_n\}$$
$$= \{(x_{11}, x_{12}, \ldots, x_{1p}), \ldots, (x_{n1}, x_{n2}, \ldots, x_{np})\} \qquad (1.1)$$

of n points in p dimensions, and we want to estimate its "center." (We have decided to denote the cases by *rows*, to keep the notation consistent

248

with previous chapters.) For this purpose, we apply a multivariate location estimator, that is, a statistic T which is *translation equivariant*,

$$T(\mathbf{x}_1 + \mathbf{b}, \ldots, \mathbf{x}_n + \mathbf{b}) = T(\mathbf{x}_1, \ldots, \mathbf{x}_n) + \mathbf{b} , \qquad (1.2)$$

for any p-dimensional vector $\mathbf{b}$. This property is also referred to as *location equivariance*. Naturally, T also has to be *permutation invariant*,

$$T(\mathbf{x}_{\pi(1)}, \ldots, \mathbf{x}_{\pi(n)}) = T(\mathbf{x}_1, \ldots, \mathbf{x}_n) , \qquad (1.3)$$

for any permutation π on $\{1, 2, \ldots, n\}$.

Of course, not every such T will be useful, and additional conditions may be required, depending on the situation. The most well-known estimator of multivariate location is the arithmetic mean

$$T(X) = \bar{\mathbf{x}} = \frac{1}{n} \sum_{i=1}^{n} \mathbf{x}_i , \qquad (1.4)$$

which is the least squares estimator in this framework because it minimizes $\sum_{i=1}^{n} \|\mathbf{x}_i - T\|^2$, where $\| \cdots \|$ is the ordinary Euclidean norm. However, it is well known that $\bar{\mathbf{x}}$ is not robust, because even a single (very bad) outlier in the sample can move $\bar{\mathbf{x}}$ arbitrarily far away. To quantify such effects, we slightly adapt the finite-sample breakdown point of Section 2 in Chapter 1 to the framework of multivariate location. We consider all corrupted samples X' obtained by replacing any m of the original data points by arbitrary values, and we define the maximal bias by

$$\text{bias } (m; T, X) = \sup_{X'} \| T(X') - T(X) \| \qquad (1.5)$$

so the breakdown point

$$\varepsilon_n^*(T, X) = \min \{ m/n; \text{bias } (m; T, X) \text{ is infinite} \} \qquad (1.6)$$

is again the smallest fraction of contamination that can cause T to take on values arbitrarily far away. Obviously, the multivariate arithmetic mean possesses a breakdown point of $1/n$. (Therefore, it is not well suited for the detection of outliers, as will be seen in Subsection d below.) We often consider the limiting breakdown point for $n \to \infty$, so we say that the multivariate mean has 0% breakdown.

It is clear that no translation equivariant T can have a breakdown point larger than 50%, because one could build a configuration of outliers which is just a translation image of the "good" data points, making it impossible for T to choose. In one dimension, this upper bound of 50% can easily be attained, for instance, by the sample median. Therefore, several multivariate generalizations of the median have been constructed, as well as some other proposals to achieve a certain amount of robustness.

Before listing some of these robust alternatives, we shall distinguish between two classes of location estimators: those that are affine equivariant and those that are not. Indeed, in many situations one wants the estimation to commute with linear transformations (i.e., a reparametrization of the space of the x_i should not change the estimate). We say that T is *affine equivariant* if and only if

$$T(\mathbf{x}_1\mathbf{A} + \mathbf{b}, \dots, \mathbf{x}_n\mathbf{A} + \mathbf{b}) = T(\mathbf{x}_1, \dots, \mathbf{x}_n)\mathbf{A} + \mathbf{b} \qquad (1.7)$$

for any vector $\mathbf{b}$ and any nonsingular matrix $\mathbf{A}$. (Because $\mathbf{x}_i$ and T are denoted by rows, the matrix $\mathbf{A}$ has to stand on the right.) For instance, the arithmetic mean does satisfy this property, but not all robust estimators do. We shall first consider some nonequivariant proposals.

a. Robust Estimators That Are Not Affine Equivariant

The simplest idea is to consider each variable separately. Indeed, for each variable j the numbers $x_{1j}, x_{2j}, \dots, x_{nj}$ can be considered as a one-dimensional data set with n points. One may therefore apply a univariate robust estimator to each such "sample" and combine the results into a p-dimensional estimate. This procedure inherits the breakdown point of the original estimator.

For instance, the *coordinatewise median* is defined as

$$(\operatorname*{med}_i x_{i1}, \operatorname*{med}_i x_{i2}, \dots, \operatorname*{med}_i x_{ip}) \qquad (1.8)$$

and possesses a 50% breakdown point. This estimator is easily computed, but fails to satisfy some "natural" properties. For instance, it does not have to lie in the convex hull of the sample when $p \geq 3$. As an example, consider the p unit vectors $(1, 0, \dots, 0)$, $(0, 1, \dots, 0), \dots,$ $(0, 0, \dots, 1)$, the convex hull of which is a simplex not containing the coordinatewise median $(0, 0, \dots, 0)$. (However, it does lie in the convex hull when $p \leq 2$, as can be shown by a geometrical argument.)

Nath (1971) proposed another coordinatewise technique, in which each component is investigated separately, and a certain fraction of the largest and the smallest observations are not used. This method is very simple, but many outliers may go unnoticed because multivariate outliers do not necessarily stick out in any of their components (see Figure 4 of Chapter 1).

Note that the multivariate arithmetic mean, although affine equivariant, can also be computed coordinatewise. However, this is an exception. Indeed, Donoho (1982, Proposition 4.6) showed that the only (measurable) location estimator that is both affine equivariant and computable as a vector of one-dimensional location estimators is the arithmetic mean.

Sometimes one does not wish equivariance with respect to all affine transformations, but only for those that preserve Euclidean distances, that is, transformations of the type $\mathbf{x} \to \mathbf{x}\Gamma + \mathbf{b}$ where Γ is an orthogonal matrix (this means that $\Gamma^t = \Gamma^{-1}$). This includes translations, rotations, and reflections. For instance, the L_1 location estimator, given as the solution T of

$$\underset{T}{\text{Minimize}} \sum_{i=1}^{n} \|\mathbf{x}_i - T\|$$ (1.9)

is orthogonal equivariant, because the objective function only depends on Euclidean distances. In operations research, this estimate is called the Weber point, and it corresponds to the optimal location to build a factory when the customers sit at the $\mathbf{x}_i$ and the total transportation cost is to be minimized. Several routines are available to compute T. The L_1 estimator is also a generalization of the univariate median, and its breakdown point is still 50%. (Some people call it the "spatial median" or the "median center.")

Orthogonal equivariant estimators are often used to estimate the center of a spherically symmetric density

$$f_\mu(\mathbf{x}) = g(\|\mathbf{x} - \boldsymbol{\mu}\|) .$$ (1.10)

This symmetry assumption may be reasonable when the data are points in some physical space (such as the customers in industrial location, or stars in three-dimensional space), but one should be careful when applying such procedures to variables of different types, in which case the choice of measurement units becomes important. For instance, when the variables are height and time, it makes a lot of difference whether these are expressed in feet and minutes or in centimeters and hours. (In such situations, it may be safer to apply either coordinatewise techniques or affine equivariant ones.)

When the spherically symmetric density is Gaussian, the maximum likelihood estimator becomes the arithmetic mean. This suggests a (modest) generalization of M-estimators, given by

$$\underset{T}{\text{Minimize}} \sum_{i=1}^{n} \rho(\|\mathbf{x}_i - T\|) . \qquad (1.11)$$

Huber (1967) considered the asymptotic behavior of these estimators, and Collins (1982) obtained some results on minimax variance.

b. Affine Equivariant Estimators

Let us now consider estimators that do satisfy the affine equivariance condition. Such estimators are particularly useful in a model with so-called *elliptical symmetry*, with density

$$|\det (\mathbf{A})|^{-1} g(\|(\mathbf{x} - \boldsymbol{\mu})\mathbf{A}^{-1}\|) . \qquad (1.12)$$

This model is obtained when starting from a spherically symmetric density, to which an affine transformation is applied. Unfortunately, the pair ($\boldsymbol{\mu}$, $\mathbf{A}$) is not a suitable parametrization because both $\mathbf{A}$ and $\boldsymbol{\Gamma}\mathbf{A}$ (with orthogonal $\boldsymbol{\Gamma}$) lead to the same distribution. However, the symmetric and positive definite matrix $\boldsymbol{\Sigma} = \mathbf{A}'\mathbf{A}$ is the same for all $\mathbf{A}$ leading to the same density, because $\|(\mathbf{x} - \boldsymbol{\mu})\mathbf{A}^{-1}\| = [(\mathbf{x} - \boldsymbol{\mu})(\mathbf{A}'\mathbf{A})^{-1}(\mathbf{x} - \boldsymbol{\mu})']^{1/2}$. Therefore, we parametrize the model as

$$f_{\boldsymbol{\mu}, \boldsymbol{\Sigma}}(\mathbf{x}) = (\det (\boldsymbol{\Sigma}))^{-1/2} g([(\mathbf{x} - \boldsymbol{\mu})\boldsymbol{\Sigma}^{-1}(\mathbf{x} - \boldsymbol{\mu})']^{1/2}) . \qquad (1.13)$$

The typical example is the multivariate normal distribution $N(\boldsymbol{\mu}, \boldsymbol{\Sigma})$, where $\boldsymbol{\Sigma}$ is the (variance–) covariance matrix. Therefore, $\boldsymbol{\Sigma}$ is called a (*pseudo-*) *covariance matrix*, or *scatter matrix*, even in the general situation. Both $\boldsymbol{\mu}$ and $\boldsymbol{\Sigma}$ must be estimated. For covariance estimators, affine equivariance means that

$$\mathbf{C}(\{\mathbf{x}_1\mathbf{A} + \mathbf{b}, \dots, \mathbf{x}_n\mathbf{A} + \mathbf{b}\}) = \mathbf{A}'\mathbf{C}(\{\mathbf{x}_1, \dots, \mathbf{x}_n\})\mathbf{A} , \qquad (1.14)$$

where $\mathbf{A}$ is any nonsingular p-by-p matrix and $\mathbf{b}$ is any vector. At $N(\boldsymbol{\mu}, \boldsymbol{\Sigma})$, maximum likelihood yields the equivariant estimators

$$T(X) = \bar{\mathbf{x}} \quad \text{and} \quad \mathbf{C}(X) = \frac{1}{n} \sum_{i=1}^{n} (\mathbf{x}_i - \bar{\mathbf{x}})'(\mathbf{x}_i - \bar{\mathbf{x}}) \qquad (1.15)$$

(to obtain an unbiased estimator of $\boldsymbol{\Sigma}$, the denominator of $\mathbf{C}(X)$ must be replaced by $n - 1$).

In order to generalize the maximum likelihood approach, Hampel (1973) suggested an affine equivariant iterative procedure to estimate μ and Σ. Maronna (1976) formally introduced affine equivariant M-estimators of location and scatter, defined as simultaneous solutions of systems of equations of the form

$$
\begin{cases}
\dfrac{1}{n} \displaystyle\sum_{i=1}^{n} u_1([(\mathbf{x}_i - T)\mathbf{C}^{-1}(\mathbf{x}_i - T)']^{1/2})(\mathbf{x}_i - T) = \mathbf{0} \\[3mm]
\dfrac{1}{n} \displaystyle\sum_{i=1}^{n} u_2((\mathbf{x}_i - T)\mathbf{C}^{-1}(\mathbf{x}_i - T)')(\mathbf{x}_i - T)'(\mathbf{x}_i - T) = \mathbf{C},
\end{cases}
\tag{1.16}
$$

where u_1 and u_2 must satisfy certain assumptions, and he considered the problems of existence, uniqueness, consistency, and asymptotic normality. Huber (1977, 1981) and Stahel (1981) computed influence functions and showed that the breakdown point of all affine equivariant M-estimators is at most $1/(p+1)$, which is disappointingly low. In a numerical study, Devlin et al. (1981, p. 361) found that M-estimators could tolerate even fewer outliers than indicated by this upper bound. Recently, Tyler (1986) found that the estimated scatter matrix becomes singular (which is a form of breaking down) when the contaminated data lie in some lower-dimensional subspace.

Affine equivariant M-estimators can be computed recursively, for instance by means of ROBETH (Marazzi 1980) or COVINTER (Dutter 1983b). A survey of affine equivariant M-estimators can be found in Chapter 5 of Hampel et al. (1986), and some recent results are given by Tyler (1985a,b).

Donoho (1982) lists some other well known affine equivariant techniques and shows that they all have a breakdown point of at most $1/(p+1)$. These proposals include:

1. *Convex Peeling* (Barnett 1976, Bebbington 1978; based on an idea of Tukey). This proceeds by discarding the points on the boundary of the sample's convex hull, and this is repeated until a sufficient number of points have been peeled away. On the remaining data, classical estimators can be applied. This procedure appeals to the intuition, and it is indeed equivariant because the convex hull is preserved by affine transformations. However, the breakdown point is quite low (even when the "good" data come from a multivariate normal distribution) because each step of peeling removes at least $p+1$ points from the sample, and often only one of these is really an outlier, so the stock of "good" points is being exhausted too fast.

2. *Ellipsoidal Peeling* (Titterington 1978, Helbling 1983). This is simi-

lar to convex peeling, but removes all observations on the boundary of the minimum volume ellipsoid containing the data. (Note that the convex hull is also the smallest convex set containing all the points.) Affine transformations $x \rightarrow xA + b$ map ellipsoids onto ellipsoids, and the volume of the image is just $|\det (A)|$ times the original volume, so ellipsoidal peeling is again affine equivariant. However, its breakdown point also tends to $1/(p+1)$ for the same reasons.

3. *Classical Outlier Rejection.* The (squared) Mahalanobis distance

$$MD^2(x_i, X) = (x_i - T(X))C(X)^{-1}(x_i - T(X))' \qquad (1.17)$$

is computed for each observation, where $T(X)$ is the arithmetic mean and $C(X)$ is the classical covariance estimate (1.15) with denominator $n - 1$ instead of n. Points for which $MD^2(x_i, X)$ is large are deleted from the sample, and one processes the "cleaned" data in the usual way. (The result will be affine equivariant because the $MD^2(x_i, X)$ do not change when the data are subjected to an affine transformation.) This approach works well if only a single outlier is present (Barnett and Lewis 1978, David 1981), but suffers from the masking effect otherwise, because one far-away outlier can make all other outliers have small $MD^2(x_i, X)$. (In other words, the breakdown point is only $2/n$.) Therefore, some refinements of this technique have been proposed.

4. *Iterative Deletion.* This consists of finding the most discrepant observation according to (1.17), deleting it, recomputing $T(X)$ and $C(X)$ for the remaining data and using it to find the most discrepant x_i among those $n - 1$ points, and so on. Several rules are possible to decide how many observations are to be removed, but at any rate the breakdown point of the mean of the remainder can be no better than $1/(p+1)$.

5. *Iterative Trimming* (Gnanadesikan and Kettenring 1972, Devlin et al. 1975). This starts with $X^{(1)} = X$ and defines $X^{(k+1)}$ recursively as the set of observations with the $(1 - \alpha)n$ smallest values in $\{MD^2(x_i, X^{(k)}), x_i \in X\}$. (Note that $X^{(2)}, X^{(3)}, \ldots$ all have the same number of points.) This iterative process is halted when both $T(X^{(k)})$ and $C(X^{(k)})$ stabilize. By means of a heuristic reasoning, Donoho (1982) concluded that the breakdown point of the final $T(X)$ is at most about $1/p$.

6. *Depth Trimming.* This is based on the concept of depth (Tukey 1974), which provides a kind of "rank statistic" for multivariate data sets. The depth of x_i is the smallest number of data points in any half-space containing it. Donoho (1982) defines a depth-trimmed mean as the average of all points with depth at least k. The higher the k, the better

the resulting breakdown point will be. The trimmed mean with $k = \max_i$ depth $(\mathbf{x}_i)$, which may be called the *deepest points estimator*, generalizes the univariate sample median. Unfortunately, not all data sets contain very deep points. In fact, Donoho proves that the maximal depth (in any data set) satisfies

$$[n/(p + 1)] \le \max_i \text{ depth } (\mathbf{x}_i) \le [n/2] \, .$$

Both bounds are sharp. The upper bound is achieved when there is a point about which the sample is centrosymmetric, and then the breakdown point of the deepest points estimator is about $1/3$. On the other hand, the lower bound is reached when the data are evenly distributed over small clusters on the $p + 1$ vertices of a simplex, in which case the breakdown point of any depth-trimmed mean is at most $1/(p + 2)$.

Recently, Oja (1983) has put forward another affine equivariant proposal, which he calls the *generalized median*. It is defined by minimization of

$$\sum \Delta(\mathbf{x}_{i_1}, \ldots, \mathbf{x}_{i_p}, T) \, , \tag{1.18}$$

where the summation is over $1 \le i_1 < \cdots < i_p \le n$, and $\Delta(\mathbf{x}_{i_1}, \ldots, \mathbf{x}_{i_p}, T)$ is the volume of the simplex with vertices $\mathbf{x}_{i_1}, \ldots, \mathbf{x}_{i_p}$ and T. Figure 1 illustrates the meaning of (1.18) in two dimensions: Instead of minimizing the total length of the lines linking T to each data point [as in (1.9), the spatial median], we now minimize the total area of all triangles formed by T and any *pair* of data points.

This is already the fourth generalization of the univariate median that we encounter. Like Donoho's deepest points estimator, also Oja's generalized median is affine equivariant. Indeed, any affine transformation $\mathbf{x} \rightarrow \mathbf{x}\mathbf{A} + \mathbf{b}$ merely multiplies the volumes $\Delta(\mathbf{x}_{i_1}, \ldots, \mathbf{x}_{i_p}, T)$ by the constant factor $|\det(\mathbf{A})|$, which is immaterial to the minimization. However, this equivariance costs a lot, because the algorithm mentioned by Oja and Niinimaa (1985) for computing $T(X)$ appears to involve the computation of the objective function (1.18), which contains C_n^p terms, in each of

$$C_{C_n^p}^p$$

candidate estimates. Oja and Niinimaa (1985) showed that the generalized median is asymptotically normal, with good statistical efficiency. It was recently found that the breakdown point equals $1/3$ in the bivariate case, and it is assumed that in the p-variate case it is $1/(p + 1)$ (Oja 1986,

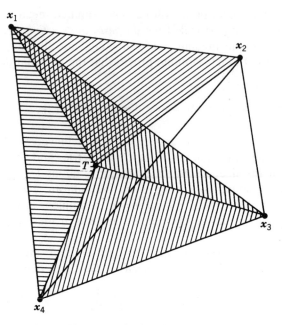

Figure 1. An illustration of Oja's generalized median, which minimizes the total area of all triangles formed with any pair of observations. (Not all such triangles are drawn.)

personal communication). Brown and Hettmansperger (1985) constructed affine invariant bivariate rank tests based on this estimator.

To conclude, all these affine equivariant estimators are facing the ubiquitous upper bound $1/(p + 1)$ on their breakdown point. Fortunately, affine equivariant estimators with a higher breakdown point have recently been introduced.

c.   Affine Equivariant Methods with High Breakdown Point

The first affine equivariant multivariate location estimator with a 50% breakdown point was obtained independently by Stahel (1981) and Donoho (1982). This estimator, called *outlyingness-weighted mean*, is defined as follows. For each observation $\mathbf{x}_i$, one looks for a one-dimensional projection leaving it most exposed. This is done by computing the following measure of the "outlyingness" of $\mathbf{x}_i$:

$$u_i = \sup_{\|\mathbf{v}\|=1} \frac{\left| \mathbf{x}_i \mathbf{v}' - \underset{j}{\mathrm{med}}\, (\mathbf{x}_j \mathbf{v}') \right|}{\underset{k}{\mathrm{med}} \left| \mathbf{x}_k \mathbf{v}' - \underset{j}{\mathrm{med}}\, (\mathbf{x}_j \mathbf{v}') \right|} , \tag{1.19}$$

where $\text{med}_j\,(x_j v^t)$ is the median of the projections of the data points x_j on the direction of the vector v, and the denominator is the median absolute deviation of these projections. To compute u_i, one must (in principle) search over all possible directions. Then one estimates location by the weighted mean

$$T(X) = \frac{\sum_{i=1}^{n} w(u_i)x_i}{\sum_{i=1}^{n} w(u_i)}, \tag{1.20}$$

where $w(u)$ is a strictly positive and decreasing function of $u \ge 0$, such that $uw(u)$ is bounded. [The latter bound is dictated by the one-dimensional case, where (1.20) reduces to a one-step W-estimator starting from the median (Hampel et al. 1986, p. 116), the influence function of which is proportional to $uw(u)$.] Analogously, these weights can also be used to compute a robust covariance matrix.

Donoho (1982) motivates (1.19) by noting that the classical (squared) Mahalanobis distance (1.17) can also be written as

$$\text{MD}^2(x_i, X) = \left(\sup_{\|v\|=1} \frac{\left| x_i v^t - \frac{1}{n} \sum_{i=1}^{n} x_i v^t \right|}{\text{SD}\,(x_1 v^t, \ldots, x_n v^t)} \right)^2, \tag{1.21}$$

which is quite vulnerable to outliers, because they affect both the mean $(1/n) \sum_{i=1}^{n} x_i v^t$ and the standard deviation $\text{SD}\,(x_1 v^t, \ldots, x_n v^t)$ of the projected data. Therefore it seems natural to replace these by robust estimators such as the median and the median absolute deviation, whereas Stahel (1981) proposed to insert also other univariate M-estimators. Points with large u_i are unlikely to belong to the "good" data, so their downweighting makes sense. It is shown that (1.20) is affine equivariant, because the u_i do not change when the x_i are transformed to $x_i A + b$. Donoho (1982) also showed that the breakdown point of (1.20) is high (supposing that n is larger than $2p + 1$) and tends to 50% for $n \to \infty$. In his proof, he assumed that X is in *general position*, which, in the context of multivariate location, means that no more than p points of X lie in any $(p - 1)$-dimensional affine subspace. (For two-dimensional data, this says that there are no more than two points of X on any line, so any three points of X determine a triangle with nonzero area.) The asymptotic behavior of this estimator has recently been investigated by Donoho (1986, personal communication).

The Stahel–Donoho estimator downweights any point that is many robust standard deviations away from the sample in some projection. Therefore, it is related to the projection pursuit principle discussed in Section 5 of Chapter 3. The projections are on one-dimensional subspaces, and their "interestingness" is measured by the objective function in (1.19). For each point, the corresponding "least favorable" projection must be found, leading to u_i.

Rousseeuw (1983, 1984) introduced a second affine equivariant estimator with maximal breakdown point, by putting

$$T(X) = \text{center of the minimal volume ellipsoid} \atop \text{covering (at least) } h \text{ points of } X, \qquad (1.22)$$

where h can be taken equal to $[n/2] + 1$. This is called the *minimum volume ellipsoid estimator* (MVE). The corresponding covariance estimator is given by the ellipsoid itself, multiplied by a suitable factor to obtain consistency. Affine equivariance of the MVE follows from the fact that the image of an ellipsoid through a nonsingular affine transformation $\mathbf{x} \to \mathbf{xA} + \mathbf{b}$ is again an ellipsoid, with volume equal to $|\det(\mathbf{A})|$ times the original volume. Because $|\det(\mathbf{A})|$ is a constant, the relative sizes of ellipsoids do not change under affine transformations.

Theorem 1. At any p-dimensional sample X in general position, the breakdown point of the MVE estimator equals

$$\varepsilon_n^*(T, X) = ([n/2] - p + 1)/n,$$

which converges to 50% as $n \to \infty$.

Proof. Without loss of generality, let $T(X) = \mathbf{0}$. We put $M :=$ volume of the smallest ellipsoid with center zero containing *all* points of X. Because X is in general position, each of its C_n^{p+1} subsets of $p + 1$ points (indexed by some $J = \{i_1, \ldots, i_{p+1}\}$) determines a simplex with nonzero volume. Therefore, for each such J there exists a bound d_J such that any ellipsoid with center $\|\mathbf{c}\| > d_J$ and containing $\{\mathbf{x}_{i_1}, \ldots, \mathbf{x}_{i_{p+1}}\}$ has a volume strictly larger than M. Put $d := \max_J d_J < \infty$.

Let us first show that $\varepsilon_n^*(T, X) \geq ([n/2] - p + 1)/n$. Take any sample X' obtained by replacing at most $[n/2] - p$ points of X. Suppose $\|T(X')\| > d$, and let E be the corresponding smallest ellipsoid containing (at least) $[n/2] + 1$ points of X'. But then E contains at least $([n/2] + 1) - ([n/2] - p) = p + 1$ points of X, so volume$(E) > M$. This is a contradiction, because the smallest ellipsoid with center zero around the

$n - ([n/2] - p) \geq [n/2] + 1$ "good" points of X' has a volume of at most M. Therefore, $\|T(X')\| \leq d$. [Even if $T(X')$ is not unique, then $\|T(X')\| \leq d$ still holds for all solutions.]

On the other hand, $\varepsilon_n^*(T, X) \leq ([n/2] - p + 1)/n$. Indeed, take any p points of X and consider the $(p - 1)$-dimensional affine subspace H they determine. Now replace $[n/2] - p + 1$ other points of X by points on H. Then H contains $[n/2] + 1$ points of the new sample X', so the minimal volume ellipsoid covering these $[n/2] + 1$ points degenerates to zero volume. Because X is in general position, no ellipsoid covering another subset of $[n/2] + 1$ points of X' can have zero volume, so $T(X')$ lies on H. Finally, we note that $T(X')$ is not bounded because the $[n/2] - p + 1$ contaminated data points on H may have arbitrarily large norms. □

In most applications it is not feasible to actually consider all "halves" of the data and to compute the volume of the smallest ellipsoid around each of them. As in the case of least median of squares (LMS) regression, we therefore resort to an approximate algorithm not unlike the one described in Section 1 of Chapter 5. We start by drawing a subsample of $(p + 1)$ different observations, indexed by $J = \{i_1, \ldots, i_{p+1}\}$. For this subsample we determine the arithmetic mean and the corresponding covariance matrix, given by

$$\bar{\mathbf{x}}_J = \frac{1}{p + 1} \sum_{i \in J} \mathbf{x}_i \quad \text{and} \quad \mathbf{C}_J = \frac{1}{p} \sum_{i \in J} (\mathbf{x}_i - \bar{\mathbf{x}}_J)'(\mathbf{x}_i - \bar{\mathbf{x}}_J), \quad (1.23)$$

where $\mathbf{C}_J$ is nonsingular whenever $\mathbf{x}_{i_1}, \ldots, \mathbf{x}_{i_{p+1}}$ are in general position. The corresponding ellipsoid should then be inflated or deflated to contain exactly h points, which corresponds to computing

$$m_J^2 = \operatorname*{med}_{i=1,\ldots,n} (\mathbf{x}_i - \bar{\mathbf{x}}_J)\mathbf{C}_J^{-1}(\mathbf{x}_i - \bar{\mathbf{x}}_J)' \quad (1.24)$$

because m_J is the right magnification factor. The volume of the resulting ellipsoid, corresponding to $m_J^2\mathbf{C}_J$, is proportional to

$$(\det (m_J^2\mathbf{C}_J))^{1/2} = (\det (\mathbf{C}_J))^{1/2}(m_J)^p . \quad (1.25)$$

This has to be repeated for many J, after which the one with the lowest objective function (1.25) is retained. We then compute

$$T(X) = \bar{\mathbf{x}}_J \quad \text{and} \quad \mathbf{C}(X) = (\chi_{p,0.50}^2)^{-1}m_J^2\mathbf{C}_J , \quad (1.26)$$

where $\chi_{p,0.50}^2$ is the median of the chi-squared distribution with p degrees of freedom. (This correction factor is for consistency at multivariate normal data.) The number of random subsamples J that are needed can be determined as in Section 1 of Chapter 5. Indeed, the probability that at least one out of m subsamples consists exclusively of "good" points is approximately

$$1 - (1 - (1 - \varepsilon)^{p+1})^m \tag{1.27}$$

when the original data contains a fraction ε of outliers. By imposing that this probability exceeds a given value, one obtains the number m of replications.

The complexity of this algorithm is similar to that of LMS regression, because for each of the m replications a p-by-p matrix $\mathbf{C}_J$ must be inverted (this can even be done somewhat faster because $\mathbf{C}_J$ is symmetric and positive definite). Then a median of the n "squared residuals" must be computed in (1.24). Note that the calculation of $\det(\mathbf{C}_J)$ does not cost extra, because it is obtained as a by-product of the computation of $\mathbf{C}_J^{-1}$. Also note that this algorithm is itself affine equivariant and that it may easily be parallelized by treating many subsamples simultaneously.

The (approximate) MVE estimates can also be used as initial solutions on which to base a one-step improvement. As in the case of reweighted least squares regression (see Section 2 of Chapter 1) we can assign a weight w_i to each observation by means of the rule

$$w_i = \begin{cases} 1 & \text{if} \quad (\mathbf{x}_i - T(X))\mathbf{C}(X)^{-1}(\mathbf{x}_i - T(X))' \le c \\ 0 & \text{otherwise} , \end{cases} \tag{1.28}$$

where the cut-off value c might be taken equal to $\chi^2_{p,0.975}$. Then one can apply the reweighted estimators

$$T_1(X) = \frac{\displaystyle\sum_{i=1}^{n} w_i \mathbf{x}_i}{\displaystyle\sum_{i=1}^{n} w_i} \tag{1.29}$$

and

$$\mathbf{C}_1(X) = \frac{\displaystyle\sum_{i=1}^{n} w_i(\mathbf{x}_i - T_1(X))'(\mathbf{x}_i - T_1(X))}{\displaystyle\sum_{i=1}^{n} w_i - 1} \tag{1.30}$$

which simply means that the classical computations are carried out on the set of points for which $w_i = 1$.

The above algorithm (both MVE and reweighted) has been implemented in a program called PROCOVIEV (Rousseeuw and van Zomeren 1987). This abbreviation stands for Program for RObust COVariance and Identification of Extreme Values, indicating its ability to discover multivariate outliers.

Example
Let us consider the Hertzsprung–Russell data of Table 3 of Chapter 2. This time we treat both variables (log temperature and log light intensity) in the same way, so we have 47 points in the plane. Figure 2 shows the classical 97.5% tolerance ellipse, obtained from the usual mean and covariance matrix (dashed line). Note that this ellipse is very big, because it is attracted by the four outliers (indeed, the tolerance ellipse tries to engulf them). On the other hand, the tolerance ellipse based on the MVE estimates (solid line) is much smaller and essentially fits the main sequence. The four giant stars, lying far away from this ellipse, can then be identified as not belonging to the same population as the bulk of the

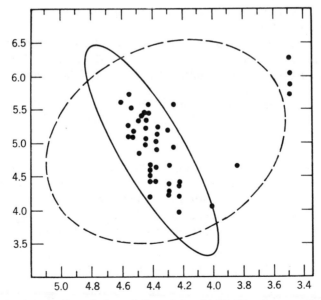

Figure 2. Hertzsprung–Russell data of Chapter 2, with 97.5% tolerance ellipse obtained from the classical estimator (dashed line) and based on the MVE estimator (solid line).

data. Note that choosing other measurement units for the variables would yield essentially the same result, because of affine equivariance. Also note that this analysis is basically different from that presented in Figure 4 of Chapter 2, because there the aim was to write the log light intensity as a linear function of the log temperature, and the LMS tried to find the narrowest *strip* covering half of the data, whereas we now made use of the smallest *ellipse* covering half of the data. However, we could construct another robust regression line of y on x based on this ellipse, by putting

$$\hat{y} = f(x) = \text{midpoint of the intersections of the vertical line}$$
$$\text{through } x \text{ with the ellipse}$$

in the same way that the LS line is derived from the classical covariance matrix. This could easily be generalized to multiple regression.

For $p = 1$ (univariate location) the minimum volume ellipsoid reduces to the shortest half, so $T(X)$ becomes the one-dimensional LMS discussed in Chapter 4. In particular, Theorem 3 in Section 4 of Chapter 4 shows that this estimator converges as $n^{-1/3}$, which is abnormally slow. It is assumed that the multivariate MVE will not have a better rate, so it is useful to perform a one-step reweighting [based on formula (1.28) above] to improve its statistical efficiency. (Alternatively, a one-step M-estimator might be applied.)

Another approach would be to generalize the least trimmed squares (LTS) estimator (which converges like $n^{-1/2}$ according to Theorem 4 in Section 4 of Chapter 4) to multivariate location. This yields

$$T(X) = \text{Mean of the } h \text{ points of } X \text{ for which the determinant}$$
$$\text{of the covariance matrix is minimal}$$

$$(1.31)$$

(Rousseeuw 1983, 1984). We call this the *minimum covariance determinant* estimator (MCD). It corresponds to finding the h points for which the classical tolerance ellipsoid (for a given level) has minimum volume, and then taking its center. This estimator is also affine equivariant because the determinant of the covariance matrix of the transformed data points equals

$$\det (\mathbf{A}'\mathbf{C}\mathbf{A}) = (\det (\mathbf{A}))^2 \det (\mathbf{C}) . \qquad (1.32)$$

The MCD has the same breakdown point as the MVE, because of the same reasoning as in Theorem 1 above. Like the MVE, the MCD also yields a robust covariance estimate at the same time: One only has to use the (classical) covariance matrix of the selected h observations, multiplied by a constant to obtain consistency in the case of multivariate normality.

Both the MVE and the MCD are very drastic, because they are intended to safeguard against up to 50% of outliers. If one is certain that the fraction of outliers is at most α (where $0 < \alpha \le \frac{1}{2}$), then one can work with the estimators MVE(α) and MCD(α) obtained by replacing h by $k(\alpha) = [n(1 - \alpha)] + 1$ in (1.22) and (1.31). The breakdown point of these estimators is equal to α (for $n \to \infty$). For $\alpha \to 0$, the MVE yields the center of the smallest ellipsoid covering all the data, whereas the MCD tends to the arithmetic mean.

In the regression context, the LMS has been generalized to S-estimators, which are described in Section 4 of Chapter 3. These S-estimators can also be extended to multivariate location and covariance, by adapting formulas (4.28)–(4.30) of Chapter 3. This means that one must find a vector T and a symmetric positive definite matrix $\mathbf{C}$ such that

$\det(\mathbf{C})$ is minimized subject to

$$\frac{1}{n} \sum_{i=1}^{n} \rho(\{(\mathbf{x}_i - T)\mathbf{C}^{-1}(\mathbf{x}_i - T)'\}^{1/2}) = K, \qquad (1.33)$$

where K is often put equal to the expected value $E[\rho(\{\mathbf{z}\mathbf{z}'\}^{1/2})]$ in which $\mathbf{z}$ follows a standard multivariate normal distribution $N(\mathbf{0}, I)$, and where ρ satisfies the same conditions as before. These S-estimators are obviously affine equivariant, and for well-chosen ρ their breakdown point is again that of Theorem 1 above (this can be shown by means of a reasoning analogous to Theorem 8 in Section 4 of Chapter 3). Note that the MVE can actually be viewed as a special case of S-estimators, obtained by inserting the *discontinuous* ρ-function

$$\rho(u) = \begin{cases} 0 & \text{if } |u| < (\chi^2_{p,0.50})^{1/2} \\ 1 & \text{otherwise} \end{cases} \qquad (1.34)$$

and putting $K = \frac{1}{2}$. As in the case of regression analysis, choosing a *smooth* ρ-function greatly improves the asymptotic behavior. Davies (1987) indeed proves that S-estimators for multivariate location and covariance, constructed from a smooth ρ-function, are asymptotically normal, which means that

$$n^{1/2}(T_n(\mathbf{x}_1, \ldots, \mathbf{x}_n) - \boldsymbol{\mu})$$

and

$$n^{1/2}(\mathbf{C}_n(\mathbf{x}_1, \ldots, \mathbf{x}_n) - \mathbf{\Sigma})$$

converge in law to multivariate normal distributions with zero mean. (He writes this down with a function κ, which is related to ρ through $\kappa(u) = [\rho(\infty) - \rho(\sqrt{u})]/\rho(\infty)$.) In actual applications, Davies (1987, Section 5) does not use S-estimators with smooth ρ-function, for which the computations become too complex. Instead, he applies the original MVE estimator, for which he proposes an algorithm that differs from the one described above in that (a) more than $p + 1$ points may be drawn at each replication and (b) reweighting is continued as long as the volume of the 50% coverage ellipsoid decreases at each step. He then runs the MVE on a five-dimensional example in which the outliers are not detectable by means of classical methods, not even on the basis of the 10 two-dimensional plots of the projections onto the eigenvectors of the ordinary covariance matrix.

REMARK. Both Davies (1987) and Lopuhaä and Rousseeuw (1987) independently found a way to increase the finite-sample breakdown point of the MVE to its best possible value. As in the case of the LMS regression estimator, this is done by letting h depend on the dimensionality. For the MVE (and the MCD), the optimal variant is to consider "halves" containing

$$h = [(n + p + 1)/2] \tag{1.35}$$

observations. The resulting breakdown value (which may be verified as in Theorem 1) is then

$$\varepsilon_n^* = \frac{[(n - p + 1)/2]}{n} . \tag{1.36}$$

Another way to construct affine equivariant estimators with high breakdown point (though usually not 50%) is to make use of the *minimum distance* principle. In this approach, one assumes that the data are sampled from an unknown member of a parametric family $\{F_\theta, \theta \in \Theta\}$ of model distributions. Let ν be some kind of distance between probability distributions. For any sample $X = \{\mathbf{x}_1, \ldots, \mathbf{x}_n\}$, one then constructs the empirical distribution F_n, and the minimum distance estimator $T(X)$ is defined as the value of θ that

$$\text{Minimize}_\theta \; \nu(F_n, F_\theta) . \tag{1.37}$$

[Sometimes one does not use F_n but other estimators of the underlying density or cumulative distribution function (cdf)]. Minimum distance estimators go back to Wolfowitz (1957), who showed that they are "automatically" consistent. By means of the half-space metric, which is affine invariant, Donoho (1982) constructed an affine equivariant minimum distance estimator with 25% breakdown point. Donoho and Liu (1986) showed that any minimum distance estimator is "automatically" robust over contamination neighborhoods defined by the metric on which the estimator is based. Tamura and Boos (1986) used the minimum Hellinger distance approach to construct affine equivariant estimators of location and covariance, which are asymptotically normal and possess a 25% breakdown point in certain situations. Unfortunately, the minimum distance approach has not yet been developed to the stage where practical algorithms are proposed for multivariate applications.

d. The Detection of Multivariate Outliers, with Application to Leverage Points

It is very important to be able to identify outliers in multivariate point clouds. Such outliers do stick out in *certain* projections, but this does not make them easy to find because most projections do not reveal anything (like the projections on the coordinate axes in Figure 4 of Chapter 1).

The classical approach to outlier detection (see, e.g., Healy 1968) has been to compute the squared Mahalanobis distance for each observation as in (1.17), based on the arithmetic mean $T(X) = (1/n) \sum_{i=1}^{n} \mathbf{x}_i$ and the unbiased covariance estimator $\mathbf{C}(X) = (1/(n-1)) \sum_{i=1}^{n} (\mathbf{x}_i - T(X))'(\mathbf{x}_i - T(X))$. Points with large $\text{MD}_i^2 = \text{MD}^2(\mathbf{x}_i, X)$ (possibly compared to some χ_p^2 quantile) are then considered outliers.

However, as already explained above, this approach suffers from the fact that it is based on exactly those statistics that are most sensitive to outliers. This is particularly acute when there are several outliers forming a small cluster, because they will move the arithmetic mean toward them and (even worse) inflate the classical tolerance ellipsoid in their direction. As in Figure 2, the tolerance ellipsoid will do its best to encompass the outliers, after which their MD_i^2 won't be large at all.

Let us look at the modified wood gravity data in Table 8 of Chapter 6 in order to have a multivariate example. By considering only the explanatory variables, we obtain a data cloud of $n = 20$ points in $p = 5$ dimensions. We know that there are four outliers, but they do not show up in any of the coordinates separately. By means of the classical mean and covariance matrix, the squared Mahalanobis distance of each case may be computed. Actually, the values of MD_i^2 were already listed in Table 9 of Chapter 6. All MD_i^2 are smaller than the 95% quantile of the χ_5^2

distribution, which equals about 11.07. The largest MD_i^2 belong to cases 7 and 16, but these are really good observations. An index plot of the $MD_i = \sqrt{MD_i^2}$ is shown in Figure 3a, which looks very regular.

In other applications it may happen that one (very bad) outlier is visible, but at the same time masks all the others. For an illustration of the masking effect, let us consider the Hawkins–Bradu–Kass data set described in Section 3 of Chapter 3. If we restrict our attention to the x-part of the data, we obtain 75 points in three dimensions. By construction, the first 14 points are outliers. However, the MD_i^2 (listed in Table 3 of Chapter 6) do not convey the whole picture. Indeed, the only MD_i exceeding $\sqrt{\chi_{3,0.95}^2} = 2.8$ are those of cases 12 and 14 (and case 13 comes close), whereas the other outliers yield quite inconspicuous values, as can be seen from Figure 4a. Only *after* the deletion of cases 12, 13, and 14 would the other outliers begin to obtain larger MD_i.

To overcome this weakness of the classical Mahalanobis distance, it is necessary to replace $T(X)$ and $C(X)$ by robust estimators of location and scatter. A first step toward this goal was the use of affine equivariant M-estimators, as advocated by Campbell (1980), who gives some interesting examples. There is no doubt that this proposal constitutes a considerable improvement over the classical approach, but the low breakdown point of M-estimators limits its applicability. (Indeed, in the modified wood gravity data we have four outliers in a sample of size 20, which is already more than $1/(p + 1) = 16.7\%$.)

To go any further, we need affine equivariant estimators with a high breakdown point. Therefore, we propose to compute the squared Mahalanobis distance relative to the MVE estimates of location and scatter, that is, to apply (1.17) where $T(X)$ and $C(X)$ are given by (1.22). In actual computations, one may use the results (1.26) of the resampling algorithm, as implemented in the program PROCOVIEV. For the explanatory part of the modified wood gravity data, this yields the robust MD_i displayed in Figure 3b, in which the four outliers are clearly visible. If we run the same program on the x-part of the Hawkins–Bradu–Kass data, we obtain Figure 4b. From this picture (constructed from a single program run!) it is at once evident that there are 14 far outliers. The fact that the robust MD_i consume more computer time than their classical counterparts is more than compensated by the reliability of the new method and the resulting gain of the statistician's time.

It is no coincidence that these examples were connected with regression data sets. Indeed, in regression analysis it is very important to discover leverage points, which are exactly those cases for which the x_i-part is outlying. In retrospect, these examples also explain why the diagonal elements h_{ii} of the hat matrix are unable to cope with multiple

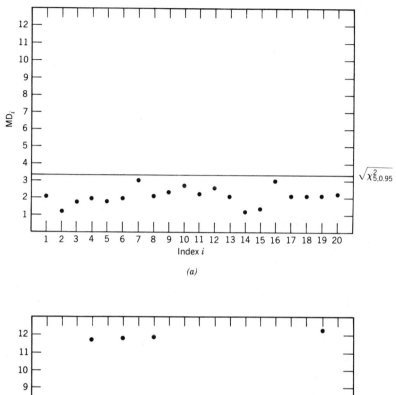

(a)

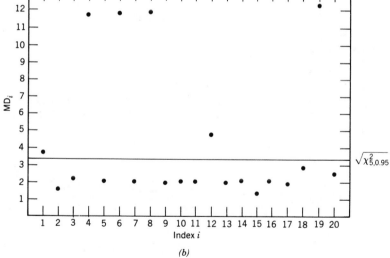

(b)

Figure 3. Index plot of Mahalanobis distances $MD_i = \sqrt{MD_i^2}$ of 20 points in five dimensions: (a) with the classical formula, (b) based on the MVE estimates.

267

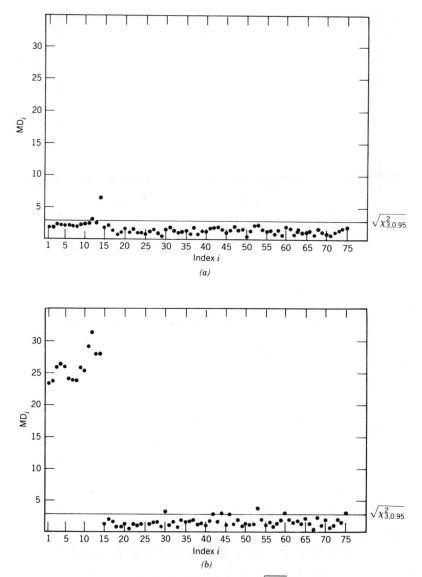

Figure 4. Index plot of Mahalanobis distances $MD_i = \sqrt{MD_i^2}$ of 75 points in three dimensions: (a) with the classical formula, (b) based on the MVE estimates.

leverage points. Indeed, these h_{ii} are related [through (2.15) of Chapter 6] to the Mahalanobis distance based on the classical estimates with zero breakdown point. An MVE analysis of the explanatory variables is very useful, because robust MD_i are more reliable *leverage diagnostics* than the h_{ii}. On the other hand, to know that x_i is a leverage point is not enough, because one also has to take the response y_i into account in order to find out if (x_i, y_i) is actually a regression outlier [i.e., whether (x_i, y_i) deviates from the linear pattern set by the majority of the data]. Hence, the question of whether x_i is a "good" or a "bad" leverage point can only be settled by means of a robust regression technique such as the LMS. Therefore, a combination of LMS and MVE gives more insight into regression data.

Also from a computational point of view, the LMS and the MVE fit together nicely. Indeed, their resampling algorithms are so similar that they could be run at the same time. The number of observations in each LMS subsample equals the number of "real" explanatory variables plus 1 (for the intercept term). This happens to be exactly the number of points that are needed in each MVE subsample. Moreover, the number of subsamples for the MVE can be taken equal to that for the LMS, because the probability of finding at least one "good" subsample is the same for both algorithms. Therefore, the whole resampling scheme is common to LMS and MVE, and the particular computations (finding a median squared residual for LMS, and the volume of a 50% ellipsoid for MVE) could be incorporated in the same program (and even run in a parallel way). Only a slight adaptation is necessary in the case of regression through the origin, because then one must restrict consideration to ellipsoids with center zero.

Gasko and Donoho (1982) provided another robustification of the MD_i to be used as a leverage diagnostic. Instead of first computing a robust estimator of location and scatter, they looked at the alternative formula (1.21) for the squared Mahalanobis distance and replaced it by the Stahel–Donoho measure of outlyingness (1.19).

e. Outlook

Multivariate location and covariance matrices are cornerstones of general multivariate statistics. Now that we have robust and affine equivariant estimators of location and scatter at our disposal, we can use them to robustify many classical techniques. There are basically two ways to do this:

(i) One may (in one way or other) insert robust estimators $T(X)$ and

$C(X)$ instead of the classical mean and empirical covariance matrices.

(ii) Otherwise, one may first compute the robust estimators in order to identify outliers, which have to be corrected or deleted. Afterwards, the usual multivariate analyses may be carried out on the "cleaned" data. For instance, the correlation matrix of these points may be computed.

Maronna (1976), Campbell (1980), and Devlin et al. (1981) proposed to robustify *principal component analysis* by inserting robust covariance matrices obtained from affine equivariant M-estimators. Campbell (1982) likewise robustified *canonical variate analysis*. Our own proposal would be to apply the MVE, because of its better breakdown properties. Other fields in which the MVE could take the place of classical normal-theory covariance matrices are *cluster analysis* and *factor analysis*.

f. Some Remarks on the Role of Affine Equivariance

We have already seen that affine equivariance (1.7) is a very natural condition. Nevertheless, it is not so easy to combine with robustness. Indeed, the only affine equivariant multivariate location/covariance estimators with high breakdown point known so far (the Stahel–Donoho weighted mean and the MVE and its relatives) need substantially more computation time than the classical estimators.

The maximal breakdown point of affine equivariant estimators (1.7) is slightly lower than that of the larger class of translation equivariant estimators (1.2). Indeed, the breakdown point of any translation equivariant estimator of location is at most

$$\frac{[(n + 1)/2]}{n} , \tag{1.38}$$

which can be proven as in Theorem 7 of Chapter 4 by replacing closed intervals by closed Euclidean balls. Note that this bound does not depend on p, the number of dimensions. Moreover, this bound is sharp because it is attained by the coordinatewise median, the L_1 estimator (1.9), and the estimator given by

$$\underset{T}{\text{Minimize}} \ \underset{i}{\text{med}} \ \|\mathbf{x}_i - T\|^2 , \tag{1.39}$$

which corresponds to the center of the smallest sphere covering at least half of the points (Rousseeuw 1984). If we restrict our attention to the smaller class of all orthogonal equivariant estimators, the bound (1.38) is still sharp because both the L_1 estimator and (1.39) are orthogonal equivariant. It is only when we switch to affine equivariance that the best possible breakdown point of any pair $(T, \mathbf{C})$ goes down to

$$\frac{[(n - p + 1)/2]}{n} ,$$

which is attained by the variants of the MVE and the MCD with $h = [(n + p + 1)/2]$.

Some people have expressed the opinion that it should be very easy to obtain high-breakdown estimators that are affine equivariant, by first rescaling the observations. They proposed the following procedure: Calculate the ordinary covariance matrix $\mathbf{C}$ given by (1.15), and take a root $\mathbf{S}$ (i.e., $\mathbf{S}'\mathbf{S} = \mathbf{C}$). Transform the data as $\tilde{\mathbf{x}}_i = \mathbf{x}_i \mathbf{S}^{-1}$, apply to these $\tilde{\mathbf{x}}_i$ an easily computable estimator $\tilde{T}$ with $\varepsilon^* = 50\%$ which is not affine equivariant, and then transform back by putting $T = \tilde{T}\mathbf{S}$. First, we observe that this construction gives a unique result if and only if $\tilde{T}$ is equivariant for orthogonal transformations, because for any orthogonal matrix Γ the product $\Gamma\mathbf{S}$ is also a root of $\mathbf{C}$. Also, $\tilde{T}$ has to be translation equivariant to ensure affine equivariance of T. Therefore, it seems that taking the L_1 estimator for $\tilde{T}$ will do the job. However, it turns out that its good breakdown behavior does not carry over.

Example

To show this, consider a two-dimensional example where a fraction $(1 - \varepsilon)$ of the data is spherically bivariate normal (we assume that both coordinates follow a standard normal distribution $N(0, 1)$ and are independent of each other), and there is a fraction ε of outliers that are concentrated at the point with coordinates $(0, u)$. Here, $u > 0$ and $0 < \varepsilon < \frac{1}{2}$. (This configuration is sketched in Figure 5a.) The (usual) covariance matrix becomes

$$\mathbf{C} = \begin{bmatrix} 1 - \varepsilon & 0 \\ 0 & (1 - \varepsilon)(1 + \varepsilon u^2) \end{bmatrix}.$$

Therefore, we can easily construct a root $\mathbf{S}$ of $\mathbf{C}$:

$$\mathbf{S} = \begin{bmatrix} \sqrt{1 - \varepsilon} & 0 \\ 0 & \sqrt{(1 - \varepsilon)(1 + \varepsilon u^2)} \end{bmatrix}.$$

For u tending to ∞, the transformed $(\tilde{x}_1, \tilde{x}_2) = (x_1, x_2)\mathbf{S}^{-1}$ are situated as in Figure 5b: The $(1 - \varepsilon)$-fraction gets concentrated on the $\tilde{x}_1$-axis with

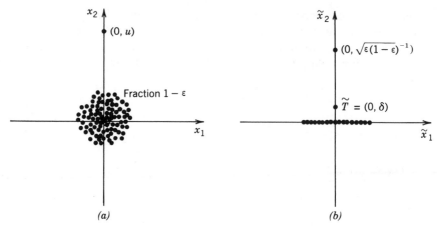

Figure 5. Sketch of the example: (*a*) original data, (*b*) after scaling by means of the classical covariance matrix.

univariate normal distribution $N(0, (1 - \varepsilon)^{-1})$, and the ε-fraction lands at the point $(0, (\varepsilon(1 - \varepsilon))^{-1/2})$.

For any finite value of u, the L_1 estimate of the transformed data lies on the vertical axis by symmetry, so we may denote it by $\tilde{T} = (0, \delta(u))$. Let us now look at the limit $\delta = \lim_{u \to \infty} \delta(u)$. By definition, δ minimizes the expected value of $\|(\tilde{x}_1, \tilde{x}_2) - (0, \delta)\|$, which is denoted by

$$D := \varepsilon((\varepsilon(1 - \varepsilon))^{-1/2} - \delta) + (1 - \varepsilon) \int_{-\infty}^{+\infty} [\delta^2 + \tilde{x}_1^2]^{1/2} \, dF(\tilde{x}_1) ,$$

where F is the normal distribution $N(0, (1 - \varepsilon)^{-1})$. By the substitution $z = (1 - \varepsilon)^{1/2} \tilde{x}_1$, the average distance D equals

$$D = \varepsilon((\varepsilon(1 - \varepsilon))^{-1/2} - \delta) + (1 - \varepsilon)^{1/2} \int_{-\infty}^{+\infty} [(1 - \varepsilon)\delta^2 + z^2]^{1/2} \, d\Phi(z) ,$$

where Φ is the standard normal cdf. Because δ minimizes D, it follows that

$$0 = \partial D / \partial \delta = -\varepsilon + (1 - \varepsilon) \int_{-\infty}^{+\infty} \{(1 - \varepsilon)\delta^2 / [(1 - \varepsilon)\delta^2 + z^2]\}^{1/2} \, d\Phi(z) ;$$

hence

$$\int_{-\infty}^{+\infty} \{(1 - \varepsilon)\delta^2 / [(1 - \varepsilon)\delta^2 + z^2]\}^{1/2} \, d\Phi(z)$$

equals the positive constant $\varepsilon/(1 - \varepsilon)$. This cannot happen for $\delta = 0$, hence $\delta > 0$. As the final estimate T equals $\tilde{T}\mathbf{S}$, we conclude that

$$\|T\| = \|\tilde{T}\mathbf{S}\| = \|(0, \delta(u))\mathbf{S}\| = \{(1 - \varepsilon)(1 + \varepsilon u^2)\}^{1/2}\delta(u)$$

tends to infinity for $u \to \infty$. (Note that this effect is caused by the explosion of $\mathbf{S}$!) This means that any fraction $\varepsilon > 0$ can make T break down, so $\varepsilon^*(T) = 0\%$.

Therefore, rescaling does not solve our problem unless we could start with a high-breakdown covariance estimator, but that is precisely what we were looking for in the first place.

2. ROBUST TIME SERIES ANALYSIS

Many people are involved in time series analysis and forecasting (e.g., in economics, engineering, and physical sciences). Software packages based on the work of Box and Jenkins (1976) are widely available, but unfortunately they are restricted to the least squares approach and do not provide for handling outliers. Indeed, the field of robust time series analysis has come into existence only fairly recently and has seen most of its activity during the last decade. (For an up-to-date survey, see Martin and Yohai 1984a.) This is partly because one had to wait for the development of robust regression techniques (of which extensive use is made) and also because of the increased difficulty inherent in dealing with dependencies between the observations. In this section, we shall first briefly describe the most commonly used time series models together with two main types of outliers and then consider robust estimation in this framework.

a. Autoregressive Moving Average Models and Types of Outliers

A time series is a sequence of n consecutive univariate observations

$$Y_1, Y_2, \ldots, Y_n$$

measured at regular intervals. Often the observations are fitted by a model

$$Y_t = \mu + X_t \qquad (t = 1, \ldots, n), \tag{2.1}$$

where μ is a location parameter, and X_t follows a zero-mean *autoregressive moving average* model, abbreviated ARMA(p, q). This means that

$$X_t - \alpha_1 X_{t-1} - \cdots - \alpha_p X_{t-p} = e_t + \beta_1 e_{t-1} + \cdots + \beta_q e_{t-q}, \quad (2.2)$$

where the e_t are distributed like $N(0, \sigma^2)$. The unknown parameters are therefore

$$\alpha = \begin{bmatrix} \alpha_1 \\ \vdots \\ \alpha_p \end{bmatrix}, \quad \beta = \begin{bmatrix} \beta_1 \\ \vdots \\ \beta_q \end{bmatrix}, \quad \mu, \text{ and } \sigma.$$

The left-hand side of (2.2) is called the *autoregressive part*, because X_t depends on its previous (lagged) values $X_{t-1}, \ldots, X_{t-p}$. The right-hand side is referred to as the *moving average part*, because the actual error term at time t is a linear combination of the original $e_t, e_{t-1}, \ldots, e_{t-q}$ which cannot be observed directly.

The general ARMA(p, q) model is rather difficult to deal with numerically because of the moving average part. Fortunately, it is often sufficient to use an AR(p) model, which is much simpler because it contains only the autoregressive part. This model is obtained by putting $q = 0$ in (2.2), which reduces to

$$X_t = \alpha_1 X_{t-1} + \cdots + \alpha_p X_{t-p} + e_t. \quad (2.3)$$

(Such AR(p) models always provide a good approximation to ARMA data if one is willing to switch to a larger value of p.) In terms of the actually observed values $Y_t = \mu + X_t$, (2.3) becomes

$$(Y_t - \mu) = \alpha_1 (Y_{t-1} - \mu) + \cdots + \alpha_p (Y_{t-p} - \mu) + e_t,$$

which can be rewritten as

$$Y_t = \alpha_1 Y_{t-1} + \cdots + \alpha_p Y_{t-p} + \gamma + e_t, \quad (2.4)$$

where the intercept term γ equals $\mu(1 - \alpha_1 - \cdots - \alpha_p)$. There are $n - p$ complete sets $(Y_t, Y_{t-1}, \ldots, Y_{t-p})$ as t ranges from $p + 1$ to n, which can be used to estimate $\alpha_1, \ldots, \alpha_p, \gamma$ in (2.4) by means of some regression estimator. Indeed, we now have the linear model

$$\begin{bmatrix} Y_{p+1} \\ Y_{p+2} \\ \vdots \\ Y_n \end{bmatrix} = \begin{bmatrix} Y_p & Y_{p-1} & \cdots & Y_1 & 1 \\ Y_{p+1} & Y_p & \cdots & Y_2 & 1 \\ \vdots & \vdots & & \vdots & \vdots \\ Y_{n-1} & Y_{n-2} & \cdots & Y_{n-p} & 1 \end{bmatrix} \begin{bmatrix} \alpha_1 \\ \alpha_2 \\ \vdots \\ \alpha_p \\ \gamma \end{bmatrix} + \begin{bmatrix} e_{p+1} \\ e_{p+2} \\ \vdots \\ e_n \end{bmatrix}.$$

We do not believe it is useful to add rows with the unobserved

$Y_0, Y_{-1}, \ldots$ put equal to zero, because this only leads to outliers as illustrated by Martin (1980, example 2).

A very simple (but frequently used) example is the AR(1) model

$$Y_t = \alpha_1 Y_{t-1} + \gamma + e_t. \tag{2.5}$$

In order for the Y_t to be stationary, the absolute value of α_1 must be less than 1. There are $n - 1$ complete pairs (Y_t, Y_{t-1}), which may conveniently be plotted in a scatterplot of Y_t (vertically) versus Y_{t-1} (horizontally). Figure 6a is a graph of a well-behaved AR(1) time series without outliers, together with the corresponding scatterplot. In such an uncontaminated situation, it is not difficult to estimate the slope α_1 and the intercept γ adequately.

Fox (1972) and Martin (1981) considered two types of outliers that may occur in time series data. The first (and relatively innocent) class is that of *innovation outliers*, which may be modeled using the original ARMA framework (2.1)–(2.2), except that the distribution of the e_t is no longer normal but has heavy tails. This implies that when a certain e_{t_0} is outlying, it immediately affects X_{t_0}. Because the model (2.2) still holds, this exceptional value of X_{t_0} will then influence X_{t_0+1}, and then X_{t_0+2}, and so on. However, after a while this effect dies out. For real-world examples of innovation outliers in the area of speech recognition, see Lee and Martin (1987). Figure 6b shows a sample path similar to Figure 6a, but with one innovation outlier at time t_0.

Let us now look at the corresponding scatterplot in Figure 6b. The innovation outlier first appears at the point $A = (Y_{t_0-1}, Y_{t_0})$ which is an outlier in the vertical direction. But afterwards it yields some large (Y_{t-1}, Y_t) that do lie close to the original line of slope α_1. In the scatterplot, the innovation outlier therefore results in one outlier in the response variable and a certain number of "good" leverage points, which have the potential of improving the accuracy of the regression estimate of α_1. When least squares is used, these good points tend to compensate for the effect of the one outlier (Whittle 1953). In fact, when the heavy-tailed distribution of e_t is symmetric, one may estimate α_1 with a better precision than in the case of Gaussian e_t.

Let us now look at the second type of outlier. *Additive outliers* occur when contamination is added to the Y_t themselves, so the resulting observations no longer obey the ARMA model. To formalize this, we must extend (2.1) to

$$Y_t = \mu + X_t + V_t. \tag{2.6}$$

For most observations V_t is zero, but we assume that $P(V_t \neq 0) = \varepsilon$ where

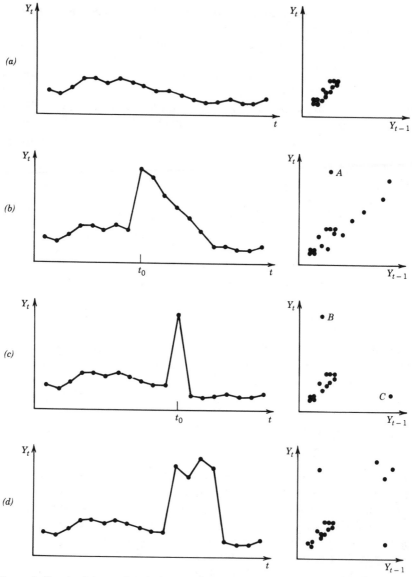

Figure 6. Sketch of time series with corresponding plot of Y_t versus Y_{t-1} for (a) no outliers, (b) innovation outlier, (c) isolated additive outlier, and (d) patch of additive outliers.

this fraction of additive outliers is positive and not too large. (This is similar to replacement contamination, which is the basis of the finite-sample breakdown point approach adopted in this book.) According to Martin (1981), many time series with outliers arising in practice have ε between 1% and 25%. Letting the V_t be i.i.d. random variables yields a model for "isolated" or "scattered" outliers, occurring independently of each other. On the other hand, certain dependency structures for the V_t could be used to describe outliers coming in "patches," "clumps," or "bursts."

Figure 6c provides an example with an isolated additive outlier, as might be caused by a keypunch or transcription error (e.g., a misplaced decimal point) or someone accidentally touching a recording instrument. This one outlying Y_{t_0} gives rise to *two* exceptional points in the scatterplot, namely $B = (Y_{t_0-1}, Y_{t_0})$, which is an outlier in the response variable, and $C = (Y_{t_0}, Y_{t_0+1})$, which is a bad leverage point. [In an AR(p) model, we obtain one outlier in the vertical direction and p leverage points.] Additive outliers are therefore a cause for much greater concern than innovative ones, because the usual least squares (LS) method cannot cope with leverage points and will yield biased parameter estimates. In Figure 6c, the LS estimate of the slope will be biased toward 0.

Figure 6d shows a whole patch of additive outliers, as is not uncommon in real data. For instance, weekly sales may be positively affected by a bargain month. Another example is that of industrial production that temporarily decreases because of a strike. The resulting effects in the scatterplot are equally unpleasant, because the leverage points will tend to make the estimated slope biased toward 1.

Additive outliers appear to occur more often than innovation outliers in actual applications, and they are more difficult to deal with because leverage points pose bigger problems than vertical outliers. This indicates the need for reliable robust estimators for time series parameters.

A research area that has recently received a lot of attention is the qualitative robustness of such estimators, generalizing Hampel's (1971) i.i.d. definition. We shall not treat this topic here, but refer the reader to Martin (1980, p. 237), Papantoni-Kazakos and Gray (1979), Bustos (1981), Cox (1981), Boente et al. (1982), and Papantoni-Kazakos (1984).

b. *M-* and GM-Estimators

If an ARMA process contains only innovation outliers, then the usual *M*-estimators for regression (which bound the influence of vertical outliers) are appropriate. Let us first restrict our attention to AR(p) models (2.4). In this case, classical least squares corresponds to

$$\text{Minimize} \sum_t (Y_t - \hat{\gamma} - \hat{\alpha}_1 Y_{t-1} - \cdots - \hat{\alpha}_p Y_{t-p})^2 . \qquad (2.7)$$

This can easily be generalized to M-estimation:

$$\underset{\hat{\gamma},\hat{\alpha}}{\text{Minimize}} \sum_t \rho \left(\frac{Y_t - \hat{\gamma} - \hat{\alpha}_1 Y_{t-1} - \cdots - \hat{\alpha}_p Y_{t-p}}{\hat{\sigma}} \right), \qquad (2.8)$$

where $\hat{\sigma}$ is a robust estimate of the error scale. This means that any computer program for regression M-estimation may be applied to such data. Finally, a natural estimate of the location parameter μ in (2.1) is given by

$$\hat{\mu} = \frac{\hat{\gamma}}{1 - \hat{\alpha}_1 - \cdots - \hat{\alpha}_p} . \qquad (2.9)$$

Under regularity conditions, $\hat{\gamma}$, $\hat{\alpha}$, and $\hat{\mu}$ are consistent and asymptotically normal (Lee and Martin 1982). Their Cramer–Rao bound and efficiency robustness can be found in Martin (1982).

In the general ARMA(p, q) framework, the least squares estimator is given by

$$\underset{\hat{\gamma},\hat{\alpha},\hat{\beta}}{\text{Minimize}} \sum_t r_t(\hat{\gamma}, \hat{\alpha}, \hat{\beta})^2 . \qquad (2.10)$$

Since the residual r_t is now a nonlinear function of the parameters, we already face a nonlinear estimation problem in the classical approach (Box and Jenkins 1976). In order to provide for innovation outliers, Martin (1981) proposed to replace (2.10) by an M-estimator

$$\underset{\hat{\gamma},\hat{\alpha},\hat{\beta}}{\text{Minimize}} \sum_t \rho \left(\frac{r_t(\hat{\gamma}, \hat{\alpha}, \hat{\beta})}{\hat{\sigma}} \right) \qquad (2.11)$$

and then to estimate μ as in (2.9). This yields essentially the same asymptotic behavior as in the autoregression case, as was proved by Lee and Martin (1982).

However, ordinary M-estimators are not robust for additive outliers, as was pointed out by Martin (1979) in the context of autoregression. Therefore, it seems natural to apply GM-estimators [formulas (2.12) and (2.13) of Chapter 1] instead, because they can also bound the influence of leverage outliers. Some computational aspects and Monte Carlo results concerning the application of GM-estimators to autoregression models can be found in Martin and Zeh (1978), Denby and Martin (1979), and Zeh (1979), and some actual analyses were performed by Martin (1980).

Bustos (1982) proved consistency and asymptotic normality. Influence functions of GM-estimators in autoregression models were defined by Martin and Yohai (1984b, 1986); see also Künsch (1984). For the computation of GM-estimators in general ARMA models, Stockinger (1985) constructed an algorithm based on nonlinear regression and carried out a small Monte Carlo study with it.

Other methods of estimation involve robust instrumental variables (Martin 1981, Section VI), approximate non-Gaussian maximum likelihood (Martin 1981, Section VII; Pham Dinh Tuan 1984), and residual autocovariance estimators (Bustos and Yohai 1986, Bustos et al. 1984). Bruce and Martin (1987) considered multiple outlier diagnostics for time series. Robust tests for time series were considered by Basawa et al. (1985). A summary of robust filtering, smoothing, and spectrum estimation can be found in Martin and Thomson (1982), and Kassam and Poor (1985) presented a survey of robust techniques for signal processing.

c. Use of the LMS

Any attempt to define breakdown points for time series parameter estimators must take the nature of the failure mechanism into account, because breakdown can occur in different ways. Martin (1985) gave breakdown points for some estimators that yield to analytical calculations. Moreover, he constructed maximal bias curves that display the global bias–robustness of an estimator.

At any rate, it appears extremely useful to have a high-breakdown regression method when fitting autoregressive models to time series with outliers. Therefore, we propose to apply the LMS to the AR(p) model (2.4). Indeed, as we saw above, an isolated additive outlier occurs in $p + 1$ subsequent regression cases $(Y_t, Y_{t-1}, \ldots, Y_{t-p})$, yielding one vertical outlier and p leverage points. This means that the fraction of additive outliers in the original time series may give rise to a much higher fraction of contaminated data when it comes to fitting the autoregression, which is all the more reason for wanting to use a very robust regression estimator. The LMS can be applied to such situations [with $n - p$ cases and $p + 1$ coefficients, as in (2.4)] as easily as to any other data set. However, further research is necessary before the LMS may be used in general ARMA(p, q) models, because of the nonlinearity caused by the moving average part.

Example
Table 1, part (a) lists the monthly time series RESX which originated at Bell Canada. It describes the installation of residential telephone exten-

Table 1. Residential Extensions Data

Jan.	Feb.	Mar.	Apr.	May	June	July	Aug.	Sep.	Oct.	Nov.	Dec.
(a) Original Series RESX$_t$, from January 1966 to May 1973											
10165	9279	10930	15876	16485	14075	14168	14535	15367	13396	12606	12932
10545	10120	11877	14752	16932	14123	14777	14943	16573	15548	15838	14159
12689	11791	12771	16952	21854	17028	16988	18797	18026	18045	16518	14425
13335	12395	15450	19092	22301	18260	19427	18974	20180	18395	15596	14778
13453	13086	14340	19714	20796	18183	17981	17706	20923	18380	17343	15416
12465	12442	15448	21402	25437	20814	22066	21528	24418	20853	20673	18746
15637	16074	18422	27326	32883	24309	24998	25996	27583	22068	75344	47365
18115	15184	19832	27597	34256							
(b) Seasonal Differences Y$_t$ = RESX$_{t+12}$ − RESX$_t$, for t = 1, . . . , 77											
380	841	947	−1124	447	48	609	408	1206	2152	3232	1227
2144	1671	894	2200	4922	2905	2211	3854	1453	2497	680	266
646	604	2679	2140	447	1232	2439	177	2154	350	−922	353
118	691	−1110	622	−1505	−77	−1446	−1268	743	−15	1747	638
−988	−644	1108	1688	4641	2631	4085	3822	3495	2473	3330	3330
3172	3632	2974	5924	7446	3495	2932	4468	3165	1215	54671	28619
2478	−890	1410	271	1373							

Source: Martin et al. (1983).

sions in a fixed geographic area from January 1966 to May 1973, so it contains 89 observations. Looking at these data, it appears that the values for November and December 1972 are extremely large. These outliers have a known cause, namely a bargain month (November) with free installation of residence extensions, and a spillover effect in December because all of November's orders could not be filled in that month.

When analyzing these data, Brubacher (1974) first performed seasonal differencing, that is, he constructed the series

$$Y_t = \text{RESX}_{t+12} - \text{RESX}_t , \qquad t = 1, \dots, 77 , \qquad (2.12)$$

which is listed in Table 1, part (b) and shown in Figure 7. The outliers are now the observations 71 and 72. Brubacher found that this new series was stationary with zero mean (except for the outliers) and that it could adequately be described by an autoregressive model without intercept. In what follows, we shall first fit an AR(1) model and then an AR(2).

The AR(1) model is given by

$$Y_t = \alpha_1 Y_{t-1} + e_t . \qquad (2.13)$$

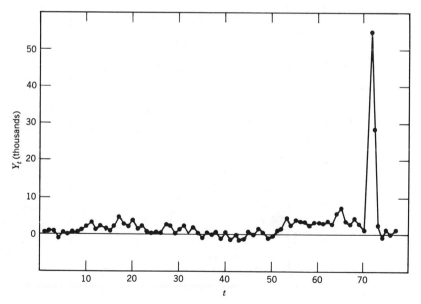

Figure 7. Plot of the series Y_t, $t = 1, \dots, 77$.

This model does not contain an intercept term because the expected value of Y_t appears to be about zero, so $\mu = 0$ in (2.1) and hence $\gamma = \mu(1 - \alpha_1) = 0$ in (2.4). The lag-one scatterplot (i.e., the plot of Y_t versus Y_{t-1}) is presented in Figure 8 and displays 76 pairs (Y_1, Y_2), $(Y_2, Y_3), \ldots, (Y_{76}, Y_{77})$. It looks a little bit like Figure 6d, indicating that a (small) patch of two additive outliers would provide a plausible description of the situation. The pair (Y_{70}, Y_{71}) is a regression outlier in the vertical direction, (Y_{71}, Y_{72}) is a leverage point that does not lie far away from the linear pattern formed by the majority, and (Y_{72}, Y_{73}) is a bad leverage point. Applying least squares to this data set yields the upper portion of column 1 in Table 2. The outliers have blown up $\hat{\sigma}$, so the only case with large standardized residual $|r_i/\hat{\sigma}|$ is (Y_{70}, Y_{71}). This means that the December 1972 value is masked in a routine LS analysis. On the other hand, the LMS obtains a higher slope $\hat{\alpha}_1$ because it is not dragged down by the bad leverage point, and its $\hat{\sigma}$ is four times smaller. Consequently, both (Y_{70}, Y_{71}) and (Y_{72}, Y_{73}) are now identified by their large $|r_i/\hat{\sigma}|$, whereas (Y_{71}, Y_{72}) has a small LMS residual, which confirms its classification as a good leverage point. The next column of Table 2 shows the LMS-based reweighted least squares (RLS) estimates, which are quite similar.

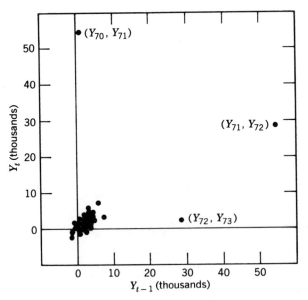

Figure 8. Plot of Y_t versus Y_{t-1}.

Table 2. Parameter Estimates for Y_t Series

Estimates	LS	LMS	RLS
AR(1) $\hat{\alpha}_1$	0.482	0.535	0.546
$\hat{\sigma}$	6585	1501	1422
AR(2) $\hat{\alpha}_1$	0.533	0.393	0.412
$\hat{\alpha}_2$	−0.106	0.674	0.501
$\hat{\sigma}$	6636	1162	1111

Let us now consider an AR(2) model

$$Y_t = \alpha_1 Y_{t-1} + \alpha_2 Y_{t-2} + e_t , \qquad (2.14)$$

which does not have an intercept term either (for the same reason as before). We try to fit the model (2.14) to the 75 triples (Y_1, Y_2, Y_3), $(Y_2, Y_3, Y_4), \ldots, (Y_{75}, Y_{76}, Y_{77})$. Now the two outliers Y_{71} and Y_{72} affect four regression cases:

Case no. 69: (Y_{69}, Y_{70}, Y_{71}) is a vertical regression outlier
Case no. 70: (Y_{70}, Y_{71}, Y_{72})
Case no. 71: (Y_{71}, Y_{72}, Y_{73}) $\Big\}$ are leverage points .
Case no. 72: (Y_{72}, Y_{73}, Y_{74})

Applying LS to all 75 cases yields the lower portion of column 1 in Table 2. Because the LS fit is attracted by the leverage points and $\hat{\sigma}$ is blown up, only case 69 may be identified through its large $|r_i/\hat{\sigma}|$, whereas for all other cases, $|r_i/\hat{\sigma}|$ is very small. The LMS fit in the next column is totally different and possesses a much lower value of $\hat{\sigma}$. The observations with large $|r_i/\hat{\sigma}|$ are now 69, 70, 71, and 72. Reweighting on the basis of the LMS yields comparable estimates.

Martin (1980) applied a GM-estimator to these data, yielding $\hat{\alpha}_1 = 0.51$ and $\hat{\alpha}_2 = 0.38$, and the robust filter used by Martin et al. (1983) gives a similar result. Both solutions are close to the estimates obtained by Brubacher (1974) by means of an interpolation technique described in Brubacher and Wilson (1976), which, however, requires that one specify the outlier positions in advance.

For the AR(2) model, all these robust estimators yield a positive $\hat{\alpha}_2$ and a small $\hat{\sigma}$, which makes them rather distinct from the naive LS fit. Nevertheless, it is somewhat discomforting to see relatively large differences between the various robust estimates (although they all identify the same outliers). A possible explanation may be the high correlation between Y_{t-1} and Y_{t-2}, at least when the outliers are taken out first (this can be seen from the lag-one plot in Figure 8), so we are almost in a collinearity situation: Whenever $\hat{\alpha}_1$ increases, $\hat{\alpha}_2$ will have to decrease. A way to avoid this problem would be to use only one explanatory variable, by sticking to the AR(1) model. However, Martin (1980) found that the AR(2) model is more suitable in this example, by means of a robustified version of Akaike's AIC function for estimating the order of the autoregression.

Also other people have used the LMS in a time series context. Heiler (1984) gave an example concerning the interest rates of a certain type of bonds. The data were recorded monthly from January 1965 to December 1982, so $n = 216$. He fitted AR(1), AR(2), and AR(3) models by means of both LS and LMS, and showed that the LMS coefficients were much easier to interpret. He further replaced the usual R^2 and partial correlation coefficients by means of Kendall's tau, which he then used to select the AR(2) fit. In his diploma thesis, Stahlhut (1985) analyzed the same example, as well as the residential extensions data and one other time series, and found that the LMS and the corresponding RLS yielded more reliable estimates and forecasts than did LS.

3. OTHER TECHNIQUES

In this final section we skim some other statistical topics in which robustification is emerging, and we provide a few supplementary references. Our selection is, of course, subjective and largely incomplete.

To begin with, there is the area of *orthogonal regression*, in which the residuals are no longer measured vertically, but instead are measured perpendicular to the regression line or hyperplane. This means that the actual (Euclidean) distances from the observations to the regression hyperplane are computed. This approach is more symmetric, in the sense that it does not matter which variable is selected as the response. On the other hand, standardization of the variables does not preserve orthogonality, so everything depends very much on the measurement units that are chosen by the user. In classical theory, the sum of the squared orthogonal distances is minimized. It is trivial to generalize the LMS by minimizing the *median* of the squared orthogonal distances. (Note that the LMS then really corresponds to the narrowest band covering half of the points, because the thickness of the band is now measured in the usual way.) Also from the computational point of view this presents no problems, because we only have to divide the original objective $\text{med}_i \, r_i^2$ (with "vertical" r_i^2) by the factor $\theta_1^2 + \cdots + \theta_{p-1}^2 + 1$, at each trial estimate. By adding a couple of lines of code, we made orthogonal versions of both PROGRESS (described in Section 1 of Chapter 5) and the simple regression program (see Section 2 of Chapter 5). The more general *errors in variables* model can be dealt with in the same way, as far as LMS is concerned. Recently, Brown (1982) constructed *w*-estimators in this context, whereas Zamar (1985) extended *M*-estimators. Kelly (1984) defined influence functions for the errors in variables situation.

Another important field is that of *nonlinear regression*, which is also useful in time series analysis and generalized linear models. The classical approach is based on the LS criterion and makes use of several iterative methods such as the Gauss–Newton, steepest descent, and Marquardt algorithms. Dutter and Huber (1981) generalized the nonlinear LS algorithm of Nagel and Wolff (1974) to nonlinear robust regression. Ways in which the LMS estimator may be applied to nonlinear problems are presently being investigated.

Nowadays there is a lot of interest in the robustification of *analysis of variance*. For the approach based on *M*-estimators we refer to Huber (1981, Section 7.10). Robust tests in general linear models (making use of GM-estimators) can be found in Hampel et al. (1986, Chapter 7). In two-way tables, the median polish provides a robust estimator of row and column effects (for a recent survey, see Hoaglin et al. 1983, 1985). By means of zero–one variables, a two-way table can also be written as a linear model, but it is not always possible to apply the LMS to it because the number of cases (cells) must be larger than twice the number of parameters. On the other hand, the LMS becomes more useful when also interval-scaled covariates occur, which may contain outliers.

Another topic is estimation on the circle and the sphere, corresponding to *directional data* (Watson 1983). On these spaces the worst kind of contamination is of an asymmetric type, when the outliers are 90° away from the "true" direction. Ko (1985) used the LMS in this framework and found that it could withstand a high percentage of contamination. For instance, he applied the LMS (which amounts to the "shortest arc" on the circle) to the well-known frog data (Ferguson et al. 1967, Collett 1980), in which the sun compass orientation of 14 frogs was investigated. In this example, the LMS comes close to the true home direction. Recently, Ko and Guttorp (1986) considered a standardized version of the gross-error-sensitivity that is tailored to estimators for directional data.

The currently used algorithms for *multidimensional scaling* are very susceptible to outlying dissimilarities, partly because these methods employ some variant of the classical Young–Householder–Torgerson scaling procedure for their initial estimator. There is a growing awareness of this problem, causing a recent interest in robustness aspects. A type of influence function was defined in this setting by de Leeuw and Meulman (1986), following a suggestion of Kruskal and Wish (1978, pp. 58–60).

Spence and Lewandowsky (1985) proposed a resistant multidimensional scaling method and implemented it in a program called TUFSCAL. Their algorithm is a median-type modification of Newton's method, and to avoid iteration toward a wrong solution they use a very robust starting configuration based on ranks. The new method was compared to some of the currently employed alternatives in a simulation study, and its break-down value turned out to be quite high (although it is probably hard to summarize in a single number because it depends on the particular pattern of outliers). Influence functions for the *principal components* problem have been computed recently by Critchley (1985), and influence functions for various forms of *correspondence analysis* and *canonical analysis* can be deduced from Gifi (1981) and de Leeuw (1984).

In biological studies of morphological change, one makes use of *shape comparison* techniques. Suppose that the first shape is characterized by the points $x_1, \ldots, x_n$ in the plane, and the second is characterized by $y_1, \ldots, y_n$. The points x_i now have to go through a rigid transformation f (composed of rotation, translation, and magnification) such that the images $f(x_i)$ come as close as possible to the y_i. The classical LS criterion is

$$\text{Minimize} \sum_{f}^{n} \| f(x_i) - y_i \|^2 . \qquad (3.1)$$

Its nonrobustness causes the effects of large deformations to become smeared out and therefore obscured. Siegel and Benson (1982) use repeated medians to obtain a robust comparison, which succeeds in detecting localized shape differences (e.g., in an example on primate skulls). The method was generalized to three dimensions by Siegel and Pinkerton (1982) and used for protein molecules. By replacing the sum in (3.1) by a median, the LMS could also be applied.

In astronomy, one often has to *estimate shifts* between stellar spectra. For instance, this happens when studying the orbits of binary star systems, in which the same star may be moving toward us at one point in time, but away from us at another time. This change of velocity may be measured by means of the Doppler effect, which causes the star spectrum to shift to the red or to the blue. To estimate the actual shift between two observed spectra in an automatic way, astronomers have typically used LS or product cross-correlation methods. However, it turns out that these techniques run into trouble in the case of "noisy" spectra, which occur frequently for hot and luminous stars. For this reason, Rousseeuw (1987a) proposed to apply the L_1 approach, that is, to choose the shift that minimizes the sum of absolute values of the differences between the intensities g and h:

$$\underset{\Delta\lambda}{\text{Minimize}} \sum_{i} |g(\lambda_i) - h(\lambda_i + \Delta\lambda)| . \tag{3.2}$$

This method is more robust against inaccuracies in the data, and it appears to be very powerful for analyzing complicated spectra, where asymmetric peaks, as well as a combination of absorption and emission features, may occur. The method was tried out on artificial profiles and on the spectrum of Beta Arietes. In all cases it compared favorably with the classical approach. The reason why L_1 does so well here is that there can be no leverage points in this one-dimensional framework. Therefore, it does not appear necessary to replace the sum in (3.2) by a median, which would be a lot more expensive in view of the size of astronomical data sets.

A similar situation occurs in cluster analysis, at least in those cases where orthogonal equivariance is sufficient (things become more difficult when affine equivariance is imposed). Indeed, in the social sciences many data sets consist of a collection of subjective distances (or dissimilarities) between entities for which no measurements or coordinates are available. Such entities are not easily represented as points in a linear space, so affine equivariance is not applicable, whereas orthogonal transformations on the original points (when the latter exist) do preserve the collection of distances. In this context, Kaufman and Rousseeuw (1988) apply L_1-type methods. For instance, the k-median approach searches for k representative objects (called "medoids") such that

$$\frac{1}{n} \sum_{i=1}^{n} d(i, m(i)) \tag{3.3}$$

is minimized, where $d(i, m(i))$ is the distance (or dissimilarity) of object i to the closest medoid, denoted by $m(i)$. Then a partitioning into k clusters is constructed by assigning each object to the nearest medoid. This method is a robust alternative to the classical k-means technique, in which sums of *squared* distances are minimized. Again, the L_1 approach is suitable because there exist no leverage points in this situation. Also, graphical methods can be used to discover outliers in such data sets (see, e.g., Rousseeuw 1987b and the references cited therein). Recently, Cooper and Milligan (1985) examined the effect of outliers on hierarchical clustering procedures.

EXERCISES AND PROBLEMS

Section 1

1. Show that the L_1 estimator is orthogonal equivariant, and construct a small two-dimensional example to illustrate that it is not affine equivariant for $p \geq 2$. What happens for $p = 1$?

2. Show that the coordinatewise median lies in the convex hull of the sample when $p \leq 2$.

3. Suppose that T is affine equivariant, $\mathbf{A}$ is a nonsingular matrix, $\mathbf{b}$ is any vector, and X is any sample of size n. Then show that $\varepsilon_n^*(T, X\mathbf{A} + \mathbf{b}) = \varepsilon_n^*(T, X)$ by using the fact that affine transformations map bounded sets on bounded sets.

4. (From Donoho 1982.) Show that convex peeling is affine equivariant, and draw the subsequent convex hulls by hand in a two-dimensional example. What is the most harmful configuration of outliers?

5. Show that the outlyingness-weighted mean given by (1.19) and (1.20) is affine equivariant.

6. Construct an approximate algorithm for the outlyingness-weighted mean, by either (a) projecting on lines through pairs of points, as done by Donoho or (b) projecting on lines orthogonal to hyperplanes through p points, as proposed by Stahel. Which algorithm is faster? Show that (b) is affine equivariant but that (a) is not.

7. What happens to the estimators (1.4), (1.8), (1.9), (1.11), (1.18), (1.20), (1.22), and (1.31) if $p = 1$? Compare these special cases by means of one of the univariate data sets of Chapter 4.

8. Show that the breakdown point of the MVE estimator becomes exactly (1.36) when $h = [(n + p + 1)/2]$.

9. Use some standard statistical program to compute the principal components of the $\mathbf{x}$-part of the modified wood gravity data, before and after deletion of the outliers identified by the MVE. Repeat this for the $\mathbf{x}$-part of the Hawkins–Bradu–Kass data.

10. Show that the breakdown point of any translation equivariant estimator of location is at most (1.38), and give an example to show that this bound is sharp.

11. Show that the breakdown point of any affine equivariant covariance estimator is at most $[(n - p + 1)/2]/n$. (Solution in Davies 1987.)

Section 2

12. (From Heiler 1984 and Stahlhut 1985.) Use PROGRESS to fit AR(1), AR(2), and AR(3) models to the time series of Table 3.

Table 3. Monthly Interest Rates of Bonds, from January 1965 to December 1982

Year	Jan.	Feb.	Mar.	Apr.	May	June	July	Aug.	Sep.	Oct.	Nov.	Dec.
1965	6.3	6.2	6.4	6.6	6.9	7.1	7.2	7.4	7.4	7.4	7.5	7.6
1966	7.6	7.6	7.6	7.7	7.8	8.0	8.3	8.6	8.5	8.1	7.9	7.6
1967	7.6	7.5	7.3	7.0	6.8	6.7	6.8	6.8	6.8	6.8	6.7	6.8
1968	6.8	6.8	6.8	6.8	6.6	6.5	6.5	6.3	6.3	6.3	6.4	6.4
1969	6.2	6.2	6.3	6.5	6.6	6.8	6.9	7.1	7.1	7.3	7.2	7.1
1970	7.4	7.8	8.2	8.2	8.1	8.4	8.6	8.5	8.5	8.6	8.6	8.3
1971	7.9	7.7	7.7	7.8	7.9	8.1	8.3	8.3	8.2	8.0	7.9	7.8
1972	7.8	7.6	7.4	7.6	8.0	8.1	8.2	8.1	8.1	8.1	8.3	8.4
1973	8.6	8.5	8.5	8.6	9.2	9.8	9.8	9.9	9.6	9.8	9.4	9.5
1974	9.5	9.6	10.3	10.6	10.6	10.5	10.6	10.6	10.3	10.3	10.2	9.7
1975	9.4	8.8	8.7	8.6	8.3	8.2	8.3	8.3	8.7	8.7	8.5	8.2
1976	8.2	8.0	7.8	7.7	7.9	8.1	8.2	8.2	8.0	7.9	7.5	7.3
1977	7.1	6.9	6.6	6.6	6.3	6.3	6.3	6.0	6.0	5.9	5.9	5.9
1978	5.8	5.7	5.5	5.4	5.8	6.0	6.2	6.5	6.3	6.3	6.4	6.5
1979	6.6	6.8	6.9	7.1	7.4	7.9	7.9	7.6	7.6	7.7	8.1	7.9
1980	7.9	8.2	9.1	9.5	8.7	8.2	8.0	7.8	8.1	8.4	8.8	8.8
1981	9.0	9.6	10.1	10.0	10.2	10.9	10.5	11.0	11.2	10.4	10.0	9.7
1982	9.8	9.7	9.5	8.9	8.7	9.1	9.3	9.0	8.7	8.3	8.2	7.9

Source: Statistische Beihefte zu den Monatsberichten der Deutschen Bundesbank: Renditen von festverzinslicher Wertpapiere, Neuemissionen.

Compare the LS and LMS coefficients. Can you identify outliers? (Also look at a plot of the data.) Which p would you select?

13. Take a time series (from your own experience) that is adequately described by an AR(p) model and analyze it by means of PROGRESS to compare the LS and LMS estimates. Then throw in a few additive outliers and run the program again to investigate their effect on both estimators.

Section 3

14. Take a simple regression data set (e.g., from Chapter 2). Is the LMS regression of y on x the same as that of x on y? What happens when the orthogonal variant of LMS is used?

15. (From Ko 1985.) Ferguson et al. (1967) described an experiment in which 14 frogs were captured, transported to another place, and then released to see if they would find their way back. Taking $0°$ to be due north, the directions taken by these frogs were

$$104°,\ 110°,\ 117°,\ 121°,\ 127°,\ 130°,\ 136°,$$
$$145°,\ 152°,\ 178°,\ 184°,\ 192°,\ 200°,\ 316°.$$

(a) Plot these data on a circle with center 0 and radius 1. Indicate the approximate position of the "mode" (in a loose sense).

(b) Compute the population mean, given by the point

$$T = \frac{\sum\limits_{i=1}^{14} \mathbf{x}_i}{\left\| \sum\limits_{i=1}^{14} \mathbf{x}_i \right\|},$$

where the $\mathbf{x}_i$ are the observations on the unit circle.

(c) Collett (1980) found that there is some evidence to indicate that $316°$ is an outlier. Does it make much difference to the population mean whether this point is deleted? Explain.

(d) Compute (approximately) the spherical median of these data, given by

$$\underset{\hat{\theta}}{\text{Minimize}} \sum_{i=1}^{14} d(\mathbf{x}_i, \hat{\theta}),$$

where the distance d is measured along the circle.

(e) Compute the LMS estimator, corresponding to the midpoint of the shortest arc containing (at least) eight points. Which points do now appear to be outliers?

(f) Which of these estimates comes closest to the true home direction 122°? Compare the various estimates with the "mode" (a). Which outliers are influential for (b)?

16. How would you generalize the LMS to data on a sphere?

17. Apply a standard multidimensional scaling program to a data set in which you replace a few dissimilarities by large values.

References

Sections in which a particular reference is cited are given in brackets. The number preceding the decimal point refers to the chapter; the number following the decimal point refers to the section.

Abrahamowicz, M., and Stegun, I. A. (1964), *Handbook of Mathematical Functions*, National Bureau of Standards No. 55, Washington, D.C. [4.4]

Adichie, J. N. (1967), Estimation of regression coefficients based on rank tests, *Ann. Math. Stat.*, **38**, 894–904. [1.2, 3.6]

Afifi, A. A., and Azen, S. P. (1979), *Statistical Analysis, A Computer Oriented Approach*, 2nd ed., Academic Press, New York. [2.6]

Aitkin, M. A. (1974), Simultaneous influence and the choice of variable subsets in multiple regression, *Technometrics*, **16**, 221–227. [3.3]

Allen, D. M. (1974), The relationship between variable selection and data augmentation and a method for prediction, *Technometrics*, **16**, 125–127. [3.3]

Andrews, D. F. (1974), A robust method for multiple linear regression, *Technometrics*, **16**, 523–531. [1.2, 2.7, 3.1, 3.6]

Andrews, D. F., and Pregibon, D. (1978), Finding the outliers that matter, *J. R. Stat. Soc. Ser. B*, **40**, 85–93. [2.4, 3.1, 6.3, 6.4]

Andrews, D. F., Bickel, P. J., Hampel, F. R., Huber, P. J., Rogers, W. H., and Tukey, J. W. (1972), *Robust Estimates of Location: Survey and Advances*, Princeton University Press, Princeton, N.J. [4.1, 4.4, 4.5, 5.4]

Atkinson, A. C. (1982), Regression diagnostics, transformations and constructed variables, *J. R. Stat. Soc. Ser. B*, **44**, 1–36. [3.1, 6.2]

Atkinson, A. C. (1983), Diagnostic regression for shifted power transformations, *Technometrics*, **25**, 23–33. [6.2]

Atkinson, A. C. (1985), *Plots, Transformations, and Regression*, Clarendon Press, Oxford. [3.1]

Atkinson, A. C. (1986), Masking unmasked, *Biometrika*, **73**, 533–541. [6.6]

292

Atkinson, A. C. (1987), Transformations unmasked, Technical Report, Department of Mathematics, Imperial College, London. [6.6]

Barnett, V. (1976), The ordering of multivariate data, *J. R. Stat. Soc. Ser. A*, **138**, 318–344. [7.1]

Barnett, V., and Lewis, T. (1978), *Outliers in Statistical Data*, John Wiley & Sons, New York. 2nd edition: 1985. [4.1, 4.3, 7.1]

Barrett, J. P. (1974), The coefficient of determination—Some limitations, *Am. Stat.*, **28**, 19–20. [3.4]

Barrodale, I., and Roberts, F. D. K. (1975), Algorithm 478: Solution of an overdetermined system of equations in the L_1 norm [F4], *Commun. ACM*, **17**(6), 319–320. [3.6]

Bartlett, M. S. (1949), Fitting a straight line when both variables are subject to error, *Biometrics*, **5**, 207–212. [1.2, 2.7]

Basawa, I. V., Huggins, R. M., and Staudte, R. G. (1985), Robust tests for time series with an application to first-order autoregressive processes, *Biometrika*, **72**, 559–571. [7.2]

Baskerville, J. C., and Toogood, J. H. (1982), Guided regression modeling for prediction and exploration of structure with many explanatory variables, *Technometrics*, **24**, 9–17. [3.3]

Beaton, A. E. (1964), The Use of Special Matrix Operators in Statistical Calculus, Research Bulletin RB-64-51, Educational Testing Service, Princeton, N.J. [3.3]

Beaton, A. E., and Tukey, J. W. (1974), The fitting of power series, meaning polynomials, illustrated on band-spectroscopic data, *Technometrics*, **16**, 147–185. [3.4]

Bebbington, A. C. (1978), A method of bivariate trimming for estimation of the correlation coefficient, *Appl. Stat.*, **27**, 221–226. [7.1]

Belsley, D. A. (1984), Demeaning conditioning diagnostics through centering (with discussion), *Am. Stat.*, **38**, 73–93. [3.3]

Belsley, D. A., Kuh, E., and Welsch, R. E. (1980), *Regression Diagnostics*, John Wiley & Sons, New York. [3.3, 6.2, 6.3]

Berk, K. N. (1978), Comparing subset regression procedures, *Technometrics*, **20**, 1–6. [3.3]

Bickel, P. J. (1973), On some analogues to linear combination of order statistics in the linear model, *Ann. Stat.*, **1**, 597–616. [1.2, 3.6]

Bickel, P. J. (1975), One-step Huber estimates in the linear model, *J. Am. Stat. Assoc.*, **70**, 428–433. [3.4, 4.1]

Bickel, P. J., and Hodges, J. L. (1967), The asymptotic theory of Galton's test and a related simple estimate of location, *Ann. Math. Stat.*, **4**, 68–85. [4.1]

Bloomfield, P., and Steiger, W. (1980), Least absolute deviations curve-fitting, *SIAM J. Sci. Stat. Comput.*, **1**, 290–301. [1.2]

Bloomfield, P., and Steiger, W. L. (1983), *Least Absolute Deviations: Theory, Applications, and Algorithms*, Birkhäuser Verlag, Boston. [1.2]

Boente, G., Fraiman, R., and Yohai, V. J. (1982), Qualitative Robustness for General Stochastic Processes, Technical Report No. 26, Department of Statistics, University of Washington, Seattle. [7.2]

Box, G. E. P. (1953), Non-normality and tests on variances, *Biometrika*, **40**, 318–335. [1.1]

Box, G. E. P., and Jenkins, G. M. (1976), *Time Series Analysis: Forecasting and Control*, 2nd ed., Holden-Day, San Francisco. [7.2]

Brown, M. L. (1982), Robust line estimation with errors in both variables, *J. Am. Stat. Assoc.*, **77**, 71–79. [7.3]

Brown, B. M., and Hettmansperger, T. P. (1985), Affine Invariant Rank Methods in the Bivariate Location Model, Technical Report No. 58, Department of Statistics, Pennsylvania State University. [7.1]

Brown, G. W., and Mood, A. M. (1951), On median tests for linear hypotheses, in *Proceedings of the 2nd Berkeley Symposium on Mathematical Statistics and Probability*, edited by J. Neyman, University of California Press, Berkeley and Los Angeles, pp. 159–166. [1.2, 2.7]

Brownlee, K. A. (1965), *Statistical Theory and Methodology in Science and Engineering*, 2nd ed., John Wiley & Sons, New York. [3.1]

Brubacher, S. R. (1974), Time Series Outlier Detection and Modeling with Interpolation, Bell Laboratories Technical Memo. [7.2]

Brubacher, S. R., and Wilson, G. T. (1976), Interpolating time series with application to the estimation of holiday effects on electricity demand, *Appl. Stat.*, **25**, 107–116. [7.2]

Bruce, A., and Martin, R. D. (1987), Leave-k-Out Diagnostics for Time Series, Technical Report, Department of Statistics, University of Washington, Seattle. In preparation. [7.2]

Bustos, O. H. (1981), Qualitative Robustness for General Processes, Informes de Matemática, Série B, No. 002/81, Instituto de Matemática Pura e Aplicada, Rio de Janeiro, Brazil. [7.2]

Bustos, O. H. (1982). General M-estimates for contaminated p-th order autoregressive processes: Consistency and asymptotic normality, *Z. Wahrsch. Verw. Geb.*, **59**, 491–504. [7.2]

Bustos, O. H., and Yohai, V. J. (1986), Robust estimates for ARMA models, *J. Am. Stat. Assoc.*, **81**, 155–168. [7.2]

Bustos, O. H., Fraiman, R., and Yohai, V. J. (1984), Asymptotic behavior of the estimates based on residual autocovariances for ARMA models, in *Robust and Nonlinear Time Series Analysis*, edited by J. Franke, W. Härdle, and R. D. Martin, Springer Verlag, New York. pp. 26–49. [7.2]

Campbell, N. A. (1980), Robust procedures in multivariate analysis I: Robust covariance estimation, *Appl. Stat.*, **29**, 231–237. [7.1]

Campbell, N. A. (1982), Robust procedures in multivariate analysis II: Robust canonical variate analysis, *Appl. Stat.*, **31**, 1–8. [7.1]

Carroll, R. J., and Ruppert, D. (1985), Transformations in regression: A robust analysis, *Technometrics*, **27**, 1–12. [3.1, 6.6]

Chambers, J. M., Cleveland, W. S., Kleiner, B., and Tukey, P. A. (1983), *Graphical Methods for Data Analysis*, Wadsworth International Group, Belmont, CA. [3.2, 3.3]

Chatterjee, S., and Hadi, A. S. (1986), Influential observations, high leverage points, and outliers in linear regression (with discussion), *Statist. Sci.*, **1**, 379–416. [6.6]

Chatterjee, S., and Price, B. (1977), *Regression Analysis by Example*, John Wiley, & Sons, New York. [3.3]

Cheng, K. S., and Hettmansperger, T. P. (1983), Weighted least-squares rank estimates, *Commun. Stat. (Theory and Methods)*, **12**, 1069–1086. [3.6]

Chernoff, H. (1964), Estimation of the mode, *Ann. Inst. Stat. Math.*, **16**, 31–41. [4.4]

Coleman, J., et al. (1966), *Equality of Educational Opportunity*, two volumes, Office of Education, U.S. Department of Health, Washington, D.C. [3.1]

Collett, D. (1980), Outliers in circular data, *Appl. Stat.*, **29**, 50–57. [7.3, 7.exercise 15]

Collins, J. R. (1982), Robust *M*-estimators of location vectors, *J. Multivar. Anal.*, **12**, 480–492. [7.1]

Cook, R. D. (1977), Detection of influential observation in linear regression, *Technometrics*, **19**, 15–18. [6.2, 6.3]

Cook, R. D. (1979), Influential observations in regression, *J. Am. Stat. Assoc.*, **74**, 169–174. [3.1]

Cook, R. D. (1986), Assessment of local influence (with discussion), *J. R. Statist. Soc. B*, **48**, 133–169. [6.6]

Cook, R. D., and Weisberg, S. (1982), *Residuals and Influence in Regression*, Chapman & Hall, London [6.2, 6.3, 6.4]

Cooper, M. C., and Milligan, G. W. (1985), An Examination of the Effect of Outliers on Hierarchical Clustering Procedures, paper presented at the Fourth European Meeting of the Psychometric Society and the Classification Societies, July 2–5, Cambridge, United Kingdom. [7.3]

Cox, D. D. (1981), Metrics on Stochastic Processes and Qualitative Robustness, Technical Report No. 3, Department of Statistics, University of Washington, Seattle. [7.2]

Critchley, F. (1985), Influence in principal components analysis, *Biometrika*, **72**, 627–636. [7.3]

Cushny, A. R., and Peebles, A. R. (1905), The action of optical isomers. II. Hyoscines, *J. Physiol.*, **32**, 501–510. [4.exercise 10]

D^2 Software (1986), *MACSPIN: Graphical Data Analysis Software for the Macintosh*, Wadsworth, Belmont, CA. [3.5]

Daniel, C., and Wood, F. S. (1971), *Fitting Equations to Data*, John Wiley & Sons, New York. [2.1, 3.1, 3.3]

David, H. A. (1981), *Order Statistics*, John Wiley & Sons, New York. [7.1]

Davies, P. L. (1987), Asymptotic Behavior of *S*-estimates of multivariate location parameters and dispersion matrices, Technical Report, University of Essen, West-Germany. [7.1, 7.exercise 11]

De Leeuw, J. (1984), Statistical properties of multiple correspondence analysis, Technical Report, Department of Data Theory, University of Leiden, The Netherlands. [7.3]

De Leeuw, J., and Meulman, J. (1986), A special jackknife for multidimensional scaling, *J. Class.*, **3**, 97–112. [7.3]

Dempster, A. P., and Gasko–Green, M. (1981), New tools for residual analysis, *Ann. Stat.*, **9**, 945–959. [3.1, 6.3]

Denby, L., and Martin, R. D. (1979), Robust estimation of the first-order autoregressive parameter, *J. Am. Stat. Assoc.*, **74**, 140–146. [7.2]

Devlin, S. J., Gnanadesikan, R., and Kettenring, J. R. (1975), Robust estimation and outlier detection with correlation coefficients, *Biometrika*, **62**, 531–545. [7.1]

Devlin, S. J., Gnanadesikan, R., and Kettenring, J. R. (1981), Robust estimation of dispersion matrices and principal components, *J. Am. Stat. Assoc.*, **76**, 354–362. [7.1]

Devroye, L., and Gyorfi, L. (1984), *Nonparametric Density Estimation: The L_1 View*, Wiley–Interscience, New York. [1.2]

De Wit, G. W. (1982), *Informatica Statistica*, Research Nationale Nederlanden N.V., Rotterdam, The Netherlands. [2.exercise 8]

Diaconis, P., and Efron, B. (1983), Computer-intensive methods in statistics, *Sci. Amer.*, **248**, 116–130. [1.2, 5.1]

Diehr, G., and Hoflin, D. R. (1974), Approximating the distribution of the sample R^2 in best subset regressions, *Technometrics*, **16**, 317–320. [3.3]

Dixon, W. J., and Tukey, J. W. (1968), Approximate behavior of the distribution of winsorized t (Trimming/Winsorization 2), *Technometrics*, **10**, 83–93. [4.1]

Dodge, Y. (1984), Robust estimation of regression coefficients by minimizing a convex combination of least squares and least absolute deviations, *Computational Statistics Quarterly*, **1**, 139–153. [3.6]

Donoho, D. L. (1982), Breakdown Properties of Multivariate Location Estimators, qualifying paper, Harvard University, Boston, M.A. [6.6, 7.1, 7.exercise 4]

Donoho, D. L. (1984), High Breakdown Made Simple, subtalk presented at the Oberwolfach Conference on Robust Statistics, West Germany, 9–15 September. [2.2, 2.5]

Donoho, D. L., and Huber, P. J. (1983), The notion of breakdown point, in *A Festschrift for Erich Lehmann*, edited by P. Bickel, K. Doksum, and J. L. Hodges, Jr., Wadsworth, Belmont, CA. [1.2, 3.4, 3.6]

Donoho, D. L., and Liu, R. C. (1986), The automatic robustness of minimum distance functionals I, Technical Report. [7.1]

Donoho, D. L., Johnstone, I., Rousseeuw, P. J., and Stahel, W. (1985), Discussion on "Projection Pursuit" of P. Huber, *Ann. Stat.*, **13**, 496–500. [3.5]

Draper, N. R., and John, J. A. (1981), Influential observations and outliers in regression, *Technometrics*, **23**, 21–26. [2.4, 6.2, 6.3, 6.4]

Draper, N. R., and Smith, H. (1966), *Applied Regression Analysis*, John Wiley & Sons, New York [3.1, 3.3, 6.6]

Dutter, R. (1977), Numerical solution of robust regression problems: Computational aspects, a comparison, *J. Stat. Comput. Simulation*, **5**, 207–238. [1.2, 3.6]

Dutter, R. (1983a), Computer Program BLINWDR for Robust and Bounded Influence Regression, Research Report No. 8, Institute for Statistics, Technische Universität Graz, Austria. [3.6]

Dutter, R. (1983b), COVINTER: A Computer Program for Computing Robust Covariances and for Plotting Tolerance Ellipses, Research Report No. 10, Institute for Statistics, Technische Universität Graz, Austria. [7.1]

Dutter, R., and Huber, P. J. (1981), Numerical methods for the nonlinear robust regression problem, *J. Stat. Comput. Simulation*, **13**, 79–114. [7.3]

Edgeworth, F. Y. (1887), On observations relating to several quantities, *Hermathena*, **6**, 279–285. [1.2]

Efron, B. (1979), Bootstrap methods: Another look at the jackknife, *Ann. Stat.*, **7**, 1–26. [5.1]

Efroymson, M. A. (1960), Multiple regression analysis, in *Mathematical Methods for Digital Computers*, edited by A. Ralston and H. S. Wilf, John Wiley & Sons, New York, pp. 191–203. [3.3]

Ellerton, R. R. W. (1978), Is the regression equation adequate?—A generalization, *Technometrics*, **20**, 313–315. [3.3]

Emerson, J. D., and Hoaglin, D. C. (1983), Resistant lines for *y* versus *x*, in *Understanding Robust and Exploratory Data Analysis*, edited by D. Hoaglin, F. Mosteller, and J. Tukey, John Wiley & Sons, New York, pp. 129–165. [2.5, 2.7]

Emerson, J. D., and Hoaglin, D. C. (1985), Resistant multiple regression, one variable at a time, in *Exploring Data Tables, Trends, and Shapes*, edited by D. Hoaglin, F. Mosteller, and J. Tukey. John Wiley & Sons, New York, pp. 241–279. [2.7]

Ercil, A. (1986), Robust Ridge Regression, paper presented at the Joint Statistical Meetings, Chicago, August 18–21. [3.3]

Ezekiel, M., and Fox, K. A. (1959), *Methods of Correlation and Regression Analysis*, John Wiley & Sons, New York. [2.6]

Ferguson, D. E., Landreth, H. F. and McKeown, J. P. (1967), Sun compass orientation of the northern cricket frog, *Anim. Behav.*, **15**, 45–53. [7.3, 7.exercise 15]

Fill, J. A., and Johnstone, I. M. (1984), On projection pursuit measures of multivariate location and dispersion, *Ann. Stat.*, **12**, 127–141. [3.5]

Forsythe, A. B. (1972), Robust estimation of straight line regression coefficients by minimizing *p*-th power deviations, *Technometrics*, **14**, 159–166. [3.6]

Fox, A. J. (1972), Outliers in time series, *J. R. Stat. Soc. Ser. B*, **34**, 350–363. [7.2]

Friedman, J. H., and Stuetzle, W. (1980), Projection pursuit classification, unpublished manuscript. [3.5]

Friedman, J. H., and Stuetzle, W. (1981), Projection pursuit regression, *J. Am. Stat. Assoc.*, **76**, 817–823. [3.5]

Friedman, J. H., and Stuetzle, W. (1982), Projection pursuit methods for data analysis, in *Modern Data Analysis*, edited by R. Launer and A. F. Siegel, Academic Press, New York. [3.5]

Friedman, J. H., and Tukey, J. W. (1974), A projection pursuit algorithm for exploratory data analysis, *IEEE Trans. Comput.*, **C-23**, 881–889. [3.5]

Friedman, J. H., Stuetzle, W., and Schroeder, A. (1984), Projection pursuit density estimation, *J. Am. Stat. Assoc.*, **79**, 599–608. [3.5]

Furnival, G. M., and Wilson, R. W., Jr. (1974), Regressions by leaps and bounds, *Technometrics*, **16**, 499–511. [3.3]

Galilei, G. (1638), *Dialogues Concerning Two New Sciences*, published in Leiden, The Netherlands. [4.5]

Gasko, M., and Donoho, D. (1982), Influential observation in data analysis, in *American Statistical Association Proceedings of the Business and Economic Statistics Section*, ASA, Washington, D.C., pp. 104–109. [7.1]

Gentle, J. E., Kennedy, W. J., and Sposito, V. A. (1977), On least absolute values estimation, *Commun. Stat. (Theory and Methods)*, **6**, 839–845. [1.2]

Gentleman, W. M. (1965), Robust Estimation of Multivariate Location by Minimizing p-th Power Transformations, Ph.D. dissertation, Princeton University, and Bell Laboratories memorandum M65-1215-16. [3.5, 3.6]

Gifi, A. (1981), *Nonlinear Multivariate Analysis*, Department of Data Theory, University of Leiden, The Netherlands. [7.3]

Gnanadesikan, R., and Kettenring, J. R. (1972), Robust estimates, residuals, and outlier detection with multiresponse data, *Biometrics*, **28**, 81–124. [4.4, 7.1]

Gorman, J. W., and Toman, R. J. (1966), Selection of variables for fitting equations to data, *Technometrics*, **8**, 27–51. [3.3]

Gray, J. B. (1985), Graphics for Regression Diagnostics, *American Statistical Association Proceedings of the Statistical Computing Section*, ASA, Washington, D.C., pp. 102–107. [3.exercise 1, 6.exercise 6]

Gray, J. B., and Ling, R. F. (1984), K-clustering as a detection tool for influential subsets in regression, *Technometrics*, **26**, 305–330. [6.5]

Groeneboom, P. (1987), Brownian motion with a parabolic drift and airy functions, to appear in *J. Probability Theory and Related Fields*. [4.4]

Grubbs, F. E. (1969), Procedures for detecting outlying observations in samples, *Technometrics*, **11**, 1–21. [4.exercise 11]

Hampel, F. R. (1971), A general qualitative definition of robustness, *Ann. Math. Stat.*, **42**, 1887–1896. [1.2, 7.2]

Hampel, F. R. (1973), Robust estimation: A condensed partial survey, *Z. Wahrsch. Verw. Geb.*, **27**, 87–104. [7.1]

Hampel, F. R. (1974), The influence curve and its role in robust estimation, *J. Am. Stat. Assoc.*, **69**, 383–393. [Preface, 1.2, 3.6, 4.5]

Hampel, F. R. (1975), "Beyond location parameters: Robust concepts and methods, *Bull. Int. Stat. Inst.*, **46**, 375–382. [1.2, 2.exercise 13]

Hampel, F. R. (1978), Optimally bounding the gross—error sensitivity and the influence of position in factor space, in *Proceedings of the Statistical Computing Section of the American Statistical Association*, ASA, Washington, D.C., 59–64. [1.2]

Hampel, F. R. (1985), The breakdown points of the mean combined with some rejection rules, *Technometrics*, **27**, 95–107. [4.1]

Hampel, F. R., Rousseeuw, P. J., and Ronchetti, E. (1981), The change-of-variance curve and optimal redescending M-estimators, *J. Am. Stat. Assoc.*, **76**, 375, 643–648. [3.4]

Hampel, F. R., Ronchetti, E. M., Rousseeuw, P. J., and Stahel, W. A. (1986), *Robust Statistics: The Approach Based on Influence Functions*, John Wiley & Sons, New York. [1.2, 2.6, 3.6, 4.1, 4.5, 7.1, 7.3]

Harter, H. L. (1977), Nonuniqueness of least absolute values regression, *Commun. Stat. (Theory and Methods)*, **6**, 829–838. [1.2]

Hawkins, D. M. (1980), *Identification of Outliers*, Chapman & Hall, London. [4.1]

Hawkins, D. M., Bradu, D., and Kass, G. V. (1984), Location of several outliers in multiple regression data using elemental sets, *Technometrics*, **26**, 197–208. [3.3, 5.1, 6.2, 6.5, 6.6]

Healy, M. J. R. (1968), Multivariate normal plotting, *Appl. Stat.*, **17**, 157–161. [7.1]

Heiler, S. (1981), Robust estimates in linear regression—A simulation approach, in *Computational Statistics*, edited by H. Büning and P. Naeve, De Gruyter, Berlin, pp. 115–136. [3.6]

Heiler, S. (1984), Robuste Schätzung bei Zeitreihen und in dynamischen Ökonometrischen Modellen, talk presented at the Statistische Woche (Deutsche Statistische Gesellschaft), October 22–26, Augsburg, West Germany. [7.2, 7.exercise 12]

Heiler, S., and Willers, R. (1979), Asymptotic Normality of R-Estimates in the Linear Model, Forschungsbericht No. 79/6, Universität Dortmund, Germany. [3.6]

Helbling, J. M. (1983), Ellipsoïdes minimaux de couverture en statistique multivariée, Ph.D. thesis, Ecole Polytechnique Fédérale de Lausanne, Switzerland. [7.1]

Helmers, R. (1980), Edgeworth expansions for linear combinations of order statistics with smooth weight function, *Ann. Stat.*, **8**, 1361–1374. [4.1]

Henderson, H. V., and Velleman, P. F. (1981), Building multiple regression models interactively, *Biometrics*, **37**, 391–411. [6.2]

Hill, R. W. (1977), Robust Regression When There Are Outliers in the Carriers, unpublished Ph.D. dissertation, Harvard University, Boston, MA. [1.2]

Hill, W. J., and Hunter, W. J. (1974), Design of experiments of subsets of parameters, *Technometrics*, **16**, 425–434. [3.3]

Hintze, J. L. (1980), On the use of elemental analysis in multivariate variable selection, *Technometrics*, **22**, 609–612. [3.3]

Hoaglin, D. C., and Welsch, R. E. (1978), The hat matrix in regression and ANOVA, *Am. Stat.*, **32**, 17–22. [6.2]

Hoaglin, D. C., Mosteller, F., and Tukey, J. W. (1983), *Understanding Robust and Exploratory Data Analysis*, John Wiley & Sons, New York. [2.5, 2.7, 7.3]

Hoaglin, D. C., Mosteller, F., and Tukey, J. W. (1985), *Exploring Data Tables, Trends, and Shapes*, John Wiley & Sons, New York. [2.7, 3.1, 7.3]

Hocking, R. R. (1976), The analysis and selection of variables in linear regression, *Biometrics*, **32**, 1–49. [3.3]

Hocking, R. R. (1983), Developments in linear regression methodology: 1959–1982, *Technometrics*, **25**, 219–249. [3.3, 6.2]

Hocking, R. R., and Pendleton, O. J. (1983), The regression dilemma, *Commun. Stat.* (*Theory and Methods*), **12**, 497–527. [6.2]

Hodges, J. L., Jr. (1967), Efficiency in normal samples and tolerance of extreme values for some estimates of location, *Proc. Fifth Berkeley Symp. Math. Stat. Probab.*, **1**, 163–168. [1.2]

Hodges, J. L., Jr., and Lehmann, E. L. (1963), Estimates of location based on rank tests, *Ann. Math. Stat.*, **34**, 598–611. [3.6, 4.1]

Hoerl, A. E., and Kennard, R. W. (1970), Ridge regression: Biased estimation for non-orthogonal problems, *Technometrics*, **12**, 55–67. [3.3]

Hoerl, A. E., and Kennard, R. W. (1981), Ridge regression—1980: Advances, algorithms, and applications, *Am. J. Math. Manage. Sci.*, **1**, 5–83. [3.3]

Holland, P. W., and Welsch, R. E. (1977), Robust regression using iteratively reweighted least squares, *Commun. Stat.* (*Theory and Methods*), **6**, 813–828. [1.2]

Huber, P. J. (1964), Robust estimation of a location parameter, *Ann. Math. Stat.*, **35**, 73–101. [Preface, 3.6, 4.1]

Huber, P. J. (1967), The behavior of maximum likelihood estimates under non-standard conditions, *Proc. Fifth Berkeley Symp. Math. Stat. Probab.*, **1**, 221–233. [7.1]

Huber, P. J. (1972), Robust statistics: A review, *Ann. Math. Stat.*, **43**, 1041–1067. [1.1, 3.4]

Huber, P. J. (1973), Robust regression: Asymptotics, conjectures and Monte Carlo, *Ann. Stat.*, **1**, 799–821. [1.2, 3.6]

Huber, P. J. (1977), Robust covariances, in *Statistical Decision Theory and Related Topics*, Vol. 2., edited by S. S. Gupta and D. S. Moore, Academic Press, New York, pp. 165–191. [7.1]

Huber, P. J. (1981), *Robust Statistics*, John Wiley & Sons, New York. [1.2, 3.4, 7.1, 7.3]

Huber, P. J. (1985), Projection pursuit, *Ann. Stat.*, **13**, 435–475. [3.5]

Huber, P. J., and Dutter, R. (1974), Numerical solutions of robust regression problems, in: *COMPSTAT 1974, Proceedings in Computational Statistics*, edited by G. Bruckmann, Physika Verlag, Vienna. [1.2]

Humphreys, R. M. (1978), Studies of luminous stars in nearby galaxies. I. Supergiants and O stars in the milky way, *Astrophys. J. Suppl. Ser.*, **38**, 309–350. [2.1]

Jaeckel, L. A. (1971), Some flexible estimates of location, *Ann. Math. Stat.*, **42**, 1540–1552. [4.1]

Jaeckel, L. A. (1972), Estimating regression coefficients by minimizing the dispersion of residuals, *Ann. Math. Stat.*, **5**, 1449–1458. [1.2, 3.6]

Jerison, H. J. (1973), *Evolution of the Brain and Intelligence*, Academic Press, New York. [2.4]

Johnstone, I. M., and Velleman, P. F. (1985a), Efficient scores, variance decompositions, and Monte Carlo swindles, *J. Am. Stat. Assoc.*, **80**, 851–862. [5.4]

Johnstone, I. M., and Velleman, P. F. (1985b), The resistant line and related regression methods, *J. Am. Stat. Assoc.*, **80**, 1041–1059. [1.2, 2.7, 5.4]

Jureckovà, J. (1971), Nonparametric estimate of regression coefficients, *Ann. Math. Stat.*, **42**, 1328–1338. [1.2, 3.6]

Jureckovà, J. (1977), Asymptotic relations of M-estimates and R-estimates in linear regression model, *Ann. Stat.*, **5**, 364–372. [3.6]

Kamber, R. (1985), Bruchpunktuntersuchungen in Robuster Regression, Diploma Thesis, ETH Zürich, Switzerland. [3.6]

Kassam, S. A., and Poor, H. V. (1985), Robust techniques for signal processing: A survey, *Proc. IEEE*, **73**, 433–481. [7.2]

Kaufman, L., and Rousseeuw, P. J. (1988), *Finding Groups in Data: Cluster Analysis with Computer Programs*, John Wiley & Sons, New York. (in preparation) [7.3]

Kaufman, L., Leroy, A., Plastria, P., De Donder, M. (1983), Linear Regression in a Non-Homogeneous Population, Research Report No. 183, Centre for Statistics and O.R., University of Brussels, Belgium. [6.6]

Kelly, G. (1984), The influence function in the errors in variables problems, *Ann. Stat.*, **12**, 87–100. [7.3]

Ko, D. (1985), Robust Statistics on Compact Metric Spaces, Ph.D. Thesis, Department of Statistics, University of Washington, Seattle. [7.3, 7.exercise 15]

Ko, D., and Guttorp, P. (1986), Robustness of Estimators for Directional Data, preprint, Department of Biostatistics, Virginia Commonwealth University, Richmond. [7.3]

Koenker, R., and Bassett, G. J. (1978), Regression quantiles, *Econometrica*, **46**, 33–50. [1.2, 3.6]

Krasker, W. S. (1980), Estimation in linear regression models with disparate data points, *Econometrica*, **48**, 1333–1346. [1.2]

Krasker, W. S., and Welsch, R. E. (1982), Efficient bounded-influence regression estimation, *J. Am. Stat. Assoc.*, **77**, 595–604. [1.2]

Kruskal, J. B. (1969), Toward a practical method which helps uncover the structure of a set of multivariate observations by finding the linear transformation which optimizes a new index of condensation, in *Statistical Computation*, edited by R. C. Milton and J. A. Nelder, Academic Press, New York. [3.5]

Kruskal, J. B., and Wish, M. (1978), *Multidimensional Scaling*. Sage University Paper Series on Quantitative Applications in the Social Sciences, Series No. 07-011, Sage Publications, Beverly Hills and London. [7.3]

Kühlmeyer, N. (1983), Das Verhalten Robuster Schätzer in Linearen Modell und bei Simultanen Gleichungssystemen—Eine Simulationsstudie, Ph.D. Thesis, University of Dortmund, West Germany. [3.6, 5.4]

Künsch, H. (1984), Infinitesimal Robustness for Autoregressive Processes, *Ann. Stat.*, **12**, 843–863. [7.2]

Kvalseth, T. O. (1985), Cautionary note about R^2, *Am. Stat.*, **39**, 279–285. [2.3]

Lackritz, J. R. (1984), Exact p-values for F and t tests, *Am. Stat.*, **38**, 312–314. [2.3]

Le Cam, L. (1986), The central limit theorem around 1935, *Stat. Sci.*, **1**, 78–96. [1.1]

Lecher, K. (1980), Untersuchung von R-Schätzern im Linearen Modell mit Simulationen, Diplomarbeit, Abt. Statistik, University of Dortmund, West Germany. [3.6]

Lee, C. H., and Martin, R. D. (1982), M-Estimates for ARMA Processes, Technical Report No. 23, Department of Statistics, University of Washington, Seattle. [7.2]

Lee, C. H., and Martin, R. D. (1987), Applications of robust time series analysis, Technical Report, Department of Statistics, University of Washington, Seattle. In preparation. [7.2]

Lenstra, A. K., Lenstra, J. K., Rinnooy Kan, A. H. G., and Wansbeek, T. J. (1982), Two lines least squares, Preprint, Mathematisch Centrum, Amsterdam. [6.6]

Leroy, A., and Rousseeuw, P. J. (1984), PROGRESS: A Program for Robust Regression Analysis, Technical Report 201, Center for Statistics and O.R., University of Brussels, Belgium. [2.2, 5.1]

Li, G. (1985), Robust regression, in *Exploring Data Tables, Trends, and Shapes*, edited by D. Hoaglin, F. Mosteller, and J. Tukey, John Wiley & Sons, New York. pp. 281–343. [3.1]

303

Lieblein, J. (1952), Properties of certain statistics involving the closest pair in a sample of three observations, *J. Res. NBS*, **48**(3), 255–268. [4.2]

Little, J. K. (1985), Influence and a quadratic form in the Andrews–Pregibon statistic, *Technometrics*, **27**, 13–15. [6.4]

Lopuhaä, H. P., and Rousseeuw, P. J. (1987), Breakdown points of affine equivariant estimators of multivariate location and covariance matrices, Technical Report, Faculty of Mathematics and Informatics, Delft University of Technology, The Netherlands. [7.1]

Madsen, R. W. (1982), A selection procedure using a screening variate, *Technometrics*, **24**, 301–306. [3.3]

Mallows, C. L. (1973), Some comments on C_p, *Technometrics*, **15**, 661–678. [3.3]

Mallows, C. L. (1975), On some topics in robustness, unpublished memorandum, Bell Telephone Laboratories, Murray Hill, NJ. [1.2]

Marazzi, A. (1980), ROBETH, A subroutine library for robust statistical procedures, in *COMPSTAT 1980, Proceedings in Computational Statistics*, Physica Verlag, Vienna. [1.2, 2.7, 7.1]

Marazzi, A. (1986), On the numerical solutions of bounded influence regression problems, in *COMPSTAT 1986*, Physica-Verlag, Heidelberg, pp. 114–119. [3.6]

Marazzi, A. (1987), Solving bounded influence regression problems with ROB-SYS, to appear in *Statistical Data Analysis Based on the L_1 Norm and Related Methods*, edited by Y. Dodge, North-Holland, Amsterdam/New York. [3.6]

Maronna, R. A. (1976), Robust *M*-estimators of multivariate location and scatter, *Ann. Stat.*, **4**, 51–67. [7.1]

Maronna, R. A., and Yohai, V. J. (1981), Asymptotic behavior of general *M*-estimates for regression and scale with random carriers, *Z. Wahrsch. Verw. Geb.*, **58**, 7–20. [3.4]

Maronna, R. A., Bustos, O., and Yohai, V. (1979), Bias- and efficiency-robustness of general *M*-estimators for regression with random carriers, in *Smoothing Techniques for Curve Estimation*, edited by T. Gasser and M. Rosenblatt, Springer Verlag, New York, pp. 91–116. [1.2, 3.6]

Marquardt, D. W., and Snee, R. D. (1975), Ridge regression in practice, *Am. Stat.*, **29**, 3–20. [3.3]

Martin, R. D. (1979), Robust estimation for time series autoregressions, in *Robustness in Statistics*, edited by R. L. Launer and G. N. Wilkinson, Academic Press, New York. [7.2]

Martin, R. D. (1980), Robust estimation of autoregressive models (with discussion), in *Directions in Time Series*, edited by D. R. Brillinger and G. C. Tiao, Institute of Mathematical Statistics Publications, Hayward, CA, pp. 228–254. [7.2]

Martin, R. D. (1981), Robust methods for time series, in *Applied Time Series Analysis II*, edited by D. F. Findley, Academic Press, New York, pp. 683–759. [7.2]

Martin, R. D. (1982), The Cramér–Rao bound and robust M-estimates for autoregressions, *Biometrika*, **69**, 437–442. [7.2]

Martin, R. D. (1985), Influence, bias and breakdown points for time series parameter estimates, abstract in *Inst. Math. Stat. Bull.*, **14**, 289. [7.2]

Martin, R. D., and Thomson, D. J. (1982), Robust-resistant spectrum estimation, *Proc. IEEE*, **70**, 1097–1115. [7.2]

Martin, R. D., and Yohai, V. J. (1984a), Robustness in time series and estimating ARMA models, in *Handbook of Statistics*, Vol. 5, edited by E. J. Hannan, P. R. Krishnaiah, and M. M. Rao, Elsevier, Amsterdam/New York, pp. 119–155. [7.2]

Martin, R. D., and Yohai, V. J. (1984b), Gross-error sensitivities of GM and RA-estimates, in *Robust and Nonlinear Time Series Analysis*, edited by J. Franke, W. Härdle, and D. Martin, Springer Verlag, New York, pp. 198–217. [7.2]

Martin, R. D., and Yohai, V. J. (1986), Influence functions for time series (with discussion), *Ann. Stat.*, **14**, 781–855. [7.2]

Martin, R. D., Samarov, A. and Vandaele, W. (1983), Robust methods for ARIMA models, in *Applied Time Series Analysis of Economic Data*, edited by A. Zellner, Economic Research Report ER-5, Bureau of the Census, Washington, D.C., pp. 153–177. [7.2]

Martin, R. D., and Zeh, J. E. (1978), Robust Generalized M-estimates for Autoregressive Parameters, Including Small-Sample Behavior, Technical Report No. 214, Department of Electrical Engineering, University of Washington, Seattle. [7.2]

Martin, R. D., Yohai, V., and Zamar, R. (1987), Minimax bias robust regression estimation, Technical Report, in preparation. [3.4, 3.6]

Mason, R. L., and Gunst, R. F. (1985), Outlier-induced collinearities, *Technometrics*, **27**, 401–407. [3.3]

Massart, D. L., Kaufman, L., Rousseeuw, P. J., and Leroy, A. (1986), Least median of squares: a robust method for outlier and model error detection in regression and calibration, *Anal. Chim. Acta*, **187**, 171–179. [2.6]

McKay, R. J. (1979), The adequacy of variable subsets in multivariate regression, *Technometrics*, **21**, 475–479. [3.3]

Mickey, M. R., Dunn, O. J., and Clark, V. (1967), Note on the use of stepwise regression in detecting outliers, *Comput. Biomed. Res.*, **1**, 105–111. [2.4, 6.3]

Montgomery, D. C., and Peck, A. E. (1982), *Introduction to Linear Regression Analysis*, John Wiley & Sons, New York. [3.exercise 2, 6.2, 6.3]

Mosteller, F., and Tukey, J. W. (1977), *Data Analysis and Regression*, Addison-Wesley, Reading, MA. [3.1]

Nagel, G., and Wolff, W. (1974), Ein Verfahren zur Minimierung einer Quadratsumme nichtlinearer Funktionen, *Biometrische Z.*, **16**, 431–439. [7.3]

Nair, K. R., and Shrivastava, M. P. (1942), On a simple method of curve fitting, *Sankhyā*, **6**, 121–132. [1.2, 2.7]

Narula, S. C., and Wellington, J. F. (1977), Prediction, linear regression and the minimum sum of relative errors, *Technometrics*, **19**, 185–190. [3.3]

Narula, S. C., and Wellington, J. F. (1979), Selection of variables in linear regression using the minimum sum of weighted absolute errors criterion, *Technometrics*, **21**, 299–306. [3.3]

Narula, S. C., and Wellington, J. F. (1982), The minimum sum of absolute errors regression: A state of the art survey, *Int. Stat. Rev.*, **50**, 317–326. [1.2]

Nath, G. B. (1971), Estimation in truncated bivariate normal distributions, *Appl. Stat.*, **20**, 313–318. [7.1]

Oja, H. (1983), Descriptive statistics for multivariate distributions, *Stat. Probab. Lett.*, **1**, 327–333. [3.6, 7.1]

Oja, H., and Niinimaa, A. (1984), On Robust Estimation of Regression Coefficients, Research Report, Department of Applied Mathematics and Statistics, University of Oulu, Finland. [3.6, 5.1]

Oja, H., and Niinimaa, A. (1985), Asymptotic properties of the generalized median in the case of multivariate normality, *J. R. Stat. Soc. Ser. B*, **47**, 372–377. [7.1]

Papantoni-Kazakos, P. (1984), Some aspects of qualitative robustness in time series, in *Robust and Nonlinear Time Series Analysis*, edited by J. Franke, W. Härdle, and R. D. Martin, Springer Verlag, New York, pp. 218–230. [7.2]

Papantoni-Kazakos, P., and Gray, R. M. (1979), Robustness of estimators on stationary observations, *Ann. Probab.*, **7**, 989–1002. [7.2]

Paul, S. R. (1983), Sequential detection of unusual points in regression, *The Statistician*, **32**, 417–424. [2.4, 6.2, 6.3]

Pearson, E. S. (1931), The analysis of variance in cases of non-normal variation, *Biometrika*, **23**, 114–133. [1.1]

Peters, S. C., Samarov, A., and Welsch, R. E. (1982), Computational Procedures for Bounded-Influence and Robust Regression (TROLL: BIF and BIF-MOD), Technical Report 30, Center for Computational Research in Economics and Management Science, Massachusetts Institute of Technology, Cambridge MA. [3.6]

Pham Dinh Tuan (1984), On robust estimation of parameters for autoregressive moving average models, in *Robust and Nonlinear Time Series Analysis*, edited by J. Franke, W. Härdle, and R. D. Martin, Springer Verlag, New York, pp. 273–286. [7.2]

Plackett, R. L. (1972), Studies in the history of probability and statistics XXIX: The discovery of the method of least squares, *Biometrika*, **59**, 239–251. [Epigraph, 1.1, 3.4]

Polasek, W. (1984), Regression diagnostics for general linear regression models, *J. Am. Stat. Assoc.*, **79**, 336–340. [6.3]

Portnoy, S. (1983), Tightness of the sequence of empiric c.d.f. processes defined from regression fractiles, Research Report, Division of Statistics, University of Illinois. [3.6]

Prescott, P. (1975), An approximate test for outliers in linear models, *Technometrics*, **17**, 129–132. [3.exercise 3]

Rencher, A. C., and Pun, F. C. (1980), Inflation of R^2 in best subset regression, *Technometrics*, **22**, 49–53. [3.3]

Rey, W. J. J. (1983), *Introduction to Robust and Quasi-Robust Statistical Methods*, Springer-Verlag, Berlin. [3.6]

Ronchetti, E. (1979), Robustheitseigenschaften von Tests, Diploma Thesis, ETH Zürich. [3.6]

Ronchetti, E. (1982), Robust Testing in Linear Models: The Infinitesimal Approach, Ph.D. Thesis, ETH Zürich. [3.6]

Ronchetti, E. (1985), Robust model selection in regression, *Stat. Probab. Lett.*, **3**, 21–23 [3.6]

Ronchetti, E., and Rousseeuw, P. J. (1985), Change-of-variance sensitivities in regression analysis, *Z. Wahrsch. Verw. Geb.*, **68**, 503–519. [1.2]

Rosenbrock, H. H. (1960), An automatic method for finding the greatest or lowest value of a function, *Comput. J.*, **3**, 175–184. [3.6]

Rosner, B. (1977), Percentage points for the RST many outlier procedure, *Technometrics*, **19**, 307–313. [4.2]

Rousseeuw, P. J. (1979), Optimally robust procedures in the infinitesimal sense, Proceedings of the 42nd Session of the ISI, Manila, pp. 467–470. [3.6]

Rousseeuw, P. J. (1981), A new infinitesimal approach to robust estimation, *Z. Wahrsch. verw. Geb.*, **56**, 127–132. [4.5]

Rousseeuw, P. J. (1982), Most robust M-estimators in the infinitesimal sense, *Z. Wahrsch. verw. Geb.*, **61**, 541–555. [4.5]

Rousseeuw, P. J. (1983), Multivariate Estimation With High Breakdown Point, paper presented at Fourth Pannonian Symposium on Mathematical Statistics and Probability, Bad Tatzmannsdorf, Austria, September 4–9, 1983. Abstract in *IMS Bull.* (1983), **12**, p. 234. Appeared in (1985), *Mathematical Statistics and Applications*, Vol. B, edited by W. Grossmann, G. Pflug, I. Vincze, and W. Wertz, Reidel, Dordrecht, The Netherlands, pp. 283–297. [1.2, 3.4, 3.5, 7.1]

Rousseeuw, P. J. (1984), Least median of squares regression, *J. Am. Stat. Assoc.*, **79**, 871–880. [Preface, 1.2, 2.5, 2.7, 3.4, 3.5, 6.6, 7.1]

Rousseeuw, P. J. (1985), A regression diagnostic for multiple outliers and leverage points, abstract in *IMS Bull.*, **14**, p. 399. [6.6]

Rousseeuw, P. J. (1987a), An application of L_1 to astronomy, to appear in *Statistical Data Analysis Based on the L_1 Norm and Related Methods*, edited by Y. Dodge, North-Holland, Amsterdam/New York. [7.3]

Rousseeuw, P. J. (1987b), Silhouettes: A graphical aid to the interpretation and validation of cluster analysis, to appear in *J. Comput. Appl. Math.* [7.3]

Rousseeuw, P. J., and Leroy, A. (1987), A robust scale estimator based on the shortest half, Technical Report, Faculty of Mathematics and Informatics, Delft University of Technology, The Netherlands. [4.3, 5.4]

Rousseeuw, P. J., and van Zomeren, B. C. (1987), Identification of multivariate outliers and leverage points by means of robust covariance matrices, Technical Report, Faculty of Mathematics and Informatics, Delft University of Technology, The Netherlands. [7.1]

Rousseeuw, P. J., and Yohai, V. (1984), Robust regression by means of S-estimators, in *Robust and Nonlinear Time Series Analysis*, edited by J. Franke, W. Härdle, and R. D. Martin, Lecture Notes in Statistics No. 26, Springer Verlag, New York, pp. 256–272. [1.2, 3.4]

Rousseeuw, P. J., Daniels, B., and Leroy, A. (1984a), Applying robust regression to insurance, *Insur. Math. Econ.*, **3**, 67–72. [1.exercise 8]

Rousseeuw, P. J., Leroy, A., and Daniels, B. (1984b), Resistant line fitting in actuarial science, in *Premium Calculation in Insurance*, edited by F. de Vylder, M. Goovaerts, and J. Haezendonck, Reidel, Dordrecht, The Netherlands, pp. 315–332. [2.7, 5.2]

Roy, S. N. (1953), On a heuristic method of test construction and its use in a multivariate analysis, *Ann. Math. Stat.*, **24**, 220–238. [3.5]

Ruppert, D., and Carroll, R. J. (1980), Trimmed least squares estimation in the linear model, *J. Am. Stat. Assoc.*, **75**, 828–838. [3.1, 3.4, 3.6]

Ruymgaart, F. H. (1981), A robust principal component analysis, *J. Multivar. Anal.*, **11**, 485–497. [3.5]

Sadovski, A. N. (1974), Algorithm AS74. L_1-norm fit of a straight line, *Appl. Stat.*, **23**, 244–248. [3.6]

Samarov, A. M. (1985), Bounded-influence regression via local minimax mean square error, *J. Am. Stat. Assoc.*, **80**, 1032–1040. [1.2]

Samarov, A., and Welsch, R. E. (1982), Computational procedures for bounded-influence regression, in *Compstat 1982, Proceedings in Computational Statistics*, Physica Verlag, Vienna. [3.6]

Seber, G. A. F. (1977), *Linear Regression Analysis*, John Wiley & Sons, New York. [3.3]

Sen, P. K. (1968), Estimates of the regression coefficient based on Kendall's tau, *J. Am. Stat. Assoc.*, **63**, 1379–1389. [1.2, 2.7]

Serfling, R. A. (1980), *Approximation Theorems of Mathematical Statistics*, John Wiley & Sons, New York. [4.1]

Sheynin, O. B. (1966), Origin of the theory of errors, *Nature*, **211**, 1003–1004. [3.4]

Shorack, G. R., and Wellner, J. A. (1986), *Empirical Processes with Applications to Statistics*, John Wiley & Sons, New York. [4.4]

Siegel, A. F. (1982), Robust regression using repeated medians, *Biometrika*, **69**, 242–244. [1.2, 2.5, 2.7, 3.4, 3.6]

Siegel, A. F., and Benson, R. H. (1982), A robust comparison of biological shapes, *Biometrics*, **38**, 341–350. [7.3]

Siegel, A. F., and Pinkerton, J. R. (1982), Robust Comparison of Three-Dimensional Shapes with an Application to Protein Molecule Configurations,

Technical Report No. 217, Series 2, Department of Statistics, Princeton University. [7.3]

Simkin, C. G. F. (1978), Hyperinflation and Nationalist China, *Stability and Inflation, A Volume of Essays to Honour the Memory of A. W. H. Phillips*, John Wiley & Sons, New York. [2.4]

Simon, S. D. (1986), The Median Star: an Alternative to the Tukey Resistant Line, paper presented at the Joint Statistical Meetings, Chicago, 18–21 August. [2.exercise 13]

Snee, R. D. (1977), Validation of regression models: Methods and examples, *Technometrics*, **19**, 415–428. [3.3]

Souvaine, D. L., and Steele, J.M. (1986), Time and space efficient algorithms for least median of squares regression, Technical Report, Princeton University. [5.2]

Spence, I., and Lewandowsky, S. (1985), Robust Multidimensional Scaling, Technical Report, Department of Psychology, University of Toronto. [7.3]

Sposito, V. A., Kennedy, W. J., and Gentle, J. E. (1977), "L_p norm fit of a straight line, *Appl. Stat.*, **26**, 114–116. [3.5, 3.6]

Stahel, W. A. (1981), Robuste Schätzungen: Infinitesimale Optimalität und Schätzungen von Kovarianzmatrizen, Ph.D. Thesis, ETH Zürich, Switzerland. [5.1, 6.6, 7.1]

Stahlhut, R. (1985), Robuste Schätzungen in Zeitreihenmodellen, Diploma Thesis, Department of Statistics, University of Dortmund, West Germany. [7.2, 7.exercise 12]

Steele, J. M., and Steiger, W. L. (1986), Algorithms and complexity for least median of squares regression, *Discrete Appl. Math.*, **14**, 93–100. [2.2, 3.4, 5.2]

Stefanski, L. A., and Meredith, M. (1986), Robust estimation of location in samples of size three, *Commun. Stat.* (*Theory and Methods*), **15**, 2921–2933. [4.2]

Stevens, J. P. (1984), Outliers and influential data points in regression analysis, *Psychol. Bull.*, **95**, 334–344. [6.2]

Stigler, S. M. (1973), Simon Newcomb, Percy Daniell, and the history of robust estimation 1885–1920, *J. Am. Stat. Assoc.*, **68**, 872–879. [Epigraph]

Stigler, S. M. (1977), Do robust estimators work with real data?, *Ann. Stat.*, **5**, 1055–1098. [Epigraph]

Stigler, S. M. (1981), Gauss and the invention of least squares, *Ann. Stat.*, **9**, 465–474. [1.1]

Stockinger, N. (1985), Generalized maximum likelihood type estimation of autoregressive moving average models, Ph.D. Thesis, Technical University Graz, Austria. [7.2]

Student (1908), The probable error of a mean, *Biometrika*, **6**, 1–25. [4.exercise 10]

Student (1927), Errors of routine analysis, *Biometrika*, **19**, 151–164. [1.1]

Suich, R., and Derringer, G. C. (1977), Is the regression equation adequate?—One criterion, *Technometrics*, **19**, 213–216. [3.3]

Suich, R., and Derringer, G. C. (1980), Is the regression equation adequate?—A further note, *Technometrics*, **22**, 125–128. [3.3]

Switzer, P. (1970), Numerical classification, in *Geostatistics*, Plenum, New York. [3.5]

Tamura, R. N., and Boos, D. D. (1986), Minimum Hellinger distance estimation for multivariate location and covariance, *J. Am. Stat. Assoc.*, **81**, 223–229. [7.1]

Theil, H. (1950), A rank-invariant method of linear and polynomial regression analysis (Parts 1–3), *Ned. Akad. Wetensch. Proc. Ser. A*, **53**, 386–392, 521–525, 1397–1412. [1.2, 2.7, 3.6]

Titterington, D. M. (1978), Estimation of correlation coefficients by ellipsoidal trimming, *Appl. Stat.*, **27**, 227–234. [7.1]

Tukey, J. W. (1960), A survey of sampling from contaminated distributions, in *Contributions to Probability and Statistics*, edited by I. Olkin, Stanford University Press, Stanford, CA. [1.1]

Tukey, J. W. (1970/71), *Exploratory Data Analysis* (Limited Preliminary Edition), Addison-Wesley, Reading, MA. [1.2, 2.7]

Tukey, J. W. (1974), T6: Order Statistics, in mimeographed notes for Statistics 411, Princeton University. [7.1]

Tyler, D. E. (1985a), A distribution-free *M*-estimator of multivariate scatter, Technical Report. [7.1]

Tyler, D. E. (1985b), Existence and uniqueness of the *M*-estimators of multivariate location and scatter, Technical Report. [7.1]

Tyler, D. E. (1986), Breakdown properties of the *M*-estimators of multivariate scatter, Technical Report. Abstract in *Inst. Math. Stat. Bull.*, **15**, 116. [7.1]

Vansina, F., and De Greve, J. P. (1982), Close binary systems before and after mass transfer, *Astrophys. Space Sci.*, **87**, 377–401. [2.1]

van Zomeren, B. C. (1986), A comparison of some robust regression estimators, Technical Report, Delft University of Technology. The Netherlands. [5.3]

van Zwet, W. R. (1985), Van de Hulst and robust statistics: A historical note, *Stat. Neerland.*, **32**, 81–95. [3.6]

Velleman, P. F., and Hoaglin, D. C. (1981), *Applications, Basics, and Computing of Exploratory Data Analysis*, Duxbury Press, Boston. [1.2, 2.7]

Velleman, P. F., and Welsch, R. E. (1981), Efficient computing of regression diagnostics, *Am. Stat.*, **35**, 234–242. [6.2, 6.3]

Venter, J. H. (1967), On estimation of the mode, *Ann. Math. Stat.*, **38**, 1446–1455. [4.4]

Wald, A. (1940), The fitting of straight lines if both variables are subject to error, *Ann. Math. Stat.*, **11**, 284–300. [1.2, 2.7]

Watson, G. S. (1983), Optimal and robust estimation on the sphere, paper presented at the 44th Session of the ISI (Madrid, September 12–22). [7.3]

Weisberg, S. (1980), *Applied Linear Regression*, John Wiley & Sons, New York. [2.4, 3.3]

Weisberg, S. (1981), A statistic for allocating C_p to individual cases, *Technometrics*, **23**, 27–31. [3.3]

Whittle, P. (1953), Estimation and information in stationary time series, *Arch. Math.*, **2**, 423–434. [7.2]

Wilkinson, L., and Dallal, G. E. (1981), Tests of significance in forward selection regression with an F-to-enter stopping rule, *Technometrics*, **23**, 377–380. [3.3]

Wilson, H. C. (1978), Least squares versus minimum absolute deviations estimation in linear models, *Decision Sci.*, 322–335. [3.6]

Wolfowitz, J. (1957), The minimum distance method, *Ann. Math. Stat.*, **28**, 75–87. [7.1]

Yale, C., and Forsythe, A. B. (1976), Winsorized Regression, *Technometrics*, **18**, 291–300. [2.1]

Yohai, V. J. (1985), High breakdown-point and high efficiency robust estimates for regression, to appear in *Ann. Stat.* [3.6]

Yohai, V., and Maronna, R. (1976), Location estimators based on linear combinations of modified order statistics, *Commun. Stat.* (*Theory and Methods*), **5**, 481–486. [4.4]

Yohai, V. and Zamar, R. (1986), High breakdown-point estimates of regression by means of the minimization of an efficient scale, Technical Report No. 84, Department of Statistics, University of Washington, Seattle. [3.6]

Young, A. S. (1982), The bivar criterion for selecting regressors, *Technometrics*, **24**, 181–189. [3.3]

Zamar, R. H. (1985), Orthogonal regression M-estimators, Technical Report No. 62, Department of Statistics, University of Washington, Seattle. [7.3]

Zeh, J. E. (1979), Efficiency Robustness of Generalized M-Estimates for Autoregression and Their Use in Determining Outlier Type, Ph.D. Thesis, Univerisity of Washington, Seattle. [7.2]

Table of Data Sets

The following list is an inventory of the data sets contained in this book, along with the page numbers where they occur, the parameters n and p, and the file name under which they are stored on floppy disk.

Index

(continued from front)